Wnt Signaling
in Embryonic
Development

T0348744

ADVANCES IN DEVELOPMENTAL BIOLOGY

Volume 17

WNT SIGNALING IN EMBRYONIC DEVELOPMENT

Editor

Sergei Y. Sokol

Mount Sinai School of Medicine
Department of Molecular, Cell,
and Developmental Biology
New York, New York

2007

ELSEVIER

AMSTERDAM • BOSTON • HEIDELBERG • LONDON • NEW YORK • OXFORD
PARIS • SAN DIEGO • SAN FRANCISCO • SINGAPORE • SYDNEY • TOKYO

Elsevier
525 B Street, Suite 1900, San Diego, California 92101-4495, USA
84 Theobald's Road, London WC1X 8RR, UK

This book is printed on acid-free paper.

For information on all Academic Press publications
visit our Web site at www.books.elsevier.com

ISBN-13: 978-0-444-52874-2
ISBN-10: 0-444-52874-1

07 08 09 10 9 8 7 6 5 4 3 2 1
Printed and bound by CPI Antony Rowe, Eastbourne

Contents

List of Contributors

Timothy Blauwkamp   Department of Molecular, Cellular and Developmental Biology, University of Michigan, Ann Arbor, Michigan

Ken M. Cadigan   Department of Molecular, Cellular and Developmental Biology, University of Michigan, Ann Arbor, Michigan

Gretchen L. Dollar   Department of Molecular, Cell, and Developmental Biology, Mount Sinai School of Medicine, New York, New York

Leah Etheridge   Departments of Pediatrics and Medicine, School of Medicine, University of California, San Diego, California

Elizabeth Heeg-Truesdell   Department of Biochemistry, Molecular Biology, and Cell Biology, Northwestern University, Evanston, Illinois

Jan Kitajewski   Department of Pathology, OB/GYN and Institute of Cancer Genetics, Columbia University Medical Center, New York, New York

Hendrik C. Korswagen   Hubrecht Laboratory, Netherlands Institute for Developmental Biology and Center for Biomedical Genetics, 3584 CT, Utrecht, The Netherlands

Almut Köhler   Universitaet Karlsruhe (TH), Zoologisches Institut II, D-76131 Karlsruhe, Germany

Carole LaBonne   Department of Biochemistry, Molecular Biology, and Cell Biology, Northwestern University, Evanston, Illinois; Robert H. Lurie Comprehensive Cancer Center, Northwestern University, Evanston, Illinois

T. Néstor H. Masckauchán   Department of Pathology, OB/GYN and Institute of Cancer Genetics, Columbia University Medical Center, New York, New York

David S. Parker   Department of Molecular, Cellular and Developmental Biology, University of Michigan, Ann Arbor, Michigan

Alexandra Schambony   Universitaet Karlsruhe (TH), Zoologisches Institut II, D-76131 Karlsruhe, Germany

Sergei Y. Sokol   Department of Molecular, Cell, and Developmental Biology, Mount Sinai School of Medicine, New York, New York

Jianbo Wang   Departments of Pediatrics and Medicine, School of Medicine, University of California, San Diego, California

Doris Wedlich   Universitaet Karlsruhe (TH), Zoologisches Institut II, D-76131 Karlsruhe, Germany

Anthony Wynshaw-Boris   Departments of Pediatrics and Medicine, School of Medicine, University of California, San Diego, California

Preface

In multicellular organisms, embryonic patterning and cell fate specification depend on precise spatial and temporal regulation of gene expression. This regulation is mediated by maternally deposited localized determinants and by inductive interactions between embryonic cells. The developmental program is known to involve only a small number of signaling pathways that are used repeatedly in different developmental processes. One of these is the Wnt pathway, which is becoming increasingly more important both in embryos and in adult organisms. This book, while not necessarily comprehensive, attempts to overview a number of modes of Wnt signaling in distinct developmental processes and in different species. We apologize to those authors in the Wnt field, whose work was not covered by this book.

Common principles and species differences

The pathway is commonly initiated by a Wnt ligand via a seven transmembrane domain receptor of the Frizzled family. There are 19 known Wnt ligands and 10 Frizzled genes in the mouse genome. LRP5 and LRP6 have been shown to function in association with Frizzled as Wnt coreceptors. Inside the cell, Wnt signals are transduced by Disheveled to different molecular targets, including glycogen synthase kinase 3 (GSK3), β-catenin, small Rho GTPases, Jun N-terminal kinases, and yet uncharacterized cytoskeletal proteins. The prevailing view is that signal transduction occurs via stabilization of a cytoplasmic pool of β-catenin, which then translocates to the nucleus to activate target genes in complex with TCF-transcription factors. A large number of other molecular components are known to regulate and fine tune these basic signaling events in different species. Many extracellular modulators of Wnt ligands in vertebrates, including secreted Frizzled-related proteins (sFRPs), dickkopfs, and WIF, have not been found in fly and worm embryos. The involvement of a Wnt ligand in β-catenin-independent planar polarity signaling in *Drosophila* has not been established; however, Wnt7a has been implicated in a similar polarization process in mammalian inner ear cells (Dabdoub, A., Donohue, M.J., Brennan, A., Wolf, V., Montcouquiol, M., Sassoon, D.A., Hseih, J.-C., Rubin, J.S., Salinas, P.C., Kelley, M.W. 2003. Wnt signaling mediates reorientation of outer hair cell stereociliary bundles in the mammalian cochlea. *Development* 130, 2375–2384). In contrast to a single mammalian β-catenin homologue, in *Caenorhabditis elegans* embryos, cell adhesion and signaling functions are carried out separately by different β-catenin homologues, suggesting more diversified intracellular Wnt signaling.

Processes involved

Several specific Wnt pathway branches have been proposed to exist, depending on the target and molecular components. The common pathway that involves β-catenin-dependent gene regulation is known as the *canonical* pathway. This pathway typically involves TCF, which helps β-catenin to recognize and bind upstream regulatory sequences of Wnt target genes. With β-catenin in the center of this signaling mode, other regulators of its stability, including GSK3, Axin, APC, and LRP5/6, which associates with Axin at the cell membrane, are thought to be specific to the canonical pathway (a few exceptions are noted below, see also Chapter 1 by Parker et al. and Chapter 3 by Korswagen in this book).

The second intensely studied pathway is known as planar cell polarity (PCP) pathway, which coordinates cytoskeletal organization in neighboring cells in the plane of epithelial tissues in *Drosophila* embryos. This mode of signaling involves the general molecular components Frizzled and Disheveled in addition to some unique mediators such as Flamingo, VanGogh/Strabismus, and Prickle (see Chapter 2 by Dollar and Sokol and Chapter 4 by Wang et al. in this book). In *C. elegans*, this pathway remains unstudied. In vertebrate embryos, this branch is thought to regulate convergent extension movements during gastrulation and neurulation, and controls inner ear cell polarity in mammals, presumably by regulating actin cytoskeleton. Whereas many reports show that Wnt ligands regulate cell shape, migration and morphogenetic events in mammalian cells, zebrafish, and *Xenopus* embryos (see Chapter 4 by Wang et al., Chapter 5 by Köhler et al., and Chapter 6 by Heeg-Truesdell and LaBonne in this book), the connection of these processes to PCP, to the regulation of gene targets and molecular mechanisms involved, remain poorly understood.

Mainly from studies in *C. elegans*, it became clear that Wnt ligands can regulate *cell polarity* and *asymmetric cell division* in the apparent absence of gene transcription (see Chapter 2 by Dollar and Sokol and Chapter 3 by Korswagen in this book). Asymmetric cell division may be modulated by localized cell polarity determinants and by mitotic spindle orientation. These activities may represent novel molecular pathways, which are likely to exist in many species beyond worms. In fact, Frizzled and Disheveled have been shown to influence mitotic spindle orientation in *Drosophila* sensory organ precursors. Among other candidate molecules that participate in noncanonical Wnt pathways are G proteins, different forms of protein kinase C, intracellular calcium ions, NFAT and Jun transcription factors, small GTPases, vesicular-trafficking machinery, apical–basal polarity regulators, and unknown second messengers.

Pathway or network: Variations of different steps in signal transduction

The basic composition of Wnt-signaling intermediates can vary to a significant degree in different physiological situations. Besides Wnt proteins, novel ligands that stimulate Frizzled receptors, such as Norrin, have been discovered (see Chapter 7 by Masckauchán and Kitajewski in this book). Conversely, Wnt proteins have been shown to use receptors other than Frizzled, for example, Ryk/Derailed and Ror-receptor protein kinases.

Wnt signaling leads to different outcomes in different cell types, depending on the signaling machinery that is available in particular cells. These findings demonstrate that Wnt pathways, similar to another major developmental pathway, the Notch pathway, operate in a *context-dependent* manner. This property has been demonstrated for Wnt ligands or extracellular Wnt antagonists, which have different (sometimes opposite) effects depending on receptor availability (Mikels, A.J., Nusse, R. 2006. Purified Wnt5a protein activates or inhibits beta-catenin-TCF signaling depending on receptor context. *PLoS Biol.* 4, e115; and Brott, B.K., Sokol, S.Y. 2002. Regulation of Wnt/LRP signaling by distinct domains of Dickkopf proteins. *Mol. Cell. Biol.* 22, 6100–6110). Thus, the same ligand may signal to different cellular targets in a different manner. For example, Wnt11 has been claimed to stimulate both canonical and noncanonical pathway branches, indicating that the outcome of signaling does not exclusively depend on Wnt-ligand structure.

Intracellular cytoplasmic mediators of signaling are also involved in multiple pathway branches. This is best studied in Disheveled, which functions in most known Wnt-signaling modes. In contrast, GSK3 has been proposed to be specific for the canonical pathway, initially as a negative regulator of β-catenin, and more recently, as a positive regulator of LRP5/6 phosphorylation (Zeng, X., Tamai, K., Doble, B., Li, S., Huang, H., Habas, R., Okamura, H., Woodgett, J., He, X. 2005. A dual-kinase mechanism for Wnt coreceptor phosphorylation and activation. *Nature* 438, 873–877). Despite this, additional evidence implicated GSK3 in the control of cell shape and migratory behavior as well as microtubule organization, possibly indicating its role for noncanonical (β-catenin independent) Wnt signaling. The involvement of GSK3 in noncanonical Wnt signaling is further substantiated by its role in the nuclear export of NFAT, a β-catenin-independent target of Wnt signaling. For other proteins proposed to transduce Wnt signals, such as Rho GTPases, casein kinases I and II, ROK, and JNK, the challenge will be to distinguish their specific roles in Wnt signal transduction from their general participation in cell shape/motility changes triggered by many growth factors. Accumulating genetic and biochemical evidence points to endocytosis playing an important role for Wnt signaling (Coudreuse, D.Y., Roel, G., Betist, M.C., Destree, O., Korswagen, H.C. 2006. Wnt gradient formation requires retromer function in Wnt-producing cells. *Science* 312, 921–924; Prasad, B.C., Clark, S.G. 2006. Wnt signaling establishes antero-posterior neuronal polarity and requires retromer in *C. elegans*. *Development* 133, 1757–1766; and Blitzer, J.T., Nusse, R. 2006. A critical role for endocytosis in Wnt signaling. *BMC Cell Biol.* 7, 28); however, the specific mechanistic understanding of this requirement is still lacking. Establishing specific molecular readout assays for different Wnt pathway branches in different cell types is essential to understand common and diverged roles of individual pathway components.

Wnt signaling in the nucleus is equally complex and is likely to involve multiple regulators of chromatin such as transcriptional coactivators and corepressors (see Chapter 1 by Parker et al. in this book). Although the essential role for TCF in signaling is commonly accepted, the specific mechanism for TCF involvement appears to vary in different species. In the worm, POP1/TCF is a repressor, which is exported from the nucleus in response to the MOM-2/Wnt signal. This is unlikely to be a general mechanism, as other TCFs have not been reported to undergo signal-dependent nuclear export (see Chapter 1 by Parker et al. and Chapter 3 by Korswagen in this book). TCF3 in vertebrates also functions as a signaling repressor, but how this repression is overcome during signal transduction is unknown. Whereas Wnt target genes usually contain consensus TCF-binding sites, some gene targets require additional transcriptional regulators besides TCF and β-catenin, and some may be TCF independent (Olson, L.E., Tollkuhn, J., Scafoglio, C., Krones, A., Zhang, J., Ohgi, K.A., Wu, W., Taketo, M.M., Kemler, R., Grosschedl, R., Rose, D., Li, X., Rosenfeld, M.G. 2006. Homeodomain-mediated beta-catenin-dependent switching events dictate cell-lineage determination. *Cell* 125, 593–605). Additional experimental data may allow to develop new general concepts of Wnt signaling in the nucleus.

Looking ahead

I hope that this book illustrates the vast number of questions that remain open in the Wnt field. Persistent efforts in different experimental systems are necessary to allow further progress in our understanding of this complex network of interacting molecules.

There is much to be learned about the ways of how Wnt signals are delivered across the cell membrane and how specific cellular targets are selected. Genetic targets of specific Wnt-signaling branches, such as PCP, are unknown, and the regulation of known canonical targets is not as simple as it seemed just a few years ago. Although artificial reporters that contain multimerized TCF sites are commonly used to measure canonical Wnt responses, further understanding of tissue-specific target gene regulation would be important. The analysis of available microarray data for genes activated by Wnt signals in different tissues may help identify new direct Wnt targets, both dependent and independent of β-catenin.

SERGEI Y. SOKOL
Mount Sinai School of Medicine

Wnt/β-catenin-mediated transcriptional regulation

David S. Parker, Timothy Blauwkamp and Ken M. Cadigan

Department of Molecular, Cellular and Developmental Biology, University of Michigan, Ann Arbor, Michigan

Contents

Advances in Developmental Biology
Volume 17 ISSN 1574-3349
DOI: 10.1016/S1574-3349(06)17001-5

Many Wnts act by stabilizing β-catenin and promoting its nuclear localization, where it can influence gene expression by associating with DNA-binding proteins. The best understood of these transcription factors are members of the TCF/LEF1 (TCF) family, which recognize specific DNA sequences via a high mobility group (HMG) domain. There is strong support for a model where TCFs repress target gene expression in the absence of Wnt signaling, through recruitment of corepressors. Upon Wnt stimulation, high levels of nuclear $β$-catenin bind to TCF and promote a switch to transcriptional activation. This is achieved by the recruitment of transcriptional coactivators and the subsequent modification of the chromatin surrounding Wnt regulated enhancers (WREs). While invertebrate TCFs can both repress and activate transcription of Wnt targets, genetic evidence in vertebrate systems suggests that these activities may have become separated among specific TCFs. In addition, recent findings demonstrate that $β$-catenin/TCF complex can also directly repress some Wnt target genes and $β$-catenin can also regulate transcription independently of TCFs, by associating with other DNA- binding proteins. Wnt-signaling components previously thought to act only in the cytoplasm or as nuclear shuttles for $β$-catenin have been found to associate directly with WREs. This added complexity to the already elaborate model of Wnt-mediated transcriptional regulation may help account for the extraordinary diversity of transcriptional responses to the Wnt/$β$-catenin pathway.

1. Introduction

The Wnt/$β$-catenin pathway regulates gene expression in numerous developmental contexts throughout the animal kingdom (Cadigan and Nusse, 1997; Moon et al., 1997; Logan and Nusse, 2004). In this evolutionarily conserved pathway, Wnts act through stabilization and nuclear translocation of $β$-catenin. In many cases, it is thought that $β$-catenin influences gene expression by interacting with members of the TCF/LEF1 (TCF) family of DNA-binding proteins. The current model for TCF/$β$-catenin action proposes that in the absence of signal, TCF binds to specific sequences in the regulatory region of Wnt target genes, where it interacts with corepressors to inhibit the transcription of these genes. After Wnt signaling is activated, $β$-catenin enters the nucleus and binds to TCF, which displaces/inactivates corepressors and recruits transcriptional coactivators, turning on target gene expression. This "transcriptional switch" has been proposed to be a general feature of signal-mediated gene activation (Barolo and Posakony, 2002; see Fig. 1).

This chapter will lay out the evidence for the transcriptional switch model, summarizing the transcriptional regulators that have been proposed to modulate the transcriptional switch. Wnt/$β$-catenin gene regulation in *Caenorhabditis elegans*, *Drosophila*, *Xenopus*, zebrafish, and mammalian systems will be reviewed, recognizing the strong degree of conservation between these systems. The ability to study the process in such a broad

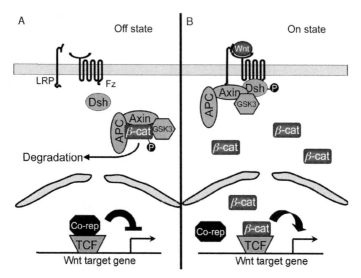

Fig. 1. Simplified model of the Wnt/β-catenin pathway. (A) In the absence of Wnt signal, a cytosolic complex containing APC, Axin, and GSK3 phosphorylate β-catenin, targeting it for ubiquitination and proteasomal degradation. In the nucleus, TCF bound by transcriptional corepressors silences Wnt target genes. (B) Upon Wnt binding to a receptor complex containing Frizzleds (Fz) and lipoprotein related binding protein (LRP), the APC/Axin/GSK3 complex is inhibited, allowing the accumulation and nuclear translocation of β-catenin, which then binds to TCF in the regulatory region of the Wnt target. This promotes transcriptional activation of the gene.

array of organisms has clearly been a huge advantage in uncovering new mechanisms of regulation.

It is well known that TCF/β-catenin regulates different genes in different tissues, a distinction between general mediators of Wnt target gene regulation and those that may act in a cell or gene-specific manner will be made. Adding to the diversity of Wnt target gene regulation are the findings that β-catenin can regulate transcription through DNA-binding proteins besides TCFs. Several examples of β-catenin-mediated gene repression that cannot be explained by the transcriptional switch model will also be discussed. Finally, evidence that several components of the pathway have been found to cycle between the nucleus and cytoplasm will be discussed, raising the possibility that components of the β-catenin-degradation complex, known for its ability to phosphorylate and destabilize β-catenin in the cytoplasm (Polakis, 2000; Ding and Dale, 2002), may also act to inhibit Wnt/β-catenin signaling in the nucleus.

2. The TCF family as mediators of Wnt/β-catenin signaling

2.1. Historical perspective and biochemical properties of TCFs

TCFs were first identified as sequence specific DNA-binding proteins expressed in T and B cell lymphocytes (Laudet et al., 1993; Clevers and van de Wetering, 1997). There are four TCFs in mammalian genomes, but only one in *Drosophila*, *TCF/pangolin* (*TCF/pan*) and one in *C. elegans*, *POP-1* (see Table 1 for list). Although TCF family members are conserved to some extent throughout their primary protein sequences, they are most similar in their high mobility group (HMG) domains (Laudet et al., 1993; Clevers and van de Wetering, 1997).

While some HMG domains bind DNA nonspecifically, the TCF subfamily recognizes DNA in a more specific manner. Initial studies with the HMG domains of LEF1 and TCF1 determined that the sequence CCTTTGWW (where W = A or T) constituted a high-affinity binding site (Giese et al., 1991; van de Wetering et al., 1991). A PCR-based screen for preferred binding sites for the fly TCF/Pan revealed that 11 nucleotides could influence binding, CCTTTGATCTT being the optimal site (van de Wetering et al., 1997). A screen with biotinylated oligonucleotides and the HMG domain of mouse TCF4 found that CCTTTGATG bound with highest affinity (Hallikas et al., 2006). Consistent with these *in vitro* studies, similar DNA sequences have been demonstrated to be functionally important in many Wnt regulated enhancers (WREs; Brannon et al., 1997; Riese et al., 1997; He et al., 1998; Galceran et al., 1999; Tetsu and McCormick, 1999; Yamaguchi et al., 1999; Lee and Frasch, 2000; Knirr and Frasch, 2001).

In addition to binding DNA in a sequence-specific fashion, LEF1 has also been shown to induce a sharp bend in the double helix (Giese et al., 1992; Love et al., 1995). The HMG domain plus several basic residues located immediately C-terminal to it are sufficient for maximal bending (Love et al., 1995; Lnenicek-Allen et al., 1996). Although DNA bending has only been shown for LEF1, the HMG domain and basic stretch are highly conversed with other TCFs. The DNA bend has been proposed to facilitate interaction among various enhancer/promoter elements (Riese et al., 1997), but this has not been demonstrated experimentally.

The connection between Wnt signaling and TCFs became apparent after several groups independently discovered that LEF1 and TCF3 could bind to β-catenin (Behrens et al., 1996; Huber et al., 1996; Molenaar et al., 1996). The importance of β-catenin in Wnt signaling was already firmly established, in large part due to the observation that loss of *armadillo* (*arm*, the fly β-catenin) resulted in phenotypes nearly identical to that of *wingless* (*wg*), the best-characterized fly Wnt (Riggleman et al., 1989; Peifer et al., 1991; Noordermeer et al., 1994; Siegfried et al., 1994). This research led to the

Table 1
Summary of TCF mutant phenotypes displaying positive or negative effects on the Wnt/β-catenin pathway

TCF gene	Positive role in Wnt/β-catenin signaling	Negative role in Wnt/β-catenin signaling	References
TCF/pan	Segment polarity phenotype in embryo; loss of Wg signaling in imaginal discs	Suppresses the cuticle phenotype of wg mutants	Brunner et al., 1997; van de Wetering et al., 1997; Schwiezer et al., 2003
POP-1	QL neuroblast migration; DTC specification in the somatic gonad; E cell-specific gene activation	Represses mesoderm (MS) cell fate in early embryo	Hermann 1991; Rocheleau et al., 1997; Thorpe et al., 1997; Kidd et al., 2005; Shetty et al., 2005; Lam et al., 2006
TCF1	Mouse limb development caudal somitogenesis (with LEF1); mesoderm induction and patterning in *Xenopus*	Tumor suppressor in intestinal and mammary epithelia; repressing dorsal cell fate in ventral blastomeres in *Xenopus*	Galceran et al., 1999; Roose et al., 1999; Liu et al., 2005; Standley et al., 2006
LEF1	Mouse limb development caudal somitogenesis (with TCF1), mesenchymal/epithelial signaling, hair follicle stem cells and defects in pro-B cell proliferation; mesoderm patterning in *Xenopus*	None reported	van Genderen et al., 1994; Galceran et al., 1999; Reya et al., 2000; Kratochwil et al., 2002; Liu et al., 2005; Lowry et al., 2005
TCF3	None reported	Anterior structures in fish and mice; A/P axis formation in mice; mesoderm induction and repressing dorsal cell fate in ventral blastomere in *Xenopus*	Kim et al., 2000; Merrill et al., 2004; Liu et al., 2005; Houston et al., 2002
TCF4	Stem cell fate in mouse intestine epithelia; dorsal cell fate in *Xenopus*	Repression of CD24 expression	Korinek et al., 1998; Shulewitz et al., 2006; Standley et al., 2006

discovery that maternal *β*-catenin is required for primary axis formation in *Xenopus* embryos, and misexpression of *β*-catenin in ventral blastomeres can induce a secondary axis (Heasman et al., 1994) as had previously been shown for several Wnts (McMahon and Moon, 1989; Smith and Harland, 1991; Sokol et al., 1991).

The finding that *β*-catenin, an essential mediator of Wnt signaling, could physically interact with specific DNA-binding proteins greatly accelerated understanding of Wnt signaling in the nucleus. The initial reports identified the N-terminal portion of TCFs as being necessary and sufficient for *β*-catenin binding (Behrens et al., 1996; Huber et al., 1996; Molenaar et al., 1996). Deletion of this portion of TCFs results in potent dominant negative inhibitors of the pathway (Behrens et al., 1996; Molenaar et al., 1996; van de Wetering et al., 1997; Kratochwil et al., 2002), consistent with the notion that TCF and *β*-catenin must physically associate to promote Wnt signaling. When TCFs are coexpressed with reporter genes containing multimerized high-affinity TCF sites upstream of a proximal promoter and reporter ORF, no activation of expression is observed. However, addition of *β*-catenin results in a marked increase in reporter gene expression (Molenaar et al., 1996; Korinek et al., 1997; van de Wetering et al., 1997), supporting the model summarized in Fig. 1B. As will be described below in more detail, the identification of WREs has been greatly facilitated by searching for the presence of TCF-binding sites in the regulatory regions of Wnt target genes.

2.2. TCF mutant phenotypes confirm a positive role in Wnt/β-catenin signaling

The analysis of *TCF* loss-of-function mutations confirms that TCFs play a physiologically important role in the activation of Wnt targets. These studies also revealed a negative role in the pathway for some TCFs in some contexts, which will be discussed in Section 2.2.1. As will be described, some TCFs seem to be predominately involved in activation and some in repression of Wnt targets, while some readily display both activities.

2.2.1. Invertebrate TCFs

The single TCF homologues in *Drosophila* and *C. elegans* are required for Wnt/*β*-catenin signaling in many contexts. Fly embryos lacking *TCF/pan* display a segment polarity phenotype consistent with a strong reduction in Wg/Arm signaling (Brunner et al., 1997; van de Wetering et al., 1997). Reduction of *TCF/pan* gene activity at later developmental stages also revealed phenotypes consistent with a block in the Wg/Arm pathway (Brunner et al., 1997; Schweizer et al., 2003). Loss-of-function mutations in *POP-1* in *C. elegans* have a block in MAB-5 expression and defects in QL neuroblast migration (Herman, 2001) and fail to activate CEH-22 expression in distal tip cell (DTC) progenitors in the somatic gonad (Lam et al., 2006).

Both events are known to require Wnt signaling (Maloof et al., 1999; Siegfried et al., 2004), indicating that worm TCF plays a positive role in the Wnt pathway.

2.2.2. Mouse TCF knockouts

Targeted disruption of murine *TCF* genes also results in phenotypes best explained by a reduction in Wnt/β-catenin signaling. Expression of the secreted Wnt antagonist Dickkopf-1 blocked follicle formation (Andl et al., 2002), and conditional knockouts of *β-catenin*, where the gene was removed in the skin epidermis, resulted in an absence of hair follicle formation due to lack of follicle stem cell proliferation (Huelsken et al., 2001; Lowry et al., 2005). Similarly, *LEF1* mutants lack hair follicles, as well as teeth and mammary glands (van Genderen et al., 1994). A mutation in *LEF1* lacking five amino acids required for β-catenin binding results in mice lacking teeth, indicating that LEF1/β-catenin association is important for this induction (Kratochwil et al., 2002). *LEF1* mutants also exhibit a reduction in pro-B cell proliferation and survival, due to a reduced ability of these lymphocytes to respond to Wnt3a (Reya et al., 2000). When *TCF* gene activity is further reduced by combining *LEF1* and *TCF1* gene disruptions, a lack of paraxial mesoderm and caudal somites is observed (Galceran et al., 1999), a phenotype strikingly similar to that observed in *Wnt3a* mutations (Takada et al., 1994). *LEF1;TCF1* double mutants also have defects in limb development associated with a loss of the apical epidermal ridge (AER; Galceran et al., 1999), which has also been shown to require *Wnt3* and *β-catenin* for its formation (Barrow et al., 2003; Hill et al., 2006).

Activation of the Wnt/β-catenin pathway in the intestinal epithelia through biallelic loss of *APC* or point mutations stabilizing β-catenin plays an important role in polyp formation and colorectal cancer (Polakis, 2000; Gregorieff and Clevers, 2005). Wnt/β-catenin signaling is thought to promote a proliferative, self-renewing population of stem cells in the crypts and pits of the small and large intestine, respectively (Pinto and Clevers, 2005). This population of progenitors is necessary to replenish the various specialized cell types found at the top of the villi (Gregorieff and Clevers, 2005; Pinto and Clevers, 2005), as well as the Paneth cells at the base of the crypts (van Es et al., 2005). Constitutive activation of Wnt/β-catenin signaling through mutation of the *APC* gene results in overproliferation of these progenitor cells (van de Wetering et al., 2002; Sansom et al., 2004). Conversely, expression of the Wnt antagonist Dickkopf-1 results in a time-dependent reduction of differentiated cell types and a general degeneration of the intestinal lining (Pinto et al., 2003; Kuhnert et al., 2004). This phenotype is very similar to that observed in mice homozygous for a *TCF4* gene disruption (Korinek et al., 1998), consistent with TCF4 being the major mediator of Wnt/β-catenin signaling in maintaining this stem cell population (van de Wetering et al., 2002; Gregorieff and Clevers, 2005; Pinto and Clevers, 2005).

Several other phenotypes reported in *TCF* knockouts are possibly related to defects in Wnt/β-catenin signaling. The defect in mammary gland formation observed in *LEF1* mutants is consistent with the proposed role of Wnt/β-catenin in mammary development (Brennan and Brown, 2004). *TCF1* and *LEF1* were first identified as lymphocyte-specific transcription factors, and disruption of *TCF1* results in a dramatic reduction in T cells (Verbeek et al., 1995), which is even more extensive in *TCF1, LEF1* double knockouts (Held et al., 2003). *TCF4;TCF1* double mutants have a severe reduction in hindgut development and at later stages an anterior transformation of the remaining gastrointestinal tract (Gregorieff et al., 2004). Whether these phenotypes are related to defective Wnt signaling is not clear, though it is a reasonable assumption based on the intimate connection of TCFs with β-catenin-mediated transcriptional activation.

2.3. TCF mutant phenotypes also reveal negative roles in Wnt/β-catenin signaling

The genetic analysis of *TCF*s was crucial in recognizing the role they play in repressing Wnt targets. Here we review the evidence that some TCFs both activate and repress transcription, while others appear dedicated to one or the other activity.

2.3.1. Drosophila *TCF/Pan*

In *Drosophila*, the phenotype of *TCF/pan* in the embryonic epidermis supports a dual role for in Wnt target gene regulation. While the *TCF/pan* phenotype clearly indicates a reduction in Wg signaling (Brunner et al., 1997; van de Wetering et al., 1997), the cuticle defects are not as severe as that of *wg* or *arm* mutants, even when both maternal and zygotic contributions of TCF are eliminated (Schweizer et al., 2003). Furthermore, a *wg, TCF/pan* double mutant is similar to *TCF/pan* alone, indicating that the loss of *TCF/pan* suppresses the *wg* segment polarity phenotype (Cavallo et al., 1998). These data are consistent with TCF/Pan acting as a repressor in the absence of Wg signaling (Fig. 1A).

The analysis of WRE-reporter gene constructs in transgenic *Drosophila* also supports a role for TCF/Pan as a repressor of Wnt targets in the absence of signaling, though the results are enhancer specific. Mutation of TCF sites in a visceral mesoderm *decapentapleigic* (*dpp*) WRE caused a large expansion of the expression domain, with no obvious reduction in expression levels in the cells where *dpp* is normally expressed (Yang et al., 2000). Knocking out TCF sites in the *wg*-dependent pericardial *even-skipped* (*eve*) WRE resulted in a reduction of expression amplitude, along with a derepression of expression in adjacent cells where the enhancer is not normally active (Knirr and

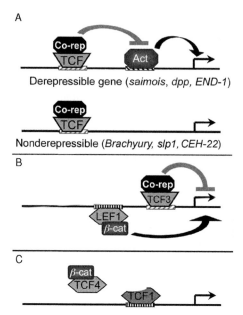

Fig. 2. Several models to explain the dual role of TCF in regulating Wnt target gene expression. TCF mutant analysis and site-directed mutagenesis of WREs indicate that TCFs both repress and activate WREs. (A) In this model, a gene will be derepressed if other positive elements are present in the regulatory region. WREs, such as *saimois* (vertebrates), *dpp* (flies), or *END-1* (worms), are derepressed when TCF-binding sites are destroyed. Other WREs, such as *Brachyury* (mice), *Slp1* (flies), or *CEH-22* (worms), may lack positive activators, so no derepression is observed. (B) In this model, different vertebrate TCFs favor activation in conjunction with β-catenin (e.g., LEF1 or TCF4) or β-catenin-independent repression (TCF3). In the case of TCF3, it is possible that β-catenin binding will block its repressive activity, though this is not clear. (C) In this model, truncated forms of TCF (e.g., TCF1) cannot bind β-catenin, and thus can act in a dominant negative way to repress TCF/β-catenin gene activation.

Frasch, 2001). Similarly mutating TCF sites in a visceral mesoderm *Ultrabithorax* (*Ubx*) WRE caused a sharp decrease in expression, with mild derepression observed in adjacent cells (Riese et al., 1997). In contrast, eliminating TCF sites in a WRE driving expression of *sloppy paired 1* (*slp1*) in the epidermis and mesoderm abolished activation, and no derepression was observed (Lee and Frasch, 2000). These studies support a model where the context of the TCF sites (i.e., whether the regulatory element contains any other Wg-independent activator elements) influences whether the WRE will be subject to derepression if TCF/Pan binding to the element is compromised (Fig. 2A).

2.3.2. C. elegans POP-1

POP-1 is thought to act largely as a repressor in early embryogenesis in the worm. At the four-cell blastomere stage, the P2 cell signals to the EMS cell to promote an asymmetric cell division. The daughter cell adjacent to the P2 cell (the E cell) develops along an endodermal cell lineage. The other daughter (the MS cell) adopts a mesodermal lineage. Mutations in a *Wnt* (*MOM-2*) or a worm *β-catenin* (*WRM-1*) result in additional mesoderm (Rocheleau et al., 1997; Thorpe et al., 1997), but mutations in the worm *TCF* (*POP-1*) cause the MS cells to adopt an E cell fate (Lin et al., 1995). Double mutants between *MOM-4* or *WRM-1* with *POP-1* result in pheno-types identical to *POP-1* alone (Rocheleau et al., 1997; Thorpe et al., 1997). Thus, in this context, POP-1 appears to be a transcriptional repressor which is inactivated by Wnt/β-catenin signaling.

Analysis of E cell-specific gene expression (i.e., genes expressed in the E cell but not the MS cell) revealed that POP-1 not only acts as a repressor in the absence of signaling, but also as an activator of transcription in Wnt-stimulated cells. In *POP-1* mutants, E cell genes are expressed at equal levels in both MS and E cells (Calvo et al., 2001; Maduro et al., 2002; Shetty et al., 2005). However, a careful examination of gene expression levels revealed that gene expression was not as high in *POP-1* mutants as in wild-type E cells (Shetty et al., 2005). Mutation of a single predicted TCF-binding site in the promoter of one of these E cell-specific genes (*END-1*) caused both a depres-sion in MS cells and a reduction in expression in E cells (Shetty et al., 2005). These data suggest that in this context, POP-1 can both activate and repress genes from the same site, depending on the activity of the Wnt pathway.

In contrast to E cells genes, POP-1 has no detectable repressive effect on CEH-22 expression in the DTC lineage of the somatic gonad. For this target, mutation of two POP-1-binding sites in the *CEH-22* WRE dramati-cally reduces expression of a reporter construct, with no derepression observed (Lam et al., 2006). Thus, like TCF/Pan, POP-1 has specific effects on different WREs, presumably due to other elements in the target gene (Fig. 2A).

2.3.3. Vertebrate TCFs

In contrast to fly and worm TCF, it appears that different vertebrate TCFs are more dedicated to either *β*-catenin-dependent activation of Wnt targets or repression of targets in the absence of signaling. As outlined above, *LEF1* and *TCF4* knockouts in the mouse demonstrate a positive role in Wnt signaling (see Table 1 for summary). The opposite is the case for *TCF3*, where loss of this TCF causes phenotypes consistent with an increase in Wnt signaling. In zebrafish, mutants in the *TCF3* orthologue *headless* (*hdl*) lack eyes, forebrain, and part of the midbrain (Kim et al., 2000). This phenotype is very similar to that obtained by overexpression of Wnts or mutation of Wnt antagonists (Mukhopadhyay et al., 2001; Niehrs et al., 2001). The *hdl* phenotype could be rescued by injection of mRNA encoding TCF3 or a chimeric TCF3 fused to a transcriptional repressor. When TCF3 was fused of VP16, a potent transcriptional activator, no rescue of the *hdl* phenotype was observed (Kim et al., 2000). These data are consistent with TCF3-repressing Wnt targets in anterior structures.

A similar loss of forebrain is also seen in mouse *TCF3* knockouts (Merrill et al., 2004). In more strongly affected embryos, an expansion and duplication of nodes and notochord is very reminiscent of mutants overexpressing Wnt (Popperl et al., 1997) or mutations in the Wnt negative regulators *Axin* (Zeng et al., 1997) or *Apc* (Ishikawa et al., 2003). A reporter gene containing multimerized TCF-binding sites (TOPGal) was expressed at wild-type levels in the *TCF3* knockouts, consistent with TCF3 only acting as a repressor (Merrill et al., 2004).

The study of Wnt-dependent early inductive events in amphibian embryogenesis has also revealed differences in TCF family member functions. Wnt/β-catenin signaling is required for mesoderm induction (Schohl and Fagotto, 2003) and then subsequently to promote ventral/lateral and repress dorsal mesodermal cell fate (Christian and Moon, 1993; Hoppler and Moon, 1998; Hamilton et al., 2001). Using morpholinos to deplete endogenous gene expression combined with rescue by injected mRNAs, it was shown that *TCF1* and *TCF3* are required for mesoderm induction, but they act non-redundantly (Liu et al., 2005). A TCF3-VP16 fusion could rescue the *TCF1* morphant but not that of *TCF3*, suggesting that TCF3 acts as a transcriptional repressor and TCF1 as an activator. Conserved motifs found in the central region of TCF3, which are absent in TCF1, were required for this repressor activity (Liu et al., 2005). In the subsequent patterning of the mesoderm, *TCF1* and *LEF1* were found to act redundantly and depleted embryos of either gene could be rescued by TCF3-VP16, suggesting that they act as transcriptional activators (Liu et al., 2005).

Before *Xenopus* mesoderm induction, *Xwnt11/β-catenin* is required for the formation of the Spemann organizer and dorsal cell fates (Heasman et al., 1994; Tao et al., 2005). Depletion of maternal *TCF3* expression using antisense oligonucleotides revealed that it functions to repress Spemman organizer markers in the ventral blastomeres (Houston et al., 2002; Yang et al., 2002). Depletion of maternal *TCF1* had a similar effect, and it was suggested that TCF1 and TCF3 act cooperatively (Standley et al., 2006). In contrast, *TCF4* was required for activation of organizer markers dorsally (Standley et al., 2006). Taken together, the amphibian data supports a model where TCF4 and LEF1 are activators of Wnt targets, TCF3 represses Wnt targets, and TCF1 has the ability to do both.

2.3.4. Vertebrate WREs reveal dual regulation by TCFs

Analysis of WREs in vertebrate developmental contexts provides support for the partitioning of TCF function. For example, WREs have been identified in the regulatory regions of *Brachyury* and *Delta-like1*, both of which are expressed in the caudal somites and are likely targets of TCF1/LEF1 (Galceran et al., 1999). Mutation of the TCF sites within these enhancers abolished expression, and no derepression was observed (Yamaguchi et al., 1999; Galceran et al., 2004). *saimois*, a marker for the Spemmann organizer, is a target which is activated by TCF4/β-catenin in the dorsal blastomeres and repressed by TCF3/TCF1 in ventral ones (Standley et al., 2006). A *siamois*-reporter gene is expressed at much higher levels when injected in dorsal blastomeres, compared to ventral blastomeres, where β-catenin is low. Mutational analysis of the TCF sites in the reporter demonstrated their requirement for both activation and repression, but some of the TCF sites are more important for repression and some for activation (Brannon et al., 1997). This raises the possibility that TCF4 and TCF3/TCF1 interact with different sites in the *siamois* WRE to mediate regulation (Fig. 2B).

One caveat with the notion that some TCFs activate and some repress Wnt targets is that it is based largely on the examination of developmental phenotypes, not direct transcriptional targets. Therefore, the genetics summarized above may only be relevant for the most developmentally important targets and not truly reflect strict mechanistic boundaries between the TCFs. For example, CD24 is activated by β-catenin in human cultured cells. In the absence of signal, RNA interference (RNAi) depletion of *TCF4*, caused a modest derepression of the gene and an inhibitory TCF site was identified in the *CD24* promoter (Shulewitz et al., 2006). It is also worth pointing out that TCF3 can efficiently activate TCF-reporter genes when overexpressed with β-catenin (Molenaar et al., 1996; Brannon et al., 1997; Fan et al., 1998), suggesting that it might have activating function in conjunction with β-catenin in some circumstances.

2.3.5. Naturally occurring dominant negative forms of TCFs

In addition to transcriptional repression, elevation of Wnt target gene expression in *TCF* mutants could also be due to reduction of inhibitory forms of TCFs. In the mouse small intestine, *TCF1* knockouts have a phenotype consistent with an elevation of Wnt signaling. As outlined above, Wnt/β-catenin/TCF4 signaling is thought to promote stem cell proliferation in the intestinal epithelium (Gregorieff and Clevers, 2005; Pinto and Clevers, 2005). Mutations in *APC*, which result in elevated β-catenin levels and increase TCF4-dependent transcriptional activation (Polakis, 2000), lead to polyp formation. Inappropriate activation of Wnt/β-catenin signaling can also cause mammary tumors in mice (Nusse and Varmus, 1992; Brennan and Brown, 2004). Disruption of the mouse *TCF1* gene caused spontaneous tumors in the intestine and mammaries, which was accelerated in an *APC* heterozygous background (Roose et al., 1999). Interestingly, TCF1 expression was activated in the intestinal epithelia by Wnt/β-catenin signaling in a TCF4-dependent manner, suggesting that TCF1 is a negative feedback regulator of the pathway. It was suggested that a truncated isoform of TCF1, lacking the β-catenin-binding domain, could be responsible for the repression, acting as a dominant negative (Roose et al., 1999; see Fig. 2C).

Such isoforms have been shown to be fairly abundant for *TCF1* and *LEF1* (Castrop et al., 1995; van de Wetering et al., 1996; Hovanes et al., 2001). Rescue experiments with a prespliced isoforms of *TCF1* will be needed to resolve whether this form is responsible for the inhibitory effect on the pathway.

TCF isoforms that are competent for β-catenin binding can also influence Wnt target genes differently. For example, while one splice form of TCF4 (TCF4C) cannot rescue *TCF3* morphants, the TCF4A isoform can, suggesting that TCF4A, but not TCF4C, can act as a transcriptional repressor (Liu et al., 2005). Other isoforms of TCF4 containing different C-terminal tails will be discussed in Section 4, as they have the ability to activate gene expression in a context-dependent manner (Atcha et al., 2003; Hecht and Stemmler, 2003).

2.4. Identification of Wnt targets through in silico detection of TCF-binding sites

Several lines of evidence argue that binding sites for TCF are necessary and sufficient for Wnt-dependent transcriptional activation. The identification of functionally important sites based on *in vitro* DNA-binding assays with the TCF-HMG domains has become the standard for characterizing WREs (Brannon et al., 1997; He et al., 1998; Tetsu and McCormick, 1999; Yamaguchi et al., 1999; Chamorro et al., 2005). Simple reporter genes containing multiple copies of a predicted high-affinity site (CCTTTGAT CTT), such as TOPFLASH (three sites; Korinek et al., 1997) or SUPER-TOPFLASH (12 sites; Lum et al., 2003; DasGupta et al., 2005), are greatly activated by Wnt signaling in many cell types (Kolligs et al., 1999; Hecht et al., 2000; Civenni et al., 2003). Multimerized TCF sites can also drive expression of reporter genes in patterns consistent with responding to Wnt signaling in transgenic fish and mice (DasGupta and Fuchs, 1999; Dorsky et al., 2002; Maretto et al., 2003; Nakaya et al., 2005).

Based on these studies, Taipale and coworkers (Hallikas et al., 2006) have developed an *in silico* method for identifying enhancers in the mammalian genomes, including WREs. To identify WREs, these researchers determined the relative affinities of oligonucleotides with all possible single-base substitutions to the consensus site to a TCF fusion protein. This information was used to create a computer algorithm [called enhancer element locator (EEL)] that aligns TCF sites on two orthologous DNA sequence from different species. EEL then scores a particular stretch of DNA based on the distance between conserved TCF sites (greater distance lowers the score) and the affinity of the sites (higher affinity increases the score). EEL has no bias based on the location of potential enhancers in relation to the transcription start site, recognizing that important enhancers can be found far upstream or downstream of proximal promoters. It also has the ability to search for multiple transcription factors in a potential enhancer, which could facilitate the search for elements regulated in a combinatorial manner. To validate EEL, several potential WREs were tested for activity in mouse embryos using reporter gene analysis. Several enhancers were expressed in the tail bud and/or AER of limb buds, consistent with possible regulation by Wnt/β-catenin signaling (Hallikas et al., 2006). Mutational analysis of the identified TCF sites will be needed to determine whether these patterns truly reflect WREs.

EEL offers an exciting and relatively quick way to search for new Wnt/β-catenin targets, and further examination of WREs should provide additional information to increase its predictive power. The detailed examination of two WREs in *Drosophila* illustrates this point. The *eve* and *slp1* WREs contain two general classes of TCF sites, based on DNase foot printing. Higher affinity sites shared a sequence of STTTGW (S = G or C), but the lower affinity sites were further removed from the consensus (e.g., ACTGT-GAA). Mutation of the higher affinity sites only caused a reduction in Wg responsiveness of these enhancers and additional mutation of the lower affinity sites was required to completely block Wg activation (Lee and Frasch, 2000; Knirr and Frasch, 2001). These studies indicate that TCF can bind to a broader array of sites then is typically appreciated. Whether such lower affinity TCF sites are important in other WREs and whether this information can help algorithms like EEL identify *bona fide* WREs remain to be seen. Integrating such computational approaches with experimental approaches combining chromatin immunoprecipitation (ChIP) and genomic microarrays (Ren et al., 2000; Wei et al., 2006) will also no doubt refine these search protocols.

2.5. *TCF-independent Wnt/β-catenin signaling*

The focus on TCFs as mediators of Wnt/β-catenin target gene regulation is justified, based on the extensive data outlined above. However, there is also growing evidence that other DNA-binding proteins mediate recruitment of β-catenin to Wnt target elements. In most instances, the case for these interactions is not as complete as for TCFs. This may simply indicate the need for further investigation but may also reflect the fact that these TCF-independent mechanisms may be utilized in a more cell- and gene-specific manner than TCFs.

In mouse heart development, the homeodomain protein Pitx2 utilizes β-catenin to specify cardiac cell fates. The development of cardiac neural crest cells requires Disheveled 2 (Dvl2), a component of the Wnt pathway, and β-catenin (Hamblet et al., 2002; Kioussi et al., 2002). The cardiac outflow tract abnormalities observed in these mutants resemble knockouts of *Pitx2* (Kioussi et al., 2002). It was found that *Pitx2* expression was activated by LEF1 and β-catenin, which could explain the similarity of the phenotypes. However, it was also shown that Pitx2 controls cell proliferation through the activation of cell cycle genes such as *cyclin D2*. β-Catenin stabilization (though LiCl treatment) caused activation of *cyclin D2* expression in a *Pitx2*-dependent manner. Pitx2 and β-catenin were found to associate through coimmunoprecipitation (even at endogenous levels), and ChIP analysis indicates that both Pitx2 and β-catenin bind to the *cyclin D2* promoter. Finally, a simple reporter gene containing multimerized Pitx2 sites could be efficiently activated by lithium treatment (Kioussi et al., 2002). Thus, it appears that like TCFs, Pitx2 can recruit β-catenin to target genes to activate transcription.

In *C. elegans*, it has recently been shown that the forkhead box O (FOXO) transcription factor orthologue DAF-16 and the worm β-catenin BAR-1 cooperate to regulate dauer formation and longevity. Like the Pitx2 story, the model is supported by genetics, physical association between DAF-16 and BAR-1 (as well as vertebrate FOXO and β-catenin), and FOXO-reporter genes coordinately regulated by FOXO and β-catenin (Essers et al., 2005).

Other examples of DNA-binding proteins that are thought to recruit β-catenin to target genes included SOX17 (Zorn et al., 1999; Sinner et al., 2004) and two members of the nuclear receptor family, androgen receptor (Song et al., 2003; Yang et al., 2006) and estrogen receptor (Kouzmenko et al., 2004). These studies demonstrate physical interactions between these transcription factors and β-catenin, as well as synergistic activation of simple reporter genes. Whether these interactions are important under physiological conditions requires further study as do many other functional interactions between Wnt/β-catenin and nuclear receptor signaling (Mulholland et al., 2005).

The analysis of some Wnt target genes has also linked TCF/β-catenin to transforming growth factor β (TGFβ) signaling. *Cis*-regulatory sequences of *twin*, a marker for the Spemann organizer and the mouse *gastrin* gene can be activated synergistically by β-catenin and Smads, the cytosolic mediators of TGFβ signaling (Nishita et al., 2000; Labbe et al., 2000; Lei et al., 2004). Both regulatory elements contain binding sites for TCF and Smads which contribute to the activation. LEF1 or TCF4 was found to bind to Smad 3 or Smad 4 (Labbe et al., 2000; Nishita et al., 2000; Lei et al., 2004), and this interaction was localized to the HMG domain in the case of LEF1 (Labbe et al., 2000). Simple reporter constructs containing either TCF- or Smad-binding sites were found to be activated by TCF/Smad combinations (Nishita et al., 2000; Lei et al., 2004). These results argue that Wnt/β-catenin and TGFβ signaling can be integrated by direct interactions on enhancers.

3. The "Off state": Repression of Wnt targets in the absence of signaling

As can be imagined for a signaling cascade that plays so many critical roles in development, the Wnt/β-catenin pathway is highly regulated at every level. In this section, we will review three general mechanisms that ensure that Wnt transcriptional targets are not expressed inappropriately. First, as described in the previous section, it is clear that TCFs can repress Wnt targets, and this involves TCF-binding transcriptional corepressors. Second, several factors that inhibit TCF/β-catenin association have been reported, and their role as TCF/β-catenin "buffers" will be explored. Finally, a variety of other TCF and β-catenin antagonists that are important negative modulators of TCF/β-catenin transcriptional activation will be discussed (see Fig. 3 for a summary).

3.1. Transcriptional corepressors and chromatin remodelers

The ability of TCF to repress Wnt targets in the absence of signaling is thought to require transcriptional corepressors. These proteins have no intrinsic ability to recognize specific DNA sequences but rather are recruited to regulatory elements through their association with DNA-binding proteins (Fisher and Caudy, 1998; Courey and Jia, 2001; Chinnadurai, 2002). In the case of TCFs, the best-characterized corepressors are a group of proteins referred to as transducin-like enhancer of split (TLE) or Groucho-related proteins (Grgs), defined by several motifs, including a C-terminal WD40 domain (Fisher and Caudy, 1998). In flies, mutations in the *groucho* (*gro*) gene suppress a *wg* cuticle phenotype, similar to that observed for *TCF/pan* mutants (Cavallo et al., 1998). TCF and Gro were found to physically

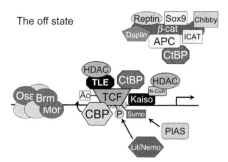

Fig. 3. Summary of nuclear factors repressing TCF/β-catenin transcriptional activation. This figure contains several general classes of inhibitory factors. Some factors act through binding directly to β-catenin or to TCF. Others modify TCF, inhibiting its function. Others (Brm/Osa) or Kaiso are thought to bind to sequences near TCF on the WRE. See Section 3 of the text for more details.

interact, consistent with Gro acting as a TCF corepressor. Vertebrate TCFs have also been shown to bind directly to several Grg/TLEs (Roose et al., 1998; Brantjes et al., 2001), and mutation of the region of TCF3 that binds to Grg/TLEs compromises its repressive activity (Liu et al., 2005; Tsuji and Hashimoto, 2005). Expression of TLEs can dramatically suppress TCF/β-catenin activation of reporter genes in cell culture and specification of a secondary axis in *Xenopus* embryos (Roose et al., 1998; Brantjes et al., 2001). Consistent with this, an inhibitory peptide that blocks Grg/TLE function causes dorsalization of ventral CNS in *Xenopus* embryos (Tsuji and Hashimoto, 2005). Similarly, a naturally occurring family member (Grg-5) lacking the WD40 domain potentiates β-catenin activation in cell culture and *Xenopus* assays (Roose et al., 1998).

The exact mechanism of how Grg/TLE repress Wnt targets is not known, but several lines of evidence point to the involvement of histone deaceytlases (HDACs). First, Gro is known to physically associate with Rpd3, a fly HDAC, and this interaction has been shown to be responsible for Gro-dependent repression of other transcription factors (Chen et al., 1999; Chen and Courey, 2000). Similarly, mammalian Grg-4 can interact with HDAC1 (Brantjes et al., 2001). This is consistent with observations that HDAC1 and Grg-1/TLE1 are physically associated with WREs in the absence of Wnt signaling (Kioussi et al., 2002; Sierra et al., 2006). Addition of the HDAC inhibitor TSA has been shown to derepress Wnt-reporter genes (Billin et al., 2000). As summarized in Fig. 3, the prevailing view is that Grg/TLE recruits HDACs to TCFs under repressive conditions, though this has never been directly demonstrated.

A second corepressor, C-terminal-binding protein (CtBP) has also been suggested to contribute to TCF-mediated repression of Wnt targets, though not without controversy. CtBP was first found to bind to the C-terminal portion of *Xenopus* TCF3, and this portion of the protein was required for repression of primary axis formation (Brannon et al., 1999). A CtBP fusion protein containing a transcriptional activation domain caused a dramatic activation of *saimois* expression in ventral blastomeres, consistent with CtBP being recruited to the *saimois* promoter by TCF3 (Brannon et al., 1999). An interaction between human TCF4 and CtBP1 was also found when both were expressed in culture cells, and overexpression of CtBP could antagonize β-catenin activation of reporters (Valenta et al., 2003). Consistent with this, disruption of both mouse *CtBP* genes in embryonic fibroblasts caused an elevation in β-catenin-reporter gene assays, but in this study, no interaction between TCF4 and CtBP could be detected (Hamada and Bienz, 2004). An alternative model for CtBP repression of TCF/β-catenin activity will be discussed below.

There is also evidence for an ATP-dependent-chromatin remodeling complex playing an important role in silencing Wg targets in *Drosophila*. The *osa* gene was originally found to play an antagonistic role to Wg signaling in several fly tissues (Treisman et al., 1997). Osa is a subunit of one of two multisubunit complexes that contain the chromatin-dependent ATPase Brahma (Collins et al., 1999; Mohrmann et al., 2004). This complex is highly conserved from yeast to humans and has been shown to modify chromatin structure *in vitro* and to be required for gene regulation in several systems (Becker and Horz, 2002; Martens and Winston, 2003). Overexpression of *osa* blocks Wg targets, and a dominant negative version of *brahma* (lacking ATPase activity) caused derepression of Wg readouts (Collins and Treisman, 2000). However, it is not clear whether a Brahma complex is directly present at WREs or if the repression is indirect.

3.2. Inhibitors of TCF/β-catenin interaction

The Wnt/β-catenin pathway is negatively regulated at several steps, including ligand–receptor interaction (Kawano and Kypta, 2003), stability of β-catenin (Polakis, 2000; Ding and Dale, 2002), as well as nuclear import of β-catenin (Tolwinski and Wieschaus, 2004). This section will focus on negative regulators which influence the association of β-catenin with TCFs. These factors can be viewed as β-catenin/TCF buffers that prevent a very low level of β-catenin from activating Wnt targets and perhaps help set the threshold by which targets are activated in a precise manner.

Several factors are thought to buffer TCF/β-catenin through binding directly to β-catenin and competitively inhibiting its association with TCF. β-Catenin contains a relatively unstructured N- and C-termini, but the majority of the protein consists of 12 Armadillo (Arm) repeats. Each repeat contains three α-helices organized in a globular manner. These domains then organize into a superhelical structure (Huber et al., 1997). Consistent with *in vitro* binding studies (Behrens et al., 1996; van de Wetering et al., 1997), cocrystals of β-catenin and various TCFs reveal extensive interactions with the superhelical groove between Arm repeats 5–10 (Graham et al., 2000, 2001; Poy et al., 2001). As discussed below, biochemical and genetic studies indicate that several factors can disrupt the formation of this TCF/β-catenin complex.

3.2.1. ICAT and Chibby bind to the C-terminal of β-catenin to inhibit its association with TCF

ICAT is a small (81 residue) protein that can bind to repeats 10–12 of β-catenin (Tago et al., 2000). Structural analysis revealed that the N-terminal helical 61 amino acids contacted repeats 11 and 12 and the C-terminal portion of ICAT extends along the groove formed by repeats 5–10, explaining how ICAT can compete with TCF for binding to β-catenin (Daniels and Weis, 2002). ICAT is a potent inhibitor of Wnt/β-catenin signaling when overexpressed, associates with β-catenin under physiological conditions (Tago et al., 2000), and its expression has been correlated with blocking Wnt signaling in the intestine (Gottardi and Gumbiner, 2004b). However, the best evidence for a physiological role for ICAT in preventing β-catenin/TCF interaction comes from a dominant negative construct analyzed in *Xenopus* embryos. This construct deletes residues 42–61 of ICAT and can still bind to β-catenin but cannot inhibit TCF/β-catenin interaction. When injected into ventral blastomeres of early embryos, this construct caused dorsalization and axis duplication with high frequency (Tago et al., 2000). This result is consistent with the truncated ICAT competing with endogenous ICAT for binding to β-catenin, though it does not rule out competition with other β-catenin-binding proteins, such as Chibby (Cby), described below.

Like ICAT, Cby was also found to interact with the C-terminal of β-catenin, with repeats 10–11 and the nonrepeat C-terminal being necessary for binding (Takemaru et al., 2003). Cby is an evolutionarily conserved protein found in the nucleus, and overexpression can inhibit β-catenin/TCF association and block activation of β-catenin-reporter genes in mammalian cell culture (Takemaru et al., 2003). Importantly, morpholino inhibition of Cby expression potentiated the ability of transfected β-catenin to activate a reporter gene in cultured cells. Likewise, RNAi-mediated knockdown of *cby* in *Drosophila* embryos results in phenotypes consistent with ubiquitous activation of Wg signaling, though this result is complicated by the fact that *wg* expression is upregulated in these embryos (Takemaru et al., 2003).

Are these TCF/β-catenin buffers important for titrating Wnt-induced-nuclear β-catenin activity, or are they also needed to keep Wnt targets off in the absence of signal? In other words, when Wnt signaling is off, is there an appreciable amount of β-catenin that reaches TCF on target genes? Evidence for the later scenario comes from experiments with Cby in the visceral mesoderm of fly embryos, using *Ubx* expression as readout for Wnt signaling (Riese, 1997). In *wg* mutants, *Ubx* expression is lost. However, *cby* depletion (by RNAi) in a *wg* mutant background displayed near normal levels of *Ubx* expression. A similar recovery of *Ubx* expression was not observed when *cby* was depleted in an *arm* mutant background (Takemaru et al., 2003). These data strongly suggest that in the absence of Wg signaling, an appreciable amount of Arm can still get into the nucleus but is prevented from associating with TCF by buffering proteins such as Cby.

3.2.2. *An additional role for APC in blocking TCF/β-catenin binding*

The product of the *adenomatous polyposis coli* (*APC*) gene is well known for its ability to bind to β-catenin and promote its degradation (Polakis, 2000). Structural analysis of β-catenin/APC binding reveals that APC binds to a similar region of β-catenin as does TCF (Spink et al., 2000; Ha et al., 2004; Xing et al., 2004). APC has been shown to shuttle between the cytoplasm and the nucleus and has been proposed to mediate efflux of β-catenin from the nucleus (Henderson, 2000; Neufeld et al., 2000a,b; Rosin-Arbesfeld et al., 2000, 2003). APC has been shown to bind to CtBP, and the APC/CtBP complex was demonstrated to compete with TCF for binding to *β*-catenin (Hamada and Bienz, 2004). As will be described in the last section of this chapter, evidence indicates that APC and CtBP act on chromatin to inhibit TCF/*β*-catenin transcriptional activation complexes (Sierra et al., 2006).

3.2.3. Members of the SOX family and Duplin also compete with TCF for binding to β-catenin

Several members of the SOX family of HMG domain proteins can bind to *β*-catenin and have been shown to repress TCF-reporter gene activity, presumably by competing with TCF for *β*-catenin (Zorn et al., 1999; Takash et al., 2001; Akiyama et al., 2004; Kan et al., 2004). In the best-studied case, *SOX9* and *β-catenin* have reciprocal phenotypes in chondrocyte differentiation (Akiyama et al., 2004), providing physiological evidence for a negative role for SOX9 in Wnt/*β*-catenin signaling. Conversely, SOX17, which can inhibit *β*-catenin activity when overexpressed (Zorn et al., 1999), has also been shown to act cooperatively with *β*-catenin in *Xenopus* development, suggesting that it acts as a *β*-catenin-dependent transcriptional factor (Sinner et al., 2004). Clearly, loss-of-function data will need to fully understand the relationship between individual SOX family members and *β*-catenin.

Overexpression of a nuclear protein called Duplin can inhibit *β*-catenin activation of reporter genes and Wnt-induced secondary axis formation in *Xenopus* embryos, consistent with its ability to compete with TCF for *β*-catenin binding (Sakamoto et al., 2000; Kobayashi et al., 2002). Nuclear localization of Duplin and its ability to bind to *β*-catenin have been correlated with its ability to block Wnt/*β*-catenin signaling (Kobayashi et al., 2002), but it is not clear yet whether this factor plays a physiological role in the pathway.

3.2.4. Groucho and CBP inhibit TCF/β-catenin association by binding or modification of TCF

In addition to the several *β*-catenin-binding antagonists just discussed, there are two proteins that prevent TCF/*β*-catenin association by interacting with TCF. Grg/TLE proteins may act as transcriptional corepressors, but they have also been shown to compete with *β*-catenin for binding to TCF (Daniels and Weis, 2005), suggesting a distinct mode of blocking the pathway. The histone acetyltransferase Nejire, the fly orthologue of Creb-binding protein (CBP), can bind to TCF and acetylate a conserved lysine in its N-terminal (Waltzer and Bienz, 1998). This acetylated TCF is less efficient at binding Arm, and consistent with this, loss of *CBP* resulted in inappropriate activation of Wg readouts, including mesodermal *Ubx* (Waltzer and Bienz, 1998).

3.3. Other antagonists of TCF and β-catenin

There are also several other nuclear factors that inhibit TCF/β-catenin transcriptional activity through mechanisms independent of TCF/β-catenin binding. For example, the BTB/POZ zinc-finger protein Kaiso has been shown to be required for repression of several Wnt targets in *Xenopus* embryogenesis (Park et al., 2005). A Kaiso-binding site was identified adjacent to two TCF sites in a *saimois* WRE, and mutation of this site resulted in significant derepression. Although not functionally tested, Kaiso sites were also found in several other WREs. Kaiso can physically associate with the N-CoR corepressor. Consistent with N-CoR acting through HDACs, morpholino knockdown of *Kaiso* caused a dramatic increase in acetylated histones at the *saimois* WRE. Finally, Kaiso and TCF3 can physically associate. Together with the close proximity of binding sites on the DNA, this suggests that Kaiso and TCF can act together to repress Wnt targets (Fig. 3). Interestingly, p120-catenin can bind to Kaiso and relieve repression (Park et al., 2005).

Another example of a Wnt signaling antagonist that acts to inhibit the TCF/β-catenin complex is the DNA-stimulated ATPase known as Reptin (sometimes called TIP49b). Reptin has been shown to inhibit TCF/β-catenin activity in cultured cells, flies, and zebrafish (Bauer et al., 2000; Rottbauer et al., 2002). Reptin can bind to β-catenin (Bauer et al., 1998, 2000), and elevation of its ATPase activity increases its inhibitory effect on β-catenin activity (Rottbauer et al., 2002). While the mechanism of Reptin action is not clear, it could involve counteracting the positive influence of a related ATPase, Pontin/TIP49a, which also binds to β-catenin as well as Reptin (Bauer et al., 1998). Reptin and Pontin act antagonistically on Wnt/β-catenin in several contexts (Bauer et al., 2000; Rottbauer et al., 2002). The histidine triad protein Hint1 may also be involved in this regulation (Weiske and Huber, 2005). In addition, both Reptin and Pontin are known to associate with several chromatin remodeling complexes (Ikura et al., 2000; Cai et al., 2003; Jin et al., 2005). It is not clear whether these complexes regulate Wnt target genes, though some have been shown to associate with β-catenin (Sierra et al., 2006).

Several other factors block TCF function through modification of TCFs. The best-studied example originally came from the genetic analysis of *C. elegans* EMS cell fate, which is known to be regulated by a Wnt-dependent pathway leading to the inhibition of POP-1 (Rocheleau et al., 1997; Thorpe et al., 1997). This Wnt pathway required two kinases, MOM-4, similar to TGFβ-activated kinase Tak-1 and LIT-1, similar to Nemo-like kinase (NLK) (Ishitani et al., 1999; Meneghini et al., 1999; Rocheleau et al., 1999; Shin et al., 1999). Biochemical studies with worm and mammalian homologues indicated that Wnt signaling activates MOM-4/TAK-1, which then activates LIT1/NLK, which then phosphorylates POP-1/TCF. This phosphorylated TCF can still bind β-catenin but cannot bind DNA (Ishitani et al., 1999, 2003;

Rocheleau et al., 1999; Shin et al., 1999). The phosphorylated POP-1 is bound by PAR-5, a 14–3–3 protein, and exported from the nucleus (Lo et al., 2004), allowing derepression of POP-1 target genes as previously discussed. In the case of the E daughter of the EMS blastomere, this nuclear export of POP-1 permits endodermal cell fate (Rocheleau et al., 1997; Thorpe et al., 1997).

In addition to worms and mammalian cell culture, overexpression of NLK can also inhibit Wnt/β-catenin signaling in flies (Zeng and Verheyen, 2004) and *Xenopus* (Ishitani et al., 1999). In zebrafish embryos, NLK was shown to act with Wnt to inhibit TCF3 isoforms (Thorpe and Moon, 2004), similar to the role LIT-1 plays in the worm embryo. Thus, this regulatory pathway of TCF function appears to be widely used throughout the animal kingdom.

Other factors have been reported to inhibit TCF function, but the *in vivo* relevance is less clear than LIT-1/NLK. A member of the protein inhibitor of activated stat (PIAS) family, PIASy, has been found to bind to inhibit LEF1-dependent transcriptional activation (Sachdev et al., 2001). Furthermore, PIASy was found to be able to promote the sumoylation of LEF1 *in vivo* and *in vitro*, suggesting that it acts as a SUMO E3 ligase. The modified LEF1 is sequestered into nuclear bodies (Sachdev et al., 2001). In contrast to its inhibitory effect on LEF1, PIASy has also been shown to sumoylate TCF4 and elevate its transcriptional activation activity (Yamamoto et al., 2003). In both cases, the need for inhibition of PIASy, through mutation or RNAi, will be necessary to further evaluate its importance in Wnt/β-catenin transcription. Likewise, other factors that bind to TCFs and inhibit their activity when overexpressed, like hydrogen peroxide-induced-clone-5 (HIC-5; Ghogomu et al., 2006), need validation via loss-of-function studies.

In contrast to these biochemical studies, two genes have been identified in *C. elegans* that genetically inhibit Wnt/β-catenin signaling (Zhang and Emmons, 2000; Moghal and Sternberg, 2003; Yoda et al., 2005). These genes (*LET-19* and *DPY-22*) encode homologues of two subunits of the mediator complex, which connects transcriptional regulators with RNA polymerase (Myers and Kornberg, 2000). While these results are interesting, it is not clear whether LET-19/DPY-22 interact directly with β-catenin or other Wnt components.

4. The "On State" in the nucleus; Arm/β-catenin triggers the switch

Wnts transduce a signal across the plasma membrane by binding to a receptor complex containing Frizzled and low-density-lipoprotein-like receptor (LRP) (He et al., 2004; Cadigan and Liu, 2006). The activated receptor complex inhibits a β-catenin-degradation complex containing APC, Axin, glycogen synthase kinase 3 (GSK3), and casein kinase I (Ding and Dale, 2002; Polakis, 2002). Blocking this complex causes an accumulation of β-catenin, which then translocates to the nucleus, binds to TCFs, and activates Wnt targets. This section will summarize our current understanding of how β-catenin crosses the nuclear pore complex (NPC) and how it recruits transcriptional coactivators to WREs, turning on target gene expression.

4.1. Nuclear import of β-catenin/Arm

In the absence of Wnt stimulation, β-catenin/Arm is primarily associated with the plasma membrane, where it is complexed with E-cadherin (Kemler, 1993), but a small cytoplasmic pool also exists (Peifer et al., 1994; Papkoff et al., 1996). Wnt-induced accumulation of nuclear β-catenin and Arm was first observed in *Xenopus* (Yost et al., 1996) and *Drosophila* embryos (Orsulic and Peifer, 1996). Nuclear β-catenin has since been associated with activation of Wnt/β-catenin signaling in a broad range of organisms, from sea urchins (Logan et al., 1999) to mammals (Park et al., 2001; van de Wetering et al., 2002). Despite the recognized importance of β-catenin nuclear import in the pathway, the mechanism by which this occurs is still largely unclear.

β-Catenin does not contain a recognizable nuclear localization signal (NLS), suggesting that it travels independently of well-characterized systems like importin-α/importin-β. With a molecular weight of approximately 90 kDa, β-catenin probably cannot enter the nucleus via diffusion, though it has a rod like structure (Huber et al., 1997), suggesting that it could diffuse through the NPC as a thread through the eye of a needle. However, β-catenin nuclear translocation is cold sensitive (Yokoya et al., 1999; Suh and Gumbiner, 2003) a condition that does not inhibit passive diffusion. When fused to β-galactosidase ($\sim$120 kDa), β-catenin can still enter the nucleus (Suh and Gumbiner, 2003), which is also inconsistent with passive diffusion as a mechanism.

β-Catenin can traverse the NPC in digitonin-permeabilized cells, even in the absence of cytosol and ATP (Fagotto et al., 1998; Yokoya et al., 1999; Suh and Gumbiner, 2003). β-Catenin shares homology with importin-α and importin-β (both are Arm repeat proteins), and there is evidence β-catenin may enter the nucleus via a similar mechanism as these proteins (Fagotto et al., 1998). However, unlike importin-β, β-catenin cannot bind to FG-rich nucleoporins and its import into the nucleus is not blocked by wheat germ agglutinin (Suh and Gumbiner, 2003). While a simple model for β-catenin nuclear import is not apparent, it appears β-catenin probably does not utilize the classic NLS/importin-α/β system.

4.1.1. Cytoplasmic and nuclear anchors regulate Arm/β-catenin subcellular localization

However, β-catenin/Arm traverses the nuclear envelope, it is clear that its distribution is heavily influenced by the presence of both cytoplasmic and nuclear anchors. There is evidence that Axin and cadherin function as cytoplasmic anchors, while TCF/Pan, Legless (Lgs), and Pygopus (Pygo) contribute to nuclear retention.

Drosophila axin mutants have constitutive Wg signaling (Hamada et al., 1999). These mutants also exhibit very high levels of nuclear Arm (Tolwinski and Wieschaus, 2001). Mutants in the fly *GSK3* (*zeste white 3/shaggy*) have a high level of Arm but did not display the dramatic increase in nuclear Arm. This argues that the nuclear Arm observed in *axin* mutants was not an indirect consequence of Arm stabilization. Consistent with a role as a cytoplasmic anchor, overexpression of Axin can trap β-catenin/Arm in the cytoplasm (Tolwinski and Wieschaus, 2001; Krieghoff et al., 2006).

Although the APC protein is largely cytoplasmic, it contains several nuclear export sequences (NESs) and has been shown to shuttle between the cytoplasm and the nucleus (Henderson, 2000; Neufeld et al., 2000a; Rosin-Arbesfeld et al., 2000). This suggested a model whereby APC could shuttle between the nucleus (to bind β-catenin) and cytoplasm (where β-catenin could be degraded), to decrease nuclear β-catenin levels. Overexpression of APC causes β-catenin degradation (Munemitsu et al., 1995), but a truncated form of APC that can bind β-catenin but not promote its degradation was found to reduce the high nuclear β-catenin found in colorectal cell lines lacking wild-type APC (Henderson, 2000; Neufeld et al., 2000a; Rosin-Arbesfeld et al., 2000). Conversely, an APC protein engineered to contain a strong NLS can trap β-catenin in the nucleus (Rosin-Arbesfeld et al., 2000).

The cell adhesion molecule E-cadherin has also been suggested to act as a cytoplasmic anchor for β-catenin. Overexpression of E-cadherin can block nuclear import of β-catenin and Wnt/β-catenin signaling (Orsulic et al., 1999; Gottardi et al., 2001). While this interaction could be nonphysiological, it has also been found that Wnt stimulation of cells modifies β-catenin so that it can bind more efficiently with TCF than to E-cadherin (Gottardi and Gumbiner, 2004a). Monomeric β-catenin bound preferentially with TCF, while β-catenin complexed with α-catenin preferred E-cadherin (Gottardi and Gumbiner, 2004a). Consistent with this, tyrosine phosphorylation of β-catenin at residue 142 lowered affinity for α-catenin and promoted the nuclear accumulation of β-catenin through binding to BCL9–2 (Brembeck et al., 2004).

In cultured fibroblasts, coexpression of LEF1 with β-catenin also causes nuclear accumulation of the normally cytoplasmic β-catenin (Behrens et al., 1996; Henderson et al., 2002). In the fly embryonic epidermis, reduced nuclear accumulation of Arm was observed when it was coexpressed with a truncated form of TCF/Pan lacking the Arm-binding domain (Tolwinski and Wieschaus, 2001). However, it has also been reported that a β-catenin mutant that cannot bind to LEF1 can still accumulate in the nucleus, presumably due to lack of interaction with APC (Prieve and Waterman, 1999). Further studies with loss-of-function alleles should clarify the importance of TCFs as a nuclear anchor for β-catenin/Arm.

In contrast to TCF, a stronger case for a nuclear anchor can be made for Lgs and Pygo. As will be described in more detail below, these genes are required for Wg signaling (Belenkaya et al., 2002; Kramps et al., 2002; Parker et al., 2002; Thompson et al., 2002). Pygo is a nuclear protein (Parker et al., 2002; Thompson, 2004; Townsley et al., 2004a; Parker and Cadigan, unpublished) and acts as a nuclear anchor for Lgs, as Lgs is no longer exclusively nuclear in *pygo* mutants (Townsley et al., 2004a). Lgs can bind to Arm (Kramps et al., 2002; Hoffmans and Basler, 2004), and consistent with this, Arm nuclear levels are decreased in *pygo* mutants where the pathway is on at high levels (Parker et al., 2002; Townsley et al., 2004a). Strikingly, the segment polarity phenotype of *lgs* and *pygo* mutants can be partially suppressed by expression of an Arm containing an NLS (Townsley et al., 2004a). In mammalian cells, inhibition of *BCL9–2*, a homologue of *lgs*, causes a shift of β-catenin from the nucleus to the plasma membrane, which is associated with a mesenchymal to epithelial transition in cell morphology (Brembeck et al., 2004).

While the studies summarized above as consistent with the cytoplasmic/nuclear retention hypothesis, the results are complicated by the fact the putative cytoplasmic anchor Axin is known to shuttle between the cytosol and nucleus (Cong and Varmus, 2004; Wiechens et al., 2004). Also, APC has been suggested to promote β-catenin nuclear efflux (Henderson, 2000; Neufeld et al., 2000a; Rosin-Arbesfeld et al., 2000). Thus, the increased nuclear β-catenin observed in *axin* or *APC* mutants could be explained by a reduction in β-catenin/Arm nuclear export. Conversely, the nuclear anchor Lgs has also been shown to undergo cytoplasmic/nuclear shuttling (Townsley et al., 2004a). Likewise some TCFs have been found in the cytoplasm (Lee et al., 2001) suggesting that they may be undergoing cytosolic/nuclear shuttling. Studies that only examine the steady state levels of β-catenin/Arm cannot distinguish between the nuclear/cytoplasmic retention or regulated import/export models.

A more detailed analysis using fluorescence recovery after photobleaching (FRAP) to kinetically measure β-catenin movement between the two compartments lends support for nuclear and cytosolic anchors. In these experiments, β-catenin was found to shuttle rapidly ($t_{1/2}$ of 4 min) between the cytoplasm and nucleus (Krieghoff et al., 2006). Overexpression of Axin, APC, TCF, or Lgs slowed the rate of shuttling and enriched β-catenin in the cytoplasm (Axin, APC) or nucleus (TCF, Lgs) (Krieghoff et al., 2006). However, none of these proteins increased the rate of β-catenin shuttling from either compartment to the other, and these proteins were found to shuttle more slowly than β-catenin (Krieghoff et al., 2006). These data suggest that none of these overexpressed proteins can increase the rate at which β-catenin moves between the two compartments, instead acting as β-catenin anchors in their appropriate subcellular compartment.

In the FRAP experiments cited above, no effect on β-catenin shuttling was found in response to Wnt stimulation (Krieghoff et al., 2006). However, several other reports raise the possibility that β-catenin/Arm nuclear accumulation can be increased by Wnt signaling through inhibition of the cytoplasmic anchor Axin. Wnt-dependent degradation of Axin has been reported in both fly and mammalian systems (Willert et al., 1999; Yamamoto et al., 1999; Mao et al., 2001; Tolwinski et al., 2003), and Wnt signaling caused the dephosphorylation of Axin, lowering its affinity for β-catenin (Willert et al., 1999). Further discussion of this potentially important mechanism can be found elsewhere (Tolwinski and Wieschaus, 2004).

4.2. β-Catenin turns the switch

Once in the nucleus at concentrations high enough to overcome inhibitors such as Cby and ICAT, β-catenin can associate with TCF. This association is thought to convert TCF from a repressor to an activator. Consistent with this, activation of the Wnt/β-catenin pathway results in a recruitment of β-catenin to WREs, and a reciprocal disappearance of HDACs (Kioussi et al., 2002; Sierra et al., 2006). How does β-catenin displace HDAC from the WRE? It is likely that HDACs are recruited to TCF through association with Grg/TLE (Chen and Courey, 2000; Brantjes et al., 2001). It has been shown that β-catenin and Grg/TLE bind competitively to TCF *in vitro* (Daniels and Weis, 2005). Therefore, high nuclear levels of β-catenin could outcompete Grg/TLE for binding to TCF, and thus also displace HDAC, which would allow histones to become acetylated and activate transcription (see below). Consistent with mutual exclusion of β-catenin and Grg/TLE association with TCF in a more physiological context, recruitment of β-catenin to the *c-Myc* WRE is correlated with a removal of TLE1 (Sierra et al., 2006). In addition to the physical displacement model, β-catenin has also been suggested to associate with HDAC and inhibit its enzymatic activity (Billin et al., 2000). This mechanism could also contribute to the inactivation of HDAC on WREs in response to Wnt/β-catenin signaling.

4.2.1. A different kind of switch in C. elegans

Although the β-catenin displacement of corepressor model has been a popular concept for several years (even before experimental data supported it!), there may be other ways to achieve the switch from TCF-associated repression to activation. The LIT-1 mediated export of POP-1 offers one such example, at least in the EMS blastomere. A variation on this mechanism has been suggested, involving a diverged worm member of the β-catenin family (Kidd et al., 2005).

Unlike fly Arm or vertebrate β-catenin, which are essential for both cadherin-mediated cell adhesion and Wnt signaling (Orsulic and Peifer, 1996; Nelson and Nusse, 2004; Brembeck et al., 2006), there are three recognizable worm β-catenins which have more specific functions. *HMP-2* is required for cell adhesion and can bind to cadherin (Costa et al., 1998;

Korswagen et al., 2000). WRM-1 binds to the LIT-1 kinase and promotes phosphorylation and nuclear export of POP-1 (Ishitani et al., 1999, 2003; Rocheleau et al., 1999; Lo et al., 2004). *BAR-1* is required for Wnt/β-catenin signaling (Eisenmann et al., 1998; Natarajan et al., 2001; Gleason et al., 2002) and can bind to POP-1 and promote activation of a TCF-reporter gene (Korswagen et al., 2000; Natarajan et al., 2001). While these three worm β-catenins appear to have specific functions, it should be noted the *HMP-2* and *WRM-1* can rescue *BAR-1* mutants when expressed at high levels (Natarajan et al., 2001).

In addition to these β-catenin homologues, worms also have an additional gene, *SYS-1*, that encodes a protein containing three Arm repeats. Despite this limited similarity, SYS-1 can bind to POP-1 and stimulate its transcriptional activity and a *SYS-1* transgene can rescue *BAR-1* mutants (Kidd et al., 2005). Thus, SYS-1 is a functional homologue of BAR-1/β-catenin and may act in Wnt/β-catenin signaling.

SYS-1 is required for an asymmetric cell division of the somatic gonadal precursor (SGP), generating a proximal and distal daughter, the later of which becomes the DTC (Miskowski et al., 2001). Mutations in SYS-1, WRM-1, LIT-1, or POP-1 result in both daughters having a proximal cell fate (Siegfried and Kimble, 2002; Siegfried et al., 2004). However, unlike *WRM-1/LIT-1* mutants, loss of *SYS-1* does not effect POP-1 distribution, that is, there is still low nuclear POP-1 levels in the presumptive distal daughter cell, but it adopts a proximal cell fate (Siegfried et al., 2004), suggesting a role for SYS-1 in modulating the activity of POP-1.

Genetic evidence suggests that the ratio between nuclear SYS-1 and POP-1 determines whether Wnt targets in the distal daughter cell are repressed or activated (Kidd et al., 2005). In the absence of Wnt signaling, high nuclear levels of POP-1 titrate out SYS-1, so that the POP-1 associated with WREs is SYS-1-free. Upon Wnt signaling, WRM-1/LIT-1 lowers nuclear POP-1 levels, increasing the chance that a POP-1/SYS-1 heterodimer will occupy the WREs and promote activation. POP-1/SYS-1 has been shown to directly activate *CEH-22* transcription, a homeodomain/tinman homologue that is critical for distal daughter cell fate (Lam et al., 2006). While important questions remain (e.g., is SYS-1 stability/nuclear localization controlled by Wnt signaling?), this data suggests that POP-1 nuclear efflux can actually increase POP-1 transcriptional activity. Whether this pathway is operational in other Wnt-dependent contexts in the worm or other organisms awaits further investigation.

4.3. β-Catenin is a transcriptional activator

When expressed alone, TCFs have no effect on reporter genes containing multiple TCF-binding sites. However, coexpression of β-catenin results in a dramatic elevation of TCF-reporter gene activity (Molenaar et al., 1996; van de Wetering et al., 1997; Korswagen et al., 2000). This suggests that transcriptional activation activity is contained in β-catenin, which was confirmed by targeting β-catenin, Arm or Bar-1 to a nave promoter via fusion with a DNA-binding domain (e.g., that of the yeast transcription factor Gal4). The resulting fusion gene could greatly activate the expression of a Gal4-reporter gene (Hsu et al., 1998; Natarajan et al., 2001; Hoffmans et al., 2005). These results indicated that β-catenin contains at least one transcriptional activation domain.

Several lines of evidence support the existence of an important transactivation domain in the C-terminal portion of β-catenin Fusion of this region of β-catenin, Arm or BAR-1 to the Gal4 DNA-binding domain resulted in high levels of transcriptional activation (van de Wetering et al., 1997; Hsu et al., 1998; Natarajan et al., 2001; Stadeli and Basler, 2005). The region C-terminal of the Arm repeats was sufficient in this assay (van de Wetering et al., 1997; Stadeli and Basler, 2005). Targeting of this domain to Wnt responsive genes *in vivo* also has physiological consequences. Overexpressing an LEF1/β-catenin C-terminal fusion protein in *Xenopus* caused axis duplication, a phenotype consistent with ectopic activation of Wnt signaling (Vleminckx et al., 1999). Consistent with the importance of the C-terminal portion of β-catenin in activation of Wnt targets is the fact that *arm* mutant alleles deleting increasingly larger portions of the C-terminal have corresponding stronger loss of Wg-signaling phenotypes (Cox et al., 1999).

In addition to the C-terminal transactivation domain just described, there is at least one more domain with similar activity in the N-terminal half of β-catenin. This has been demonstrated in mammalian β-catenin (Hsu et al., 1998), BAR-1 (Natarajan et al., 2001), and Arm (Stadeli and Basler, 2005). In the case of Arm, it has been localized to the first four Arm repeats (Stadeli and Basler, 2005). Data in the fly embryo is also consistent with an additional transcriptional activation domain besides the C-terminal of Arm. The overall levels of Arm protein in *arm*^{xm19} (this allele is truncated after the 10th Arm repeat) mutant animals are significantly lower than that of wild-type flies, perhaps due to an unstable mRNA (Peifer and Wieschaus, 1990). Unexpectedly, when the levels of Arm are manipulated to approximate those found in wild-type animals, Wg signaling is substantially (but not completely) rescued (Cox et al., 1999), strongly arguing that the C-terminal is not solely responsible for the transcriptional activity of Arm. Adding to this view are the findings the C-terminal of Arm fused to TCF/Pan is not sufficient for Wg signaling (Thompson, 2004).

As will be described in the following section, there are several factors that can bind to β-catenin and promote its ability to activate transcription. Not surprisingly, given the data with Gal4 fusions, some of these factors bind to the N-terminal portion of β-catenin, while others associate with the C-terminal portion.

4.4. Lgs and Pygo are required for activation of Wg signaling

lgs was identified in a loss-of-function genetic screen in *Drosophila* for Wg pathway components. Fly embryos genetically null for *lgs* have a segment polarity phenotype very similar to that of *wg* mutants (Kramps et al., 2002). Lgs has no recognizable domains, but has three short stretches of homology with two vertebrate genes, BCL9 and BCL9–2. As will be described, two of these domains have been shown to be crucial for Lgs function in Wg signaling.

The *pygo* gene was identified in a two-hybrid screen for Lgs-binding partners (Kramps et al., 2002) and in three independent genetic screens in *Drosophila* (Belenkaya et al., 2002; Parker et al., 2002; Thompson et al., 2002). The requirement for *pygo* in Wg signaling was examined in several embryonic and larval tissues, utilizing more than a dozen molecular readouts for the pathway. In all cases, regulation of Wg targets required *pygo* (Belenkaya et al., 2002; Kramps et al., 2002; Parker et al., 2002; Thompson et al., 2002), though it was also observed that a molecular null allele of *pygo* did not completely abolish Wg activation of Engrailed expression in the embryonic epidermis (Parker et al., 2002). Overall, the genetic analysis supports an essential role for Pygo in the Wg pathway in *Drosophila*.

Pygo has two recognizable domains, and a third domain identified by conservation between the fly and two vertebrate *pygo* genes (*pygo1* and *pygo2*). Pygos contain an NLS at the N-terminal, consistent with fly Pygo being predominately a nuclear protein (Parker et al., 2002; Thompson, 2004; Townsley et al., 2004a; Parker and Cadigan, unpublished data). In addition, Pygo has a conserved *plant homeodomain* (PHD domain) at the C-terminal. PHD domains are zinc finger-like domains that moderate protein–protein interactions in many situations. Despite their name, there is no evidence that PHD fingers can bind nucleic acids. PHD finger domains contain a unique Cys_4-His-Cys_3 pattern that distinguishes them from the similar RING finger (Cys_3-His-Cys_4) and LIM (Cys_2-His-Cys_5) domains. Like the RING and LIM domains, PHD fingers are found in many nuclear proteins such as transcription factors and chromatin remodeling agents (Aasland et al., 1995). The third domain [called the N-terminal homology domain (NHD)] is a stretch of approximately 50 amino acids encompassing the NLS (Kramps et al., 2002). This region is vital for the ability of Pygo to activate gene transcription (Hoffmans et al., 2005; Stadeli and Basler, 2005).

4.4.1. An Arm/Lgs/Pygo complex is required for Wg signaling

Pygo and Arm can be coimmunoprecipitated when overexpressed (Belenkaya et al., 2002; Thompson et al., 2002), and a tagged Pygo protein can pull down endogenous Arm and Lgs, providing evidence for an Arm/Lgs/Pygo complex *in vivo* (Thompson, 2004). However, our understanding of the Arm/Lgs/Pygo complex comes mainly from the work of Basler and coworkers, who have used a combination of high resolution protein–protein interactions experiments and elegant genetic experiments to elucidate the relationships between the three proteins.

Lgs/BCL9 have three homology domains (termed HD1, HD2, and HD3), two of which are important for Wnt gene regulation. HD2 binds to the first four Arm repeats of the Arm protein, while HD1 binds to the PHD domain of Pygo (Kramps et al., 2002). Thus Lgs acts as an adaptor between Arm and Pygo. To confirm the importance of this interaction, the HD2 domain of Lgs was fused to Pygo lacking the PHD domain. Remarkably, this Pygo-HD2 fusion was able to rescue both *lgs* and *pygo* mutants (Kramps et al., 2002). This result leads to two important conclusions about the role of Lgs and Pygo in Wg signaling. First, the only essential role of the Pygo PHD domain is to bind to Lgs. Second, the only essential role of Lgs in the pathway is to serve as an adaptor for Arm and Pygo.

To confirm the importance of the Arm/Lgs interaction in the pathway, point mutations in Arm were identified that compromised the ability of the two proteins to bind *in vitro*. Substitution of a single aspartate in the third Arm repeat (residue 164 in β-catenin, 172 in Arm) abolished binding to Lgs. This protein had no ability to activate Wg targets when overexpressed in fly tissues (Hoffmans et al., 2005). Likewise, it could rescue the cell adhesion defects of *arm* null mutants but not the Wg-signaling defect (Hoffmans and Basler, 2004). These results argue strongly that Arm/Lgs interaction is crucial for Wg signaling in *Drosophila*.

As outlined in the previous section, there is data supporting a role for Lgs/ Pygo in retaining Arm in the nucleus (Brembeck et al., 2004; Townsley et al., 2004a). While this may contribute to Wg signaling, this is clearly not the only role for Pygo in Wg signaling, as there is another convincing body of work suggesting that Pygo plays a transcriptional role in the pathway.

Three lines of evidence suggest a transcriptional coactivator role for Pygo. First, Pygo can activate the expression of a UAS-Luciferase-reporter gene when fused to the Gal4 DNA-binding domain (Belenkaya et al., 2002; Thompson, 2004; Townsley et al., 2004b). Importantly, this is not dependent on *arm* or *lgs* (Stadeli and Basler, 2005). This activity is localized to a conserved tripeptide, NPF, found in the NHD of pygo (Stadeli and Basler, 2005). Second, a Pygo-TCF fusion protein (with two deletions such that it cannot bind Lgs or Arm via Pygo or TCF) can rescue a *pygo* mutant in *Drosophila* (Thompson, 2004). This suggests that recruiting Pygo to the promoter in the absence of Arm is sufficient to activate Wg target genes. Third, and most convincingly, a *pygo* point mutant that cannot activate transcription but can bind to Lgs fails to rescue a *pygo* mutant in flies. Although this construct cannot rescue Wg signaling, it does restore Lgs nuclear levels (which are low in *pygo* mutants), suggesting that the Arm tethering function of Lgs/Pygo has also been restored (Hoffmans and Basler, 2004). These data argue convincingly that Pygo has a transcriptional coactivator role in Wg signaling.

Experiments in cultured fly cells support a model where Lgs/Pygo acts through the N-terminal transcriptional activation domain of Arm and independently of the one found at the C-terminal. As mentioned above, Lgs binds to the first four Arm repeats in Arm (Kramps et al., 2002). RNAi inhibition of *lgs* or *pygo* blocks the ability of the Arm N-terminal (containing Arm repeats 1–4) to activate a reporter gene. However, inhibition of either gene had no effect on the ability of C-terminal fragments of Arm to activate transcription (Stadeli and Basler, 2005). These data are consistent with Arm containing two functionally separate activation domains. However, a full-length Arm fusion protein also requires *lgs* and *pygo* (Stadeli and Basler, 2005), suggesting that in the context of the complete Arm protein, the two activation domains are interdependent.

In conclusion, it seems as though the Pygo/Lgs complex performs at least two roles in transducing the Wg signal, helping to tether Arm in the nucleus and also providing a transcriptional activation function.

4.4.2. Vertebrate counterparts of Lgs and Pygo

Although not as well studied as in *Drosophila*, there is some support for Pygo homologues being important for Wnt/β-catenin signaling in vertebrate systems. *hPygo1* and *hPygo2* are required for Wnt signaling in human cell culture (Thompson et al., 2002). Further, injection of morpholinos directed against *Pygo2* into *Xenopus* embryos results in embryonic brain-patterning defects, a known Wnt-dependent phenotype (Lake and Kao, 2003).

The vertebrate counterparts of Lgs also appear to be important for Wnt/β-catenin signaling. Despite the relatively low amount of similarity at the amino acid level, the mouse *BCL9* gene can rescue *lgs* mutants (Kramps et al., 2002), suggesting that the function of Lgs has been evolutionarily conserved. BCL9–2 overexpression can enhance β-catenin transcription in mammalian cells and appears to be important in converting β-catenin from a form that promotes cell adhesion to one that activates gene expression, thereby promoting an epithelial to mesenchymal transition (EMT; Brembeck, 2004). Morpholino knockdown of *BCL9–2* in *Xenopus* embryos revealed a dramatic defect in patterning of the ventral mesoderm, consistent with a block in Wnt8/β-catenin signaling (Brembeck et al., 2004).

It is interesting to note that the data for BCL9–2 differs from that of Lgs in flies in one aspect. The ability of BCL9–2 to augment β-catenin transcriptional activation does not require HD2, its presumptive Pygo-binding domain. This suggests that BCL9–2 may have functions independent of a β-catenin/BCL9/Pygo complex (Brembeck et al., 2004). Interestingly, this activity requires the HD3 of the protein, which has no known function in *Drosophila*.

BCL9 was first discovered as a gene at a recurrent breakpoint in several B-cell leukemias, (Willis et al., 1998; Dave et al., 2002). Given the growing appreciation of Wnt/β-catenin signaling in regulating B-cell lymphocyte proliferation (Reya et al., 2000; Reya and Clevers, 2005), this suggests that activation of the Wnt pathway may play an important role in these malignancies.

4.5. β-Catenin associates with several other transcriptional coactivators

In addition to Lgs/BCL9 and Pygo, there is abundant evidence that β-catenin also recruits several other transcriptional activators to regulate Wnt targets (Fig. 4). These factors have been identified in various genetic and biochemical screens. The relationships (if any) between these factors are not understood, though some (like Lgs/BCL9) bind to the N-terminal so they will be described in three general groups: (1) factors that bind to the N-terminal of β-catenin, (2) factors that bind to the C-terminal of β-catenin, and (3) tissue-specific regulators of β-catenin gene activation.

4.5.1. Pontin

The putative ATP-dependent-DNA helicase Pontin (also known as TIP49, TIP29a, Pontin52, NMP238, RUVBL1, Rvbl, TAP54α, and TIH1P) was originally identified as a protein interacting with the N-terminal half of β-catenin (Bauer et al., 1998). Pontin binds directly to β-catenin and also to the TATA box-binding protein (TBP; Bauer et al., 1998). This provided the first evidence of a direct link between Wnt signaling and the basal transcription machinery. It should also be noted that *in vitro* β-catenin directly binds TBP (Hecht et al., 1999).

When transfected in mammalian cells, Pontin can mildly activate a Wnt-reporter gene (Bauer et al., 2000). More significantly, when cotransformed with activated β-catenin, a putative dominant negative Pontin construct (TIP49DN, predicted to lack helicase activity) can inhibit the expression of

The on state

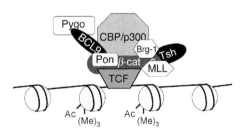

Fig. 4. Summary of nuclear factors that promote TCF/β-catenin transcriptional activation. This figure supports the view that β-catenin acts as a "landing platform" for transcriptional coactivators. The function of several of these, such as Pygo, Pontin (Pon), and Tsh, is not known. Brg-1 may act through chromatin remodeling, while CBP and MLL2 may acetylate (Ac) or trimethylate (Me3) the histones, respectively. See Section 4 of the text for more details.

endogenous Wnt target genes. Knock down of Pontin by RNAi also reduced transcript levels of β-catenin-induced endogenous genes, further suggesting a role for this protein in Wnt signaling (Feng et al., 2003).

Pontin is a member of several TRRAP/TIP60 HAT complexes (Ikura et al., 2000; Cai et al., 2003; Jin et al., 2005), implicating Pontin in chromatin remodeling. Consistent with this, ChIP data shows recruitment of Pontin to the proximal upstream region of an endogenous Wnt target gene upon pathway activation (Feng et al., 2003). This recruitment is coincident with TCF-4 and TRRAP at the same site. In addition, stable transfection of dominant negative form of Pontin (TIP49DN) reduced Wnt-pathway-dependent histone H4 acetylation at this site, suggesting a functional role for TIP49 in remodeling the chromatin at this gene. Intriguingly, TIP49DN appeared to decrease histone H4 acetylation over more widespread region of chromatin than its binding site alone, hinting that Pontin may exert some sort of long-range effect on chromatin structure (Feng et al., 2003).

As outlined in Section 3.3, Pontin directly interacts with the highly related protein Reptin (also known as TIP48, TIP49b, Reptin52, RVBL2, Rvb2, TAP54B, and TIH2p; Bauer et al., 1998; Feng et al., 2003). Reptin is also a predicted ATP-dependent-DNA helicase, and like Pontin can bind directly to β-catenin and TBP. However, in overexpression studies in mammalian cell culture and in loss-of-function experiments in flies and zebrafish, Reptin acts in a manner consistent with a repressor of Wnt signaling (Bauer et al., 2000; Rottbauer et al., 2002).

Pontin and Reptin are widely expressed in vertebrates (Bauer et al., 1998), and both are expressed ubiquitously throughout fly development (Bauer et al., 2000). However, *pontin* and *reptin* mutants are both lethal in *Drosophila* with no easily identifiable Wg defects. A more careful genetic analysis in both flies and vertebrate systems should shed more light on the role of these two proteins in the Wnt/β-catenin pathway.

4.5.2. CBP/p300

CBP and p300 are two closely related acetyltransferases that can acetylate histones H3 and H4 and several other proteins (Ogryzko et al., 1996). As described above, the fly CBP has been demonstrated to block Wg signaling by acetylation of TCF/Pan, which reduces its ability to bind Arm (Waltzer and Bienz, 1998).

In direct contrast to the study in *Drosophila*, there are several reports that indicate an important positive role for CBP/p300 in Wnt/β-catenin signaling in vertebrates. Overexpression of CBP/p300 was found to augment the ability of β-catenin to activate several Wnt-reporter genes in cultured cells (Hecht et al., 2000; Miyagishi et al., 2000; Sun et al., 2000). In *Xenopus* embryos and cultured cells, expression of the Adenovirus E1A protein, a known CBP/p300 inhibitor resulted in downregulation of the Wnt target *siamois* and TCF reporters (Hecht et al., 2000; Takemaru and Moon, 2000).

An E1A mutant that could not bind CBP/p300 had no effect on TCF/
β-catenin activity. Consistent with a direct role in the pathway, CBP and
p300 were found to bind directly to β-catenin, both in the N-terminal
(Miyagishi et al., 2000; Sun et al., 2000) and C-terminal half of β-catenin
(Hecht et al., 2000; Miyagishi et al., 2000; Takemaru and Moon, 2000;
Daniels and Weis, 2002). This suggests multiple contacts between the two
proteins, though a small molecule inhibitor that blocks CBP interaction with
β-catenin's C-terminal can inhibit activation of Wnt transcriptional targets
(Emami et al., 2004; Ma et al., 2005).

CBP has also been found to be recruited upon activation of Wnt/β-catenin
to the WREs of the *PitX2* and *c-Myc* genes and a concomitant increase in
acetylated histone H3 and H4 subunits was observed (Kioussi et al., 2002;
Sierra et al., 2006). Whether CBP was responsible for this activation was
not determined. Another study found that HAT activity of p300 was not
required for its ability to augment β-catenin-dependent activation of TCF
transcription, though this was only with reporter genes (Hecht et al., 2000).

In addition to the modification of histones, CBP/p300 has also been shown
to acetylate β-catenin (Wolf et al., 2002; Levy et al., 2004), and this acetylation appears to increase the affinity of β-catenin for TCF (Levy et al., 2004).
CBP and β-catenin can form a ternary complex with the multiple zinc-finger
protein FHL2. FHL2 is required for maximal activation of TOPFLASH
by β-catenin and CBP and may increase the ability of CBP to acetylate
β-catenin (Labalette et al., 2004).

Do the apparent differences between the role of CBP in *Drosophila* and
vertebrates reflect a true difference between phyla, or is it due to an incomplete understanding of these HATs in both systems? For example, the
activation of Wg signaling in flies observed in *nejire/CBP* mutants was
observed with a partial reduction of gene activity (Waltzer and Bienz,
1998). Perhaps a greater inactivation of *nejire/CBP* would reveal its positive
role in the pathway. Likewise the assays employed in vertebrate experiments
were designed to test for an activating function in Wnt/β-catenin transcription and the lack of loss-of-function data may have caused an inhibitory role
to be overlooked. Further study will be required to resolve this issue.

4.5.3. Chromatin, MLL/SET, and histone methylation
It is widely assumed that chromatin modification/remodeling must play a
prominent role in the regulation of Wg target genes. This assumption is based
on the rapid progress made in the last 10 years demonstrating the widespread
importance of chromatin structure on gene regulation in eukaryotes (Grewal
and Moazed, 2003), and many factors, such as CBP/p300, implicated in
chromatin modification that have been shown to play roles in Wnt signaling.

The importance of thinking about TCF-binding sites as chromatin,
rather than DNA is illustrated by the finding that while human LEF1
could bind TCF sites on naked DNA with high affinity, it does not bind

high-affinity TCF sites *in vitro* when the DNA is packaged into chromatin (Tutter et al., 2001). The addition of β-catenin to the assay resulted in much higher affinity for the chromatinized TCF sites, as did deletion of the β-catenin-binding domain of LEF1. These results suggest that β-catenin binding to LEF1 results in a conformational change that allows recognition of sites on chromatin (Tutter et al., 2001). Furthermore, binding of LEF1/β-catenin to fully assembled chromatin *in vitro* was facilitated by addition of ATP-dependent-chromatin remodeling factors, and this increase in binding correlated with an increase in transcription (Tutter et al., 2001).

In vertebrate cells with constitutive Wnt/β-catenin signaling, the C-terminal of β-catenin has been shown to associate with subunits of several known chromatin remodeling complexes. These interactions show that TRRAP/Tip60 HAT complexes, previously implicated in Wnt signaling via Pontin (Feng et al., 2003), ATPase imitation switch (ISWI1) complexes, and histone methyl transferase containing SET1 complexes can associate with β-catenin.

Consistent with these factors playing a role in Wnt target gene activation, time course ChIP experiments in synchronized cells show that 30 min following pathway activation (via LiCl treatment), β-catenin, Pygo, Bcl9, CBP, SET complex components, and RNA polymerase II (RNAPII) are recruited to the enhancer of *cMyc*, a direct Wnt target gene. Interestingly, recruitment of β-catenin, Pygo, SET complex, and CBP is transient and cyclical, alternating with TLE1 and, surprisingly, GSK3β (see below). In contrast, RNAPII recruitment is permanent, consistent with a constant increase in *cMyc* transcript levels. Similarly, despite the cyclical nature of SET complex recruitment, H3K4 trimethylation levels are constantly elevated following SET complex arrival at the enhancer (Sierra et al., 2006). In agreement with the switch model shown in Fig. 1, LEF1 remains bound to the WRE throughout the time course. The importance of this histone methylation was supported by the finding the siRNA depletion of a SET complex component, MLL2, caused a significant reduction in Wnt activation of *c-Myc* expression (Sierra et al., 2006).

β-Catenin can also influence chromatin modification when recruited to an endogenous promoter in a TCF-independent manner. Pitx2 is a bicoid-like transcription factor that interacts with β-catenin to activate the *CyclinD2* gene directly (Kioussi et al., 2002). Stabilization of β-catenin by LiCl treatment results in a rise in *CyclinD2* transcript levels concurrent with recruitment of Pitx2, β-catenin, CBP, RNAPII, and acetylated histone 3/histone 4 to the *CyclinD2* promoter (Kioussi et al., 2002).

4.6. Tissue/stage/promoter-specific nuclear factors required for Wnt signaling

The broad array of biological functions influenced by Wnt/β-catenin signaling and the large number of Wnt transcriptional targets in different cells suggests that there must be mechanisms to ensure gene/cell-specific responses to the pathway. The existence of non-TCF proteins that can utilize β-catenin to activate transcription certainly contributes to this specificity (see Section 2.5), but there are also indications that some factors help TCF/β-catenin achieve the appropriate target specificity. Several nuclear factors that interact with the Wnt/β-catenin in a tissue or gene-specific manner are described below. In addition, the role of different TCF isoforms will also be discussed.

4.6.1. TCF isoforms display different promoter specificities

Vertebrate TCFs have several different isoforms differing in sequences C-terminal to the DNA-binding HMG domain, suggesting that they may have different functions. For example, the TCF1-E and TCF4-E isoforms contain a conserved stretch of residues that has been called the E-box (Atcha et al., 2003). Other TCF family members, such as LEF1, lack this motif. Wnt/β-catenin signaling can activate LEF1 expression in intestinal cells, and a WRE in the *LEF1* control region could be activated by TCF1E and TCF4E in a β-catenin-dependent manner. Strikingly, LEF1 could not act with β-catenin to activate this WRE (Atcha et al., 2003). Of note is the fact that the TCF1-E isoform is highly expressed in colorectal cell lines compared to normal cells (Atcha et al., 2003).

Another example of gene-specific regulation by TCFs once again involves TCF4E, which can act with β-catenin and CBP to activate the *Cdx1* promoter but not that of *siamois* in cultured cells (Hecht and Stemmler, 2003). LEF1 has a reciprocal effect on these two promoters. The specificity of LEF1 can be switched by fusion with the C-terminal of TCF4E which can interact with p300 independently of β-catenin (Hecht and Stemmler, 2003). The interaction of p300 with both TCF4E and β-catenin apparently creates a transcriptional activation complex competent for activating Cdx1.

The evidence for TCF isoforms playing a role in determining Wnt target gene specificity is thus far based on overexpression/reporter gene assays. Inhibition of specific isoforms using RNAi or mouse genetics will be required to determine whether this diversity of TCFs plays an important role in normal or pathological states.

4.6.2. Brg-1 and Teashirt

Brg1 is a human homologue of the yeast ATPase-dependent chromatin remodeling protein SWI/SNF and the fly gene *brm* (Becker and Horz, 2002; Martens and Winston, 2003). While Brm has been shown to inhibit some Wg targets in *Drosophila* (Treisman et al., 1997; Collins and Treisman, 2000), Brg1 can potentiate β-catenin-dependent reporter genes, and Brg-1 can interact with β-catenin in coimmunoprecipitation assays (Barker et al., 2001; Lickert et al., 2004). Knockdown of a subunit (BAF60c) of the Brg1 chromatin remodeling complex results in cardiac defects reminiscent of a loss of Wnt/β-catenin signaling. Consistent with this, PitX2, a known LEF1/β-catenin target (Kioussi et al., 2002), is not expressed in the *BAF60c* knockdown embryos and *PitX2* expression in cultured cells is potentiated by *BAF60c* overexpression and requires *Brg-1* (Lickert et al., 2004). In flies, a twofold reduction in *brm* gene activity was found to suppress an Arm misexpression phenotype in the eye, and similar genetic interactions were observed in the wing (Barker et al., 2001). In addition, a loss of Wg-signaling phenotype in the wing was enhanced by reduction in *brm* gene activity (Barker et al., 2001). These data support a role for Brm/Brg-1 in promoting Wnt/β-catenin signaling.

In flies, Brm has been shown to have both negative and positive roles in the Wnt/β-catenin pathway. It is possible that Brm may act ubiquitously as a repressor in the absence of Wg signaling and then again as a positive regulator of Arm gene activation. Alternatively, these conflicting data may reflect gene-specific roles for this chromatin remodeler.

Another example of a gene-specific regulator of Wg signaling is the zinc-finger protein Teashirt (Tsh), which is required for Wg autoregulation and the specification of epidermal cell fate (Gallet et al., 1998, 1999). However, *tsh* is not required for Wg maintenance of Engrailed expression. Tsh can directly bind to the C-terminal of Arm in embryos, but the mechanism by which it activates Wg signaling is unclear (Gallet et al., 1998). Likewise the basis for the differential requirement in regulating Wg targets is not known.

4.6.3. Lines and Split ends

Drosophila genetics has identified other factors that are only required for a subset of Wg functions. The gene *lines* encode a protein that is required cell autonomously for Wg signaling in a subset of cells in the dorsal epidermis of the fly embryo (Hatini et al., 2000). This readout is consistent with *lin* being required for late (and not early) Wg signaling in the fly embryonic epidermis and in only a small number of cells that are responding to Wg. Split ends (Spen) is a nuclear protein that is expressed throughout embryogenesis and larval fly development, yet is only genetically required for Wg signaling in larval tissues (Lin et al., 2003). It is not yet known whether either protein directly interacts with components of the Wg pathway.

It should be pointed out that all the Arm/β-catenin-binding coactivators that have been described may act in a tissue or gene-specific manner. With the exception of Lgs and Pygo in the fly, none of these factors have been tested in a broad array of Wnt transcriptional readouts. It is possible that "core" transcriptional cofactors for β-catenin could be the exception rather than the rule.

5. β-Catenin-mediated gene repression

Although β-catenin is best known for activating transcription, there are a few reports that it can also act to repress target gene expression. Analysis of Wg signaling in flies has identified several genes that are repressed by the pathway (Cadigan et al., 1998, 2002; Alexandre et al., 1999; Payre et al., 1999; Duman-Scheel et al., 2004). In mammalian cells, genome-wide analysis of transcriptional regulation by Wnt/β-catenin signaling has revealed that the pathway both activates and represses large numbers of genes (van de Wetering et al., 2002; Jung and Kim, 2005; Naishiro et al., 2005; Ziegler et al., 2005). It is often assumed that these repressed targets are regulated indirectly, that is, Wnt/β-catenin signaling activates a repressor of these genes. Confirming whether a repressed target is due to direct or indirect regulation can be labor-intensive, and it is currently unclear at this point how common direct Wnt-mediated transcriptional repression is in biology. A handful of examples published in the last few years demonstrate that direct Wnt/β-catenin-mediated repression does occur and provides a basis on which to begin trying to identify other genes that might be repressed by Wnt signaling.

The best-characterized target of β-catenin-mediated repression is E-cadherin. Wnt/β-catenin signaling has been shown to correlate with E-cadherin downregulation in several contexts (Nelson and Nusse, 2004), including during hair follicle morphogenesis (Jamora et al., 2003). In keratinocytes this repression is direct as an *E-cadherin*-reporter gene was repressed by β-catenin expression in a manner that depended on the presence of a TCF-binding site in the enhancer. Accordingly, LEF1 was localized to the corresponding region of the endogenous gene using ChIP. This data strongly supports a model where LEF1/β-catenin acts directly on the enhancer to repress E-cadherin expression. Interestingly, E-cadherin is rendered sensitive to direct Wnt regulation only in the presence of Noggin, which caused the transcriptional activation of LEF1 (Jamora et al., 2003), suggesting a possible mechanism for conferring context-dependent repression.

In addition to being repressed by LEF1/β-catenin, E-cadherin expression is known to be a target of Snail, a DNA-binding trasnscriptional repressor (Nieto, 2002). Downregulation of E-cadherin marks the onset of the EMT during gastrulation and neural crest development and plays a key role in the progression of primary tumors to metastases (Kang and Massague, 2004). The Snail protein plays a key role in EMT during development and transformation, in large part through direct repression of E-cadherin transcription. Three E-box sequences near the E-cadherin proximal promoter bind Snail and presumably mediate repression (Batlle et al., 2000; Hajra et al., 2002). In keratinocytes, Wnt/β-catenin signaling was found to repress E-cadherin additively with Snail (Jamora et al., 2003). Interestingly, despite a direct role for LEF1/β-catenin in cultured cells, *LEF1* null mice still develop large guard hair follicles and repress E-cadherin (Jamora et al., 2003), suggesting a redundant mechanism. This other mechanism may involve Wnt regulation of Snail stability.

Snail proteins contain a conserved phosphorylation motif that causes Snail turnover to be regulated by the same GSK-3-dependent phosphorylation and ubiquitin-dependent proteasomal destruction that regulates β-catenin stablility (Zhou et al., 2004; Yook et al., 2005). Consequently, activation of the Wnt pathway stabilizes the Snail protein through the same mechanism that stabilizes β-catenin. The result of this in MCF-7 cells is that Wnt represses *E-cadherin* expression and promotes invasiveness of MCF-7 cancer cells (Yook et al., 2005). It is possible that in some cells, Wnt/β-catenin signaling represses E-cadherin in parallel through both LEF1/β-catenin and stabilized Snail.

Direct repression by TCF/Arm is also required in the fruit fly for proper differentiation of precursor cells that will form the muscle attachment at the segmental boundaries of larvae. Activity of the *stripe* (*sr*) gene appears to specify these cells as muscle attachment sites (Frommer et al., 1996). Genetic analysis of a minimal *sr*-reporter gene that recapitulated the endogenous gene expression pattern in transgenic fly embryos revealed that sr is activated by Hedgehog and repressed by Wnt signaling (Piepenburg et al., 2000). Activation required Cubitus interruptus (Ci)-binding sites, and repression required TCF-binding sites. Interestingly each Ci-binding site was closely juxtaposed to a TCF site (within four bases or overlapping), suggesting that a model in which the TCF/Arm complex represses *sr* expression by displacing the essential Ci activator.

Just as β-catenin can activate transcription through transcription factors other than TCF, work has shown that β-catenin can also repress transcription through at least one other DNA-binding protein. KAI1 is a tumor metastasis suppressor (Jackson et al., 2005) that is activated by NF-kappaB p50 homodimers in combination with the TIP60 transcriptional activation complex (Kim et al., 2005). β-Catenin has been shown to repress *KAI1* transcription through a two-pronged approach. Expression of β-catenin down regulates transcription of TIP60 through an unidentified mechanism. β-Catenin also interacts directly with p50 (Deng et al., 2002) and recruits Reptin to the *KAI1* promoter, but not the *ICAM1* promoter, both reported targets of NF-kappaB (Kim et al., 2005). ChIP assays showed that binding of the β-catenin/Reptin complex to p50 is mutually exclusive to binding of the TIP60 complex to p50 at the KAI1 promoter. This suggests a mechanism in which the β-catenin/Reptin repressor complex competes with the TIP60 activator complex, and activation of the Wnt pathway both recruits components of the repressor complex and down regulates components of the activator complex.

Why the recruitment of β-catenin to these repressed enhancers does not cause activation of transcription remains an open question. Although tissue-specific factors, such as high Reptin levels, that could change β-catenin from a transcriptional activator to a dedicated repressor are a formal possibility, they certainly cannot explain how Wnt signaling represses E-cadherin in mouse keratinocytes while simultaneously activating another putative direct Wnt target, *hK1* (Zhou et al., 1995; Jamora et al., 2003). Clearly, the answer must lie in the Wnt-responsive enhancers, at least in this case, since both the *E-cadherin* and *hK1* promoters have access to the same pool of corepressors and coactivators. With only a few examples in hand, it is difficult to make any mechanistic conclusions about Wnt-mediated repression, but even just the few examples mentioned highlight the key roles Wnt-mediated repression is likely to play in development and disease.

6. Switching off signaling: A nuclear role for the destruction complex

The current model for Wnt signaling suggests that when a cell no longer perceives Wnts, the destruction complex is no longer inhibited, β-catenin destruction resumes, and the pathway returns to the off state. This model makes no predictions as to what unfolds at the enhancers of Wnt responsive genes during this process. For example, does the subsequent decrease in β-catenin levels simply shift the equilibrium of TCF binding from β-catenin to Grg/TLE, resulting in Wnt gene repression, or is there a more proactive course of action that occurs? Recent evidence suggests the latter, as the so-called cytoplasmic destruction complex appears to be more mobile than was thought, moving between the cytoplasm and the nucleus.

Experiments using leptomycin B (LMB), a pharmacological inhibitor of chromosome maintenance region 1 (CRM1)/Exportin1-mediated nuclear export, show that APC, Axin, GSK3, and Dsh all can accumulate in the nucleus (Henderson, 2000; Rosin-Arbesfeld et al., 2000; Franca-Koh et al., 2002; Cong and Varmus, 2004; Wiechens et al., 2004; Itoh et al., 2005). These experiments suggest that all of these proteins shuttle in and out of the nucleus. Although it is unclear at this point how important this shuttling is in regulating the Wnt signal, evidence is accumulating that suggests these proteins may play a hitherto unforeseen nuclear role in the pathway. Two broad hypotheses ensue: first, that these proteins play an active role in exporting β-catenin from the nucleus, and second, that the complex plays an active role on the chromatin at WREs.

6.1. How important is chaperoned shuttling of β-catenin out of the nucleus

All of the evidence suggesting these proteins act as shuttling chaperones for β-catenin is based on overexpressing known β-catenin-binding partners. While these proteins can certainly perform this function under these conditions, β-catenin can exit the nucleus independently of CRM1 (i.e., in an LMB insensitive manner) (Yokoya et al., 1999; Eleftheriou et al., 2001; Wiechens et al., 2004; Hendriksen et al., 2005). As nuclear export of APC, Axin, and GSK3β is LMB sensitive, it is unlikely that β-catenin nucleocytoplasmic shuttling depends upon these factors, although they may play a regulatory role in this process.

One explanation for how β-catenin can exit the nucleus is based on the findings that β-catenin can interact directly with Ran-binding protein 3 (RanBP3), which could link β-catenin directly to nuclear export machinery (Hendriksen et al., 2005). RanBP3 can stimulate nuclear export of β-catenin in the presence of LMB and gain and loss-of-function experiments with RanBP3 in flies and frogs produces phenotypes consistent with β-catenin nuclear efflux (Hendriksen et al., 2005). If nuclear export of β-catenin is not dependent on APC, Axin, and GSK3β, one might expect that they have another reason for entering the nucleus. As mentioned above, recent evidence suggests this may be the case.

6.2. APC, GSK3β, and βTrcp on the chromatin

A recent report studying the dynamics of activation and repression of a WRE suggests a much more complicated model than previously suspected (Sierra et al., 2006). In a human cell line, c-Myc expression was activated within 15 min of LiCl treatment, which inhibits GSK3 activity (Klein and Melton, 1996). ChIP analysis at a WRE upstream of the c-Myc promoter revealed that LEF1, TLE1, and GSK3β occupied the DNA before LiCl treatment. LiCl caused a rapid (within 30 min) recruitment of β-catenin, Pygo, BCL9, MLL2, p300, and other factors associated with transcriptional activation. Surprisingly, APC and β-Trcp, an F-box protein that helps target β-catenin for degradation (Hart et al., 1999; Kitagawa et al., 1999) are recruited with identical kinetics as β-catenin (Sierra et al., 2006). Recruitment of β-catenin and all the other factors occurs in a transient and cyclical manner, with β-catenin, the coactivators, APC and β-Trcp alternating with GSK3β and TLE1 on the c-Myc WRE. These results suggest a cyclical switch of repression to activation to repression upon Wnt/β-catenin signaling.

In contrast to this cycling observed with transcriptional regulators, RNAPII is recruited concomitantly with β-catenin, and it remains on the DNA throughout the time course. Likewise, histone 3 trimethylation (H3K4TriMe) levels, presumably catalyzed by MLL2, remain constant. Consistent with the RNAPII distribution and epigenetic modification, c-Myc transcript levels constantly rise throughout the time course. It will be important to determine whether the cycling of β-catenin and the other factors also occurs when c-Myc is activated by Wnt stimulation, since LiCl inhibition of GSK3 may not reflect the physiological situation.

Many colorectal cell lines lack full-length APC, instead having a truncated, inactive form of the protein (Polakis, 1997, 2000). These cells have high levels of nuclear β-catenin and constitutive Wnt signaling. Consistent with this, Sierra and colleagues found that c-Myc transcription was high, and β-catenin, Pygo, BCL9, RNAPII, and MLL2, are all present at the WRE and the enhancer contains acetylated histone H4 (residue K8) and H3K4TriMe. Expression of full-length APC in such cells causes a dramatic reduction of β-catenin levels (Munemitsu et al., 1995), though this requires several hours to take effect (Sierra et al., 2006). Prior to that, there is a transient recruitment of full-length APC, CtBP, and β-Trcp to the WRE of the *c-Myc* gene. At the same time, there is a decrease in H4K8 acetylation, H3K4 trimethylation and a loss of β-catenin, Pygo, BCL9, and RNAPII and the SET complex from the WRE. Recruitment of HDAC1 and TLE1 follows and remains, presumably stabilizing the repressed state. LEF1 remains bound to the WRE both in the absence and presence of full-length APC. Importantly, full-length APC is present at the chromatin 3 hour before β-catenin protein levels begin to fall, suggesting that any role APC has in repression of *c-Myc* is distinct from the role APC plays in general β-catenin stability (Sierra et al., 2006).

These studies prompt several questions. First, the finding that GSK3β resides on the WRE in the repressed state with TLE1 and HDAC is surprising, and its roles there are currently uncertain. The fact that β-catenin can occupy the WRE at the same time as LEF1 and APC is also surprising, since APC and TCF compete for the same site on β-catenin (Spink et al., 2000; Ha et al., 2004; Xing et al., 2004; Sierra et al., 2006). Perhaps there is an additional docking site on the WRE for APC (Fig. 5).

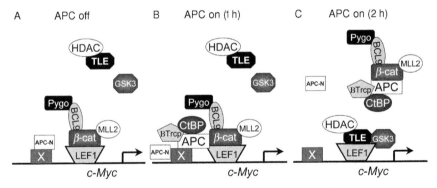

Fig. 5. Speculative model of APC-dependent inactivation of an LEF1/β-catenin transcription activation complex. This tentative model is based on the data from Sierra et al. (2006). (A) A colorectal cell line containing a nonfunction APC lacking its C-terminal half (APC-N) has a typical LEF1/β-catenin activation complex on the *c-Myc* WRE, along with APC-N-associated with the chromatin through an unknown mechanism. (B) One hour after induction of full-length APC, a CtBP/APC/β-Trcp complex occupies the site formerly bound by APC-N, where it can interact in a transient way with LEF1/β-catenin. (C) Two hours after onset of APC induction, the CtBP/APC/β-Trcp complex has removed β-catenin and coactivators from the chromatin. Possible fates of this β-catenin are nuclear efflux and/or degradation. LEF1 is now bound by TLE1/HDAC1 and GSK3β is associated with the chromatin through an unknown mechanism. *c-Myc* expression is now silenced.

6.3. What is the function of the destruction complex in the nucleus

ChIP data implies guilt by association, and as such does not equate to functionality. There are several models consistent with the presence of APC, GSK3β, and βTrcp at a Wnt responsive enhancer. For example, APC could enter the nucleus, bind CtBP and then be recruited to the c-Myc WRE, where it somehow removes Arm, allowing the TLE1 corepressor to occupy LEF1. This is consistent with a previous report, which showed that an APC/CtBP complex functionally inhibits β-catenin-mediated transcription by diverting β-catenin from TCF (Hamada and Bienz, 2004). Another possibility is that APC and β-Trcp induce localized degradation of β-catenin at the enhancer (an effect that would be masked by measuring total β-catenin levels). Both mechanisms could be important, and it will be important to determine whether GSK, APC, and β Trcp also occupy WREs in other genes.

Given the likely important role for components of the β-catenin destruction complex in the nucleus, another outstanding question is how is there function inhibited by Wnt signaling. Surprisingly (or perhaps not in this increasingly baroque pathway), recent data suggest the answer could lie with Disheveled (Dsh).

6.4. A role for Dsh in the nucleus

Like APC, Axin, and GSK3β, endogenous Dsh accumulates in the nucleus upon LMB treatment (Itoh et al., 2005). Dsh contains a functional leucine-rich NES is at the C-terminal, but no recognizable NLS. However, scanning mutagenesis revealed a novel sequence of IVLT required for the nuclear localization of Dsh. As IVLT bears no resemblance to any known NLS, the mechanism of Dsh nuclear import is currently unclear. Interestingly, mutation of this sequence affects the function of Dsh, blocking the ability of exogenous Dsh to activate β-catenin and induces a secondary axis in Xenopus embryos (Itoh et al., 2005). Although this is a negative result, this mutant protein can still bind to CK1-ε and Axin, and is still active in noncanonical signaling, suggesting the protein is not simply misfolded or unstable (Itoh et al., 2005). These data suggest that, at least in Xenopus, Dsh requires cycling through the nucleus to function in canonical Wnt signaling.

These data suggest the possibility that Dsh might enter the nucleus in order to inactivate the repressive activity of the APC complex at the level of Wnt responsive enhancers, in addition to its role in stabilizing β-catenin levels in the cytoplasm. Dsh can associate with the degradation complex through interactions with Axin (Polakis, 2000; Ding and Dale, 2002). While Axin cycles in and out of the nucleus (Cong and Varmus, 2004; Wiechens et al., 2004), it remains to be seen if it associates with WRE chromatin.

The work of Jones and colleagues (Sierra et al., 2006) has just started to elucidate this previously unsuspected level of regulation of Wnt/β-catenin transcription in the nucleus.

7. Concluding remarks

As outlined in this chapter, the regulation of TCF and TCF/β-catenin transcriptional regulation is extraordinarily complex. One concern is that cell or gene-specific factors may be mistaken for general regulators, confusing the issue of how many factors are required for a single WRE to be regulated. Clearly, combining biochemical techniques like ChIP with genetic analysis will help sort through which factors are important for which genes. The use of ChIP to analyze chromatin modifications and recruitment of factors to WREs clearly should be a priority for the field and will no doubt continue to provide surprises, such as those recounted in Section 6. Studying the role of APC and associated proteins on the chromatin is a challenging problem, since they clearly play an earlier role in the pathway. Likewise, trying to understand the mechanisms of gene-specific regulation by the pathway will be difficult. However, given the rich biology of the Wnt signaling field, there will be no shortage of investigators willing to try to solve it.

References

Aasland, R., Gibson, T.J., Stewart, A.F. 1995. The PHD finger: Implications for chromatin-mediated transcriptional regulation. Trends Biochem. Sci. 20, 56–59.

Akiyama, H., Lyons, J.P., Mori-Akiyama, Y., Yang, X., Zhang, R., Zhang, Z., Deng, J.M., Taketo, M.M., Nakamura, T., Behringer, R.R., McCrea, P.D., de Crombruqqhe, B. 2004. Interactions between Sox9 and beta-catenin control chondrocyte differentiation. Genes Dev. 18, 1072–1087.

Alexandre, C., Lecourtois, M., Vincent, J. 1999. Wingless and Hedgehog pattern *Drosophila* denticle belts by regulating the production of short-range signals. Development 126, 5689–5698.

Andl, T., Reddy, S.T., Gaddapara, T., Millar, S.E. 2002. WNT signals are required for the initiation of hair follicle development. Dev. Cell 2, 643–653.

Atcha, F.A., Munguia, J.E., Li, T.W., Hovanes, K., Waterman, M.L. 2003. A new beta-catenin-dependent activation domain in T cell factor. J. Biol. Chem. 278, 16169–16175.

Barker, N., Hurlstone, A., Musisi, H., Miles, A., Bienz, M., Clevers, H. 2001. The chromatin remodelling factor Brg-1 interacts with beta-catenin to promote target gene activation. EMBO J. 20, 4935–4943.

Barolo, S., Posakony, J.W. 2002. Three habits of highly effective signaling pathways: Principles of transcriptional control by developmental cell signaling. Genes Dev. 16, 1167–1181.

Barrow, J.R., Thomas, K.R., Boussadia-Zahui, O., Moore, R., Kemler, R., Capecchi, M.R., McMahon, A.P. 2003. Ectodermal Wnt3/beta-catenin signaling is required for the establishment and maintenance of the apical ectodermal ridge. Genes Dev. 17, 394–409.

Batlle, E., Sancho, E., Franci, C., Dominguez, D., Monfar, M., Baulida, J., Garcia De Herreros, A. 2000. The transcription factor snail is a repressor of E-cadherin gene expression in epithelial tumour cells. Nat. Cell Biol. 2, 84–89.

Bauer, A., Chauvet, S., Huber, O., Usseglio, F., Rothbacher, U., Aragnol, D., Kemler, R., Pradel, J. 2000. Pontin52 and reptin52 function as antagonistic regulators of beta-catenin signalling activity. EMBO J. 19, 6121–6130.

Bauer, A., Huber, O., Kemler, R. 1998. Pontin52, an interaction partner of beta-catenin, binds to the TATA box binding protein. Proc. Natl. Acad. Sci. USA 95, 14787–14792.

Becker, P.B., Horz, W. 2002. ATP-dependent nucleosome remodeling. Annu. Rev. Biochem. 71, 247–273.

Behrens, J., von Kries, J.P., Kuhl, M., Bruhn, L., Wedlich, D., Grosschedl, R., Birchmeier, W. 1996. Functional interaction of beta-catenin with the transcription factor LEF-1. Nature 382, 638–642.

Belenkaya, T.Y., Han, C., Standley, H.J., Lin, X., Houston, D.W., Heasman, J. 2002. Pygopus encodes a nuclear protein essential for wingless/Wnt signaling. Development 129, 4089–4101.

Billin, A.N., Thirlwell, H., Ayer, D.E. 2000. Beta-catenin-histone deacetylase interactions regulate the transition of LEF1 from a transcriptional repressor to an activator. Mol. Cell Biol. 20, 6882–6890.

Brannon, M., Brown, J.D., Bates, R., Kimelman, D., Moon, R.T. 1999. XCtBP is a XTcf-3 co-repressor with roles throughout *Xenopus* development. Development 126, 3159–3170.

Brannon, M., Gomperts, M., Sumoy, L., Moon, R.T., Kimelman, D. 1997. A beta-catenin/XTcf-3 complex binds to the siamois promoter to regulate dorsal axis specification in *Xenopus*. Genes Dev. 11, 2359–2370.

Brantjes, H., Roose, J., van de Wetering, M., Clevers, H. 2001. All Tcf HMG box transcription factors interact with Groucho-related co-repressors. Nucleic Acids Res. 29, 1410–1419.

Brembeck, F.H., Rosario, M., Birchmeier, W. 2006. Balancing cell adhesion and Wnt signaling, the key role of beta-catenin. Curr. Opin. Genet. Dev. 16, 51–59.

Brembeck, F.H., Schwarz-Romond, T., Bakkers, J., Wilhelm, S., Hammerschmidt, M., Birchmeier, W. 2004. Essential role of BCL9-2 in the switch between beta-catenin's adhesive and transcriptional functions. Genes Dev. 18, 2225–2230.

Brennan, K.R., Brown, A.M. 2004. Wnt proteins in mammary development and cancer. J. Mammary Gland Biol. Neoplasia 9, 119–131.

Brunner, E., Peter, O., Schweizer, L., Basler, K. 1997. Pangolin encodes a Lef-1 homologue that acts downstream of Armadillo to transduce the Wingless signal in *Drosophila*. Nature 385, 829–833.

Cadigan, K.M., Nusse, R. 1997. Wnt signaling: A common theme in animal development. Genes Dev. 11, 3286–3305.

Cadigan, K.M., Liu, Y.I. 2006. Wnt signaling: Complexity at the surface. J. Cell Sci. 119, 395–402.

Cadigan, K.M., Fish, M.P., Rulifson, E.J., Nusse, R. 1998. Wingless repression of *Drosophila* frizzled 2 expression shapes the Wingless morphogen gradient in the wing. Cell 93, 767–777.

Cadigan, K.M., Jou, A.D., Nusse, R. 2002. Wingless blocks bristle formation and morphogenetic furrow progression in the eye through repression of Daughterless. Development 129, 3393–3402.

Cai, Y., Jin, J., Tomomori-Sato, C., Sato, S., Sorokina, I., Parmely, T.J., Conaway, R.C., Conaway, J.W. 2003. Identification of new subunits of the multiprotein mammalian TRRAP/TIP60-containing histone acetyltransferase complex. J. Biol. Chem. 278, 42733–42736.

Calvo, D., Victor, M., Gay, F., Sui, G., Luke, M.P., Dufourcq, P., Wen, G., Maduro, M., Rothman, J., Shi, Y. 2001. A POP-1 repressor complex restricts inappropriate cell type-specific gene transcription during *Caenorhabditis elegans* embryogenesis. EMBO J. 20, 7197–7208.

Castrop, J., van Wichen, D., Koomans-Bitter, M., van de Wetering, M., de Weger, R., van Dongen, J., Clevers, H. 1995. The human TCF-1 gene encodes a nuclear DNA-binding protein uniquely expressed in normal and neoplastic T-lineage lymphocytes. Blood 86, 3050–3059.

Cavallo, R.A., Cox, R.T., Moline, M.M., Roose, J., Polevoy, G.A., Clevers, H., Peifer, M., Bejsovec, A. 1998. *Drosophila* Tcf and Groucho interact to repress Wingless signalling activity. Nature 395, 604–608.

Chamorro, M.N., Schwartz, D.R., Vonica, A., Brivanlou, A.H., Cho, K.R., Varmus, H.E. 2005. FGF-20 and DKK1 are transcriptional targets of beta-catenin and FGF-20 is implicated in cancer and development. EMBO J. 24, 73–84.

Chen, G., Courey, A.J. 2000. Groucho/TLE family proteins and transcriptional repression. Gene 249, 1–16.

Chen, G., Fernandez, J., Mische, S., Courey, A.J. 1999. A functional interaction between the histone deacetylase Rpd3 and the corepressor groucho in *Drosophila* development. Genes Dev. 13, 2218–2230.

Chinnadurai, G. 2002. CtBP, an unconventional transcriptional corepressor in development and oncogenesis. Mol. Cell 9, 213–224.

Christian, J.L., Moon, R.T. 1993. Interactions between Xwnt-8 and Spemann organizer signaling pathways generate dorsoventral pattern in the embryonic mesoderm of *Xenopus*. Genes Dev. 7, 13–28.

Civenni, G., Holbro, T., Hynes, N.E. 2003. Wnt1 and Wnt5a induce cyclin D1 expression through ErbB1 transactivation in HC11 mammary epithelial cells. EMBO Rep. 4, 166–171.

Clevers, H., van de Wetering, M. 1997. TCF/LEF factor earn their wings. Trends Genet. 13, 485–489.

Collins, R.T., Treisman, J.E. 2000. Osa-containing Brahma chromatin remodeling complexes are required for the repression of wingless target genes. Genes Dev. 14, 3140–3152.

Collins, R.T., Furukawa, T., Tanese, N., Treisman, J.E. 1999. Osa associates with the Brahma chromatin remodeling complex and promotes the activation of some target genes. EMBO J. 18, 7029–7040.

Cong, F., Varmus, H. 2004. Nuclear-cytoplasmic shuttling of Axin regulates subcellular localization of beta-catenin. Proc. Natl. Acad. Sci. USA 101, 2882–2887.

Costa, M., Raich, W., Agbunag, C., Leung, B., Hardin, J., Priess, J.R. 1998. A putative catenin-cadherin system mediates morphogenesis of the *Caenorhabditis elegans* embryo. J. Cell Biol. 141, 297–308.

Courey, A.J., Jia, S. 2001. Transcriptional repression: The long and the short of it. Genes Dev. 15, 2786–2796.

Cox, R.T., Pai, L.M., Kirkpatrick, C., Stein, J., Peifer, M. 1999. Roles of the C terminus of Armadillo in wingless signaling in *Drosophila*. Genetics 153, 319–332.

Daniels, D.L., Weis, W.I. 2002. ICAT inhibits beta-catenin binding to Tcf/Lef-family transcription factors and the general coactivator p300 using independent structural modules. Mol. Cell 10, 573–584.

Daniels, D.L., Weis, W.I. 2005. Beta-catenin directly displaces Groucho/TLE repressors from Tcf/Lef in Wnt-mediated transcription activation. Nat. Struct. Mol. Biol. 12, 364–371.

DasGupta, R., Fuchs, E. 1999. Multiple roles for activated LEF/TCF transcription complexes during hair follicle development and differentiation. Development 126, 4557–4568.

DasGupta, R., Kaykas, A., Moon, R.T., Perrimon, N. 2005. Functional genomic analysis of the Wnt-wingless signaling pathway. Science 308, 826–833.

Dave, B.J., Nelson, M., Pickering, D.L., Chan, W.C., Greiner, T.C., Weisenburger, D.D., Armitage, J.O., Sanger, W.G. 2002. Cytogenetic characterization of diffuse large cell lymphoma using multi-color fluorescence in situ hybridization. Cancer Genet. Cytogenet. 132, 125–132.

Deng, J., Miller, S.A., Wang, H.Y., Xia, W., Wen, Y., Zhou, B.P., Li, Y., Lin, S.Y., Hung, M.C. 2002. Beta-catenin interacts with and inhibits NF-kappa B in human colon and breast cancer. Cancer Cell 2, 323–334.

Ding, Y., Dale, T. 2002. Wnt signal transduction: Kinase cogs in a nano-machine? Trends Biochem. Sci. 27, 327–329.

Dorsky, R.I., Sheldahl, L.C., Moon, R.T. 2002. A transgenic Lef1/beta-catenin-dependent reporter is expressed in spatially restricted domains throughout zebrafish development. Dev. Biol. 241, 229–237.

Duman-Scheel, M., Johnston, L.A., Du, W. 2004. Repression of dMyc expression by Wingless promotes Rbf-induced G1 arrest in the presumptive *Drosophila* wing margin. Proc. Natl. Acad. Sci.USA 101, 3857–3862.

Eisenmann, D.M., Maloof, J.N., Simske, J.S., Kenyon, C., Kim, S.K. 1998. The beta-catenin homolog BAR-1 and LET-60 Ras coordinately regulate the Hox gene lin-39 during *Caenorhabditis elegans* vulval development. Development 125, 3667–3680.

Eleftheriou, A., Yoshida, M., Henderson, B.R. 2001. Nuclear export of human beta-catenin can occur independent of CRM1 and the adenomatous polyposis coli tumor suppressor. J. Biol. Chem. 276, 25883–25888.

Emami, K.H., Nguyen, C., Ma, H., Kim, D.H., Jeong, K.W., Eguchi, M., Moon, R.T., Teo, J.L., Kim, H.Y., Moon, S.H., Ha, J.R., Kahn, M. 2004. A small molecule inhibitor of beta-catenin/CREB-binding protein transcription [corrected]. Proc. Natl. Acad. Sci. USA 101, 12682–12687.

Essers, M.A., de Vries-Smits, L.M., Barker, N., Polderman, P.E., Burgering, B.M., Korswagen, H.C. 2005. Functional interaction between beta-catenin and FOXO in oxidative stress signaling. Science 308, 1181–1184.

Fagotto, F., Gluck, U., Gumbiner, B.M. 1998. Nuclear localization signal-independent and importin/karyopherin-independent nuclear import of beta-catenin. Curr. Biol. 8, 181–190.

Fan, M.J., Gruning, W., Walz, G., Sokol, S.Y. 1998. Wnt signaling and transcriptional control of Siamois in *Xenopus* embryos. Proc. Natl. Acad. Sci. USA 95, 5626–5631.

Feng, Y., Lee, N., Fearon, E.R. 2003. TIP49 regulates beta-catenin-mediated neoplastic transformation and T-cell factor target gene induction via effects on chromatin remodeling. Cancer Res. 63, 8726–8734.

Fisher, A.L., Caudy, M. 1998. Groucho proteins: Transcriptional corepressors for specific subsets of DNA-binding transcription factors in vertebrates and invertebrates. Genes Dev. 12, 1931–1940.

Franca-Koh, J., Yeo, M., Fraser, E., Young, N., Dale, T.C. 2002. The regulation of glycogen synthase kinase-3 nuclear export by Frat/GBP. J. Biol. Chem. 277, 43844–43848.

Frommer, G., Vorbruggen, G., Pasca, G., Jackle, H., Volk, T. 1996. Epidermal egr-like zinc finger protein of *Drosophila* participates in myotube guidance. EMBO J. 15, 1642–1649.

Galceran, J., Farinas, I., Depew, M.J., Clevers, H., Grosschedl, R. 1999. $Wnt3a^{-/-}$-like phenotype and limb deficiency in $Lef1^{-/-}Tcf1^{-/-}$ mice. Genes Dev. 13, 709–717.

Galceran, J., Sustmann, C., Hsu, S.C., Folberth, S., Grosschedl, R. 2004. LEF1-mediated regulation of Delta-like1 links Wnt and Notch signaling in somitogenesis. Genes Dev. 18, 2718–2723.

Gallet, A., Angelats, C., Erkner, A., Charroux, B., Fasano, L., Kerridge, S. 1999. The C-terminal domain of armadillo binds to hypophosphorylated teashirt to modulate wingless signalling in *Drosophila*. EMBO J. 18, 2208–2217.

Gallet, A., Erkner, A., Charroux, B., Fasano, L., Kerridge, S. 1998. Trunk-specific modulation of wingless signalling in *Drosophila* by teashirt binding to armadillo. Curr. Biol. 8, 893–902.

Ghogomu, S.M., van Venrooy, S., Ritthaler, M., Wedlich, D., Gradl, D. 2006. HIC-5 is a novel repressor of lymphoid enhancer factor/T-cell factor-driven transcription. J. Biol. Chem. 281, 1755–1764.

Giese, K., Amsterdam, A., Grosschedl, R. 1991. DNA-binding properties of the HMG domain of the lymphoid-specific transcriptional regulator LEF-1. Genes Dev. 5, 2567–2578.

Giese, K., Cox, J., Grosschedl, R. 1992. The HMG domain of lymphoid enhancer factor 1 bends DNA and facilitates assembly of functional nucleoprotein structures. Cell 69, 185–195.

Gleason, J.E., Korswagen, H.C., Eisenmann, D.M. 2002. Activation of Wnt signaling bypasses the requirement for RTK/Ras signaling during *C. elegans* vulval induction. Genes Dev. 16, 1281–1290.

Gottardi, C.J., Gumbiner, B.M. 2004a. Distinct molecular forms of beta-catenin are targeted to adhesive or transcriptional complexes. J. Cell Biol. 167, 339–349.

Gottardi, C.J., Gumbiner, B.M. 2004b. Role for ICAT in beta-catenin-dependent nuclear signaling and cadherin functions. Am. J. Physiol. Cell Physiol. 286, C747–C756.

Gottardi, C.J., Wong, E., Gumbiner, B.M. 2001. E-cadherin suppresses cellular transformation by inhibiting beta-catenin signaling in an adhesion-independent manner. J. Cell Biol. 153, 1049–1060.

Graham, T.A., Ferkey, D.M., Mao, F., Kimelman, D., Xu, W. 2001. Tcf4 can specifically recognize beta-catenin using alternative conformations. Nat. Struct. Biol. 8, 1048–1052.

Graham, T.A., Weaver, C., Mao, F., Kimelman, D., Xu, W. 2000. Crystal structure of a beta-catenin/Tcf complex. Cell 103, 885–896.

Gregorieff, A., Clevers, H. 2005. Wnt signaling in the intestinal epithelium: From endoderm to cancer. Genes Dev. 19, 877–890.

Gregorieff, A., Grosschedl, R., Clevers, H. 2004. Hindgut defects and transformation of the gastro-intestinal tract in $Tcf4^{-/-}/Tcf1^{-/-}$ embryos. EMBO J. 23, 1825–1833.

Grewal, S.I., Moazed, D. 2003. Heterochromatin and epigenetic control of gene expression. Science 301, 798–802.

Ha, N.C., Tonozuka, T., Stamos, J.L., Choi, H.J., Weis, W.I. 2004. Mechanism of phosphorylation-dependent binding of APC to beta-catenin and its role in beta-catenin degradation. Mol. Cell 15, 511–521.

Hajra, K.M., Chen, D.Y., Fearon, E.R. 2002. The SLUG zinc-finger protein represses E-cadherin in breast cancer. Cancer Res. 62, 1613–1618.

Hallikas, O., Palin, K., Sinjushina, N., Rautiainen, R., Partanen, J., Ukkonen, E., Taipale, J. 2006. Genome-wide prediction of mammalian enhancers based on analysis of transcription-factor binding affinity. Cell 124, 47–59.

Hamada, F., Bienz, M. 2004. The APC tumor suppressor binds to C-terminal binding protein to divert nuclear beta-catenin from TCF. Dev. Cell 7, 677–685.

Hamada, F., Tomoyasu, Y., Takatsu, Y., Nakamura, M., Nagai, S., Suzuki, A., Fujita, F., Shibuya, H., Toyoshima, K., Ueno, N., Akiyama, T. 1999. Negative regulation of wingless signaling by D-axin, a *Drosophila* homolog of axin. Science 283, 1739–1742.

Hamblet, N.S., Lijam, N., Ruiz-Lozano, P., Wang, J., Yang, Y., Luo, Z., Mei, L., Chien, K. R., Sussman, D.J., Wynshaw-Boris, A. 2002. Dishevelled 2 is essential for cardiac outflow tract development, somite segmentation and neural tube closure. Development 129, 5827–5838.

Hamilton, F.S., Wheeler, G.N., Hoppler, S. 2001. Difference in XTcf-3 dependency accounts for change in response to beta-catenin-mediated Wnt signalling in *Xenopus* blastula. Development 128, 2063–2073.

Hart, M., Concordet, J.P., Lassot, I., Albert, I., del los Santos, R., Durand, H., Perret, C., Rubinfeld, B., Margottin, F., Benarous, R., Polakis, P. 1999. The F-box protein beta-TrCP associates with phosphorylated beta-catenin and regulates its activity in the cell. Curr. Biol. 9, 207–210.

Hatini, V., Bokor, P., Goto-Mandeville, R., DiNardo, S. 2000. Tissue- and stage-specific modulation of Wingless signaling by the segment polarity gene lines. Genes Dev. 14, 1364–1376.

He, T.C., Sparks, A.B., Rago, C., Hermeking, H., Zawel, L., da Costa, L.T., Morin, P.J., Vogelstein, B., Kinzler, K.W. 1998. Identification of c-MYC as a target of the APC pathway. Science 281, 1509–1512.

He, X., Semenov, M., Tamai, K., Zeng, X. 2004. LDL receptor-related proteins 5 and 6 in Wnt/beta-catenin signaling: Arrows point the way. Development 131, 1663–1677.

Heasman, J., Crawford, A., Goldstone, K., Garner-Hamrick, P., Gumbiner, B., McCrea, P., Kintner, C., Noro, C.Y., Wylie, C. 1994. Overexpression of cadherins and underexpression of beta-catenin inhibit dorsal mesoderm induction in early *Xenopus* embryos. Cell 79, 791–803.

Hecht, A., Stemmler, M.P. 2003. Identification of a promoter-specific transcriptional activation domain at the C terminus of the Wnt effector protein T-cell factor 4. J. Biol. Chem. 278, 3776–3785.

Hecht, A., Litterst, C.M., Huber, O., Kemler, R. 1999. Functional characterization of multiple transactivating elements in beta-catenin, some of which interact with the TATA-binding protein *in vitro*. J. Biol. Chem. 274, 18017–18025.

Hecht, A., Vleminckx, K., Stemmler, M.P., van Roy, F., Kemler, R. 2000. The p300/CBP acetyltransferases function as transcriptional coactivators of beta-catenin in vertebrates. EMBO J. 19, 1839–1850.

Held, W., Clevers, H., Grosschedl, R. 2003. Redundant functions of TCF-1 and LEF-1 during T and NK cell development, but unique role of TCF-1 for Ly49 NK cell receptor acquisition. Eur. J. Immunol. 33, 1393–1398.

Henderson, B.R. 2000. Nuclear-cytoplasmic shuttling of APC regulates beta-catenin subcellular localization and turnover. Nat. Cell Biol. 2, 653–660.

Henderson, B.R., Galea, M., Schuechner, S., Leung, L. 2002. Lymphoid enhancer factor-1 blocks adenomatous polyposis coli-mediated nuclear export and degradation of beta-catenin. Regulation by histone deacetylase 1. J. Biol. Chem. 277, 24258–24264.

Hendriksen, J., Fagotto, F., van der Velde, H., van Schie, M., Noordermeer, J., Fornerod, M. 2005. RanBP3 enhances nuclear export of active (beta)-catenin independently of CRM1. J. Cell Biol. 171, 785–797.

Herman, M. 2001. *C. elegans* POP-1/TCF functions in a canonical Wnt pathway that controls cell migration and in a noncanonical Wnt pathway that controls cell polarity. Development 128, 581–590.

Hill, T.P., Taketo, M.M., Birchmeier, W., Hartmann, C. 2006. Multiple roles of mesenchymal {beta}-catenin during murine limb patterning. Development 133, 1219–1229.

Hoffmans, R., Basler, K. 2004. Identification and *in vivo* role of the Armadillo-Legless interaction. Development 131, 4393–4400.

Hoffmans, R., Stadeli, R., Basler, K. 2005. Pygopus and legless provide essential transcriptional coactivator functions to armadillo/beta-catenin. Curr. Biol. 15, 1207–1211.

Hoppler, S., Moon, R.T. 1998. BMP-2/4 and Wnt-8 cooperatively pattern the *Xenopus* mesoderm. Mech. Dev. 71, 119–129.

Houston, D.W., Kofron, M., Resnik, E., Langland, R., Destree, O., Wylie, C., Heasman, J. 2002. Repression of organizer genes in dorsal and ventral *Xenopus* cells mediated by maternal XTcf3. Development 129, 4015–4025.

Hovanes, K., Li, T.W., Munguia, J.E., Truong, T., Milovanovic, T., Lawrence Marsh, J., Holcombe, R.F., Waterman, M.L. 2001. Beta-catenin-sensitive isoforms of lymphoid enhancer factor-1 are selectively expressed in colon cancer. Nat. Genet. 28, 53–57.

Hsu, S.C., Galceran, J., Grosschedl, R. 1998. Modulation of transcriptional regulation by LEF-1 in response to Wnt-1 signaling and association with beta-catenin. Mol. Cell. Biol. 18, 4807–4818.

Huber, A.H., Nelson, W.J., Weis, W.I. 1997. Three-dimensional structure of the armadillo repeat region of beta-catenin. Cell 90, 871–882.

Huber, O., Korn, R., McLaughlin, J., Ohsugi, M., Herrmann, B.G., Kemler, R. 1996. Nuclear localization of beta-catenin by interaction with transcription factor LEF-1. Mech. Dev. 59, 3–10.

Huelsken, J., Vogel, R., Erdmann, B., Cotsarelis, G., Birchmeier, W. 2001. Beta-catenin controls hair follicle morphogenesis and stem cell differentiation in the skin. Cell 105, 533–545.

Ikura, T., Ogryzko, V.V., Grigoriev, M., Groisman, R., Wang, J., Horikoshi, M., Scully, R., Qin, J., Nakatani, Y. 2000. Involvement of the TIP60 histone acetylase complex in DNA repair and apoptosis. Cell 102, 463–473.

Ishikawa, T.O., Tamai, Y., Li, Q., Oshima, M., Taketo, M.M. 2003. Requirement for tumor suppressor Apc in the morphogenesis of anterior and ventral mouse embryo. Dev. Biol. 253, 230–246.

Ishitani, T., Ninomiya-Tsuji, J., Nagai, S., Nishita, M., Meneghini, M., Barker, N., Waterman, M., Bowerman, B., Clevers, H., Shibuya, H., Matsumoto, K. 1999. The TAK1-NLK-MAPK-related pathway antagonizes signalling between beta-catenin and transcription factor TCF. Nature 399, 798–802.

Ishitani, T., Ninomiya-Tsuji, J., Matsumoto, K. 2003. Regulation of lymphoid enhancer factor 1/T-cell factor by mitogen-activated protein kinase-related Nemo-like kinase-dependent phosphorylation in Wnt/beta-catenin signaling. Mol. Cell. Biol. 23, 1379–1389.

Itoh, K., Brott, B.K., Bae, G.U., Ratcliffe, M.J., Sokol, S.Y. 2005. Nuclear localization is required for dishevelled function in Wnt/beta-catenin signaling. J. Biol. 4, 3–10.

Jackson, P., Marreiros, A., Russell, P.J. 2005. KAI1 tetraspanin and metastasis suppressor. Int. J. Biochem. Cell Biol. 37, 530–534.

Jamora, C., DasGupta, R., Kocieniewski, P., Fuchs, E. 2003. Links between signal transduction, transcription and adhesion in epithelial bud development. Nature 422, 317–322.

Jin, J., Cai, Y., Yao, T., Gottschalk, A.J., Florens, L., Swanson, S.K., Gutierrez, J.L., Coleman, M.K., Workman, J.L., Mushegian, A., Washburn, M.P., Conaway, R.C., et al. 2005. A mammalian chromatin remodeling complex with similarities to the yeast INO80 complex. J. Biol. Chem. 280, 41207–41212.

Jung, H.C., Kim, K. 2005. Identification of MYCBP as a beta-catenin/LEF-1 target using DNA microarray analysis. Life Sci. 77, 1249–1262.

Kan, L., Israsena, N., Zhang, Z., Hu, M., Zhao, L.R., Jalali, A., Sahni, V., Kessler, J.A. 2004. Sox1 acts through multiple independent pathways to promote neurogenesis. Dev. Biol. 269, 580–594.

Kang, Y., Massague, J. 2004. Epithelial-mesenchymal transitions: Twist in development and metastasis. Cell 118, 277–279.

Kawano, Y., Kypta, R. 2003. Secreted antagonists of the Wnt signalling pathway. J. Cell Sci. 116, 2627–2634.

Kemler, R. 1993. From cadherins to catenins: Cytoplasmic protein interactions and regulation of cell adhesion. Trends Genet. 9, 317–321.

Kidd, A.R., 3rd, Miskowski, J.A., Siegfried, K.R., Sawa, H., Kimble, J. 2005. A beta-catenin identified by functional rather than sequence criteria and its role in Wnt/MAPK signaling. Cell 121, 761–772.

Kim, C.H., Oda, T., Itoh, M., Jiang, D., Artinger, K.B., Chandrasekharappa, S.C., Driever, W., Chitnis, A.B. 2000. Repressor activity of Headless/Tcf3 is essential for vertebrate head formation. Nature 407, 913–916.

Kim, J.H., Kim, B., Cai, L., Choi, H.J., Ohgi, K.A., Tran, C., Chen, C., Chung, C.H., Huber, O., Rose, D.W., Sawyer, C.L., Rosenfeld, M.G., et al. 2005. Transcriptional regulation of a metastasis suppressor gene by Tip60 and beta-catenin complexes. Nature 434, 921–926.

Kioussi, C., Briata, P., Baek, S.H., Rose, D.W., Hamblet, N.S., Herman, T., Ohgi, K.A., Lin, C., Gleiberman, A., Wang, J., Brault, V., Ruiz-Lozano, P., et al. 2002. Identification of a

Wnt/Dvl/beta-Catenin → Pitx2 pathway mediating cell-type-specific proliferation during development. Cell 111, 673–685.

Kitagawa, M., Hatakeyama, S., Shirane, M., Matsumoto, M., Ishida, N., Hattori, K., Nakamichi, I., Kikuchi, A., Nakayama, K. 1999. An F-box protein, FWD1, mediates ubiquitin-dependent proteolysis of beta-catenin. EMBO J. 18, 2401–2410.

Klein, P.S., Melton, D.A. 1996. A molecular mechanism for the effect of lithium on development. Proc. Natl. Acad. Sci. USA 93, 8455–8459.

Knirr, S., Frasch, M. 2001. Molecular integration of inductive and mesoderm-intrinsic inputs governs even-skipped enhancer activity in a subset of pericardial and dorsal muscle progenitors. Dev. Biol. 238, 13–26.

Kobayashi, M., Kishida, S., Fukui, A., Michiue, T., Miyamoto, Y., Okamoto, T., Yoneda, Y., Asashima, M., Kikuchi, A. 2002. Nuclear localization of Duplin, a beta-catenin-binding protein, is essential for its inhibitory activity on the Wnt signaling pathway. J. Biol. Chem. 277, 5816–5822.

Kolligs, F.T., Hu, G., Dang, C.V., Fearon, E.R. 1999. Neoplastic transformation of RK3E by mutant beta-catenin requires deregulation of Tcf/Lef transcription but not activation of c-myc expression. Mol. Cell. Biol. 19, 5696–5706.

Korinek, V., Barker, N., Morin, P.J., van Wichen, D., de Weger, R., Kinzler, K.W., Vogelstein, B., Clevers, H. 1997. Constitutive transcriptional activation by a beta-catenin-Tcf complex in $APC^{-/-}$ colon carcinoma. Science 275, 1784–1787.

Korinek, V., Barker, N., Moerer, P., van Donselaar, E., Huls, G., Peters, P.J., Clevers, H. 1998. Depletion of epithelial stem-cell compartments in the small intestine of mice lacking Tcf-4. Nat. Genet. 19, 379–383.

Korswagen, H.C., Herman, M.A., Clevers, H.C. 2000. Distinct beta-catenins mediate adhesion and signalling functions in *C. elegans*. Nature 406, 527–532.

Kouzmenko, A.P., Takeyama, K., Ito, S., Furutani, T., Sawatsubashi, S., Maki, A., Suzuki, E., Kawasaki, Y., Akiyama, T., Tabata, T., Kato, S. 2004. Wnt/beta-catenin and estrogen signaling converge *in vivo*. J. Biol. Chem. 279, 40255–40258.

Kramps, T., Peter, O., Brunner, E., Nellen, D., Froesch, B., Chatterjee, S., Murone, M., Zullig, S., Basler, K. 2002. Wnt/wingless signaling requires BCL9/legless-mediated recruitment of pygopus to the nuclear beta-catenin-TCF complex. Cell 109, 47–60.

Kratochwil, K., Galceran, J., Tontsch, S., Roth, W., Grosschedl, R. 2002. FGF4, a direct target of LEF1 and Wnt signaling, can rescue the arrest of tooth organogenesis in $Lef1^{-/-}$ mice. Genes Dev. 16, 3173–3185.

Krieghoff, E., Behrens, J., Mayr, B. 2006. Nucleo-cytoplasmic distribution of {beta}-catenin is regulated by retention. J. Cell Sci. 119, 1453–1463.

Kuhnert, F., Davis, C.R., Wang, H.T., Chu, P., Lee, M., Yuan, J., Nusse, R., Kuo, C.J. 2004. Essential requirement for Wnt signaling in proliferation of adult small intestine and colon revealed by adenoviral expression of Dickkopf-1. Proc. Natl. Acad. Sci. USA 101, 266–271.

Labalette, C., Renard, C.A., Neuveut, C., Buendia, M.A., Wei, Y. 2004. Interaction and functional cooperation between the LIM protein FHL2, CBP/p300, and beta-catenin. Mol. Cell. Biol. 24, 10689–10702.

Labbe, E., Letamendia, A., Attisano, L. 2000. Association of Smads with lymphoid enhancer binding factor 1/T cell-specific factor mediates cooperative signaling by the transforming growth factor-beta and wnt pathways. Proc. Natl. Acad. Sci. USA 97, 8358–8363.

Lake, B.B., Kao, K.R. 2003. Pygopus is required for embryonic brain patterning in *Xenopus*. Dev. Biol. 261, 132–148.

Lam, N., Chesney, M.A., Kimble, J. 2006. Wnt signaling and CEH-22/tinman/Nkx2.5 specify a stem cell niche in *C. elegans*. Curr. Biol. 16, 287–295.

Laudet, V., Stehelin, D., Clevers, H. 1993. Ancestry and diversity of the HMG box superfamily. Nucleic Acids Res. 21, 2493–2501.

Lee, E., Salic, A., Kirschner, M.W. 2001. Physiological regulation of [beta]-catenin stability by Tcf3 and CK1epsilon. J. Cell Biol. 154, 983–993.

Lee, H.H., Frasch, M. 2000. Wingless effects mesoderm patterning and ectoderm segmentation events via induction of its downstream target sloppy paired. Development 127, 5497–5508.

Lei, S., Dubeykovskiy, A., Chakladar, A., Wojtukiewicz, L., Wang, T.C. 2004. The murine gastrin promoter is synergistically activated by transforming growth factor-beta/Smad and Wnt signaling pathways. J. Biol. Chem. 279, 42492–42502.

Levy, L., Wei, Y., Labalette, C., Wu, Y., Renard, C.A., Buendia, M.A., Neuveut, C. 2004. Acetylation of beta-catenin by p300 regulates beta-catenin-Tcf4 interaction. Mol. Cell. Biol. 24, 3404–3414.

Lickert, H., Takeuchi, J.K., Von Both, I., Walls, J.R., McAuliffe, F., Adamson, S.L., Henkelman, R.M., Wrana, J.L., Rossant, J., Bruneau, B.G. 2004. Baf60c is essential for function of BAF chromatin remodelling complexes in heart development. Nature 432, 107–112.

Lin, H.V., Doroquez, D.B., Cho, S., Chen, F., Rebay, I., Cadigan, K.M. 2003. Splits ends is a tissue/promoter specific regulator of Wingless signaling. Development 130, 3125–3135.

Lin, R., Thompson, S., Priess, J.R. 1995. pop-1 encodes an HMG box protein required for the specification of a mesoderm precursor in early *C. elegans* embryos. Cell 83, 599–609.

Liu, F., van den Broek, O., Destree, O., Hoppler, S. 2005. Distinct roles for Xenopus Tcf/Lef genes in mediating specific responses to Wnt/{beta}-catenin signalling in mesoderm development. Development 132, 5375–5385.

Lnenicek-Allen, M., Read, C.M., Crane-Robinson, C. 1996. The DNA bend angle and binding affinity of an HMG box increased by the presence of short terminal arms. Nucleic Acids Res. 24, 1047–1051.

Lo, M.C., Gay, F., Odom, R., Shi, Y., Lin, R. 2004. Phosphorylation by the beta-catenin/MAPK complex promotes 14-3-3-mediated nuclear export of TCF/POP-1 in signal-responsive cells in *C. elegans*. Cell 117, 95–106.

Logan, C.Y., Nusse, R. 2004. The Wnt signaling pathway in development and disease. Annu. Rev. Cell Dev. Biol. 20, 781–810.

Logan, C.Y., Miller, J.R., Ferkowicz, M.J., McClay, D.R. 1999. Nuclear beta-catenin is required to specify vegetal cell fates in the sea urchin embryo. Development 126, 345–357.

Love, J.J., Li, X., Case, D.A., Giese, K., Grosschedl, R., Wright, P.E. 1995. Structural basis for DNA bending by the architectural transcription factor LEF-1. Nature 376, 791–795.

Lowry, W.E., Blanpain, C., Nowak, J.A., Guasch, G., Lewis, L., Fuchs, E. 2005. Defining the impact of beta-catenin/Tcf transactivation on epithelial stem cells. Genes Dev. 19, 1596–1611.

Lum, L., Yao, S., Mozer, B., Rovescalli, A., Von Kessler, D., Nirenberg, M., Beachy, P.A. 2003. Identification of Hedgehog pathway components by RNAi in *Drosophila* cultured cells. Science 299, 2039–2045.

Ma, H., Nguyen, C., Lee, K.S., Kahn, M. 2005. Differential roles for the coactivators CBP and p300 on TCF/beta-catenin-mediated survivin gene expression. Oncogene 24, 3619–3631.

Maduro, M.F., Lin, R., Rothman, J.H. 2002. Dynamics of a developmental switch: Recursive intracellular and intranuclear redistribution of *Caenorhabditis elegans* POP-1 parallels Wnt-inhibited transcriptional repression. Dev. Biol. 248, 128–142.

Maloof, J.N., Whangbo, J., Harris, J.M., Jongeward, G.D., Kenyon, C. 1999. A Wnt signaling pathway controls hox gene expression and neuroblast migration in *C. elegans*. Development 126, 37–49.

Mao, J., Wang, J., Liu, B., Pan, W., Farr, G.H., 3rd, Flynn, C., Yuan, H., Takada, S., Kimelman, D., Li, L., Wu, D. 2001. Low-density lipoprotein receptor-related protein-5 binds to Axin and regulates the canonical Wnt signaling pathway. Mol. Cell 7, 801–809.

Maretto, S., Cordenonsi, M., Dupont, S., Braghetta, P., Broccoli, V., Hassan, A.B., Volpin, D., Bressan, G.M., Piccolo, S. 2003. Mapping Wnt/beta-catenin signaling during mouse development and in colorectal tumors. Proc. Natl. Acad. Sci. USA 100, 3299–3304.

Martens, J.A., Winston, F. 2003. Recent advances in understanding chromatin remodeling by Swi/Snf complexes. Curr. Opin. Genet. Dev. 13, 136–142.

McMahon, A.P., Moon, R.T. 1989. Ectopic expression of the proto-oncogene int-1 in *Xenopus* embryos leads to duplication of the embryonic axis. Cell 58, 1075–1084.

Meneghini, M.D., Ishitani, T., Carter, J.C., Hisamoto, N., Ninomiya-Tsuji, J., Thorpe, C.J., Hamill, D.R., Matsumoto, K., Bowerman, B. 1999. MAP kinase and Wnt pathways converge to downregulate an HMG-domain repressor in *Caenorhabditis elegans*. Nature 399, 793–797.

Merrill, B.J., Pasolli, H.A., Polak, L., Rendl, M., Garcia-Garcia, M.J., Anderson, K.V., Fuchs, E. 2004. Tcf3: A transcriptional regulator of axis induction in the early embryo. Development 131, 263–274.

Miskowski, J., Li, Y., Kimble, J. 2001. The sys-1 gene and sexual dimorphism during gonadogenesis in *Caenorhabditis elegans*. Dev. Biol. 230, 61–73.

Miyagishi, M., Fujii, R., Hatta, M., Yoshida, E., Araya, N., Nagafuchi, A., Ishihara, S., Nakajima, T., Fukamizu, A. 2000. Regulation of Lef-mediated transcription and p53-dependent pathway by associating beta-catenin with CBP/p300. J. Biol. Chem. 275, 35170–35175.

Moghal, N., Sternberg, P.W. 2003. A component of the transcriptional mediator complex inhibits RAS-dependent vulval fate specification in *C. elegans*. Development 130, 57–69.

Mohrmann, L., Langenberg, K., Krijgsveld, J., Kal, A.J., Heck, A.J., Verrijzer, C.P. 2004. Differential targeting of two distinct SWI/SNF-related *Drosophila* chromatin-remodeling complexes. Mol. Cell. Biol. 24, 3077–3088.

Molenaar, M., van de Wetering, M., Oosterwegel, M., Peterson-Maduro, J., Godsave, S., Korinek, V., Roose, J., Destree, O., Clevers, H. 1996. XTcf-3 transcription factor mediates beta-catenin-induced axis formation in *Xenopus* embryos. Cell 86, 391–399.

Moon, R.T., Brown, J.D., Torres, M. 1997. WNTs modulate cell fate and behavior during vertebrate development. Trends Genet. 13, 157–162.

Mukhopadhyay, M., Shtrom, S., Rodriguez-Esteban, C., Chen, L., Tsukui, T., Gomer, L., Dorward, D.W., Glinka, A., Grinberg, A., Huang, S.P., Niehrs, C., Belmonte, J.C.I., et al. 2001. Dickkopf1 is required for embryonic head induction and limb morphogenesis in the mouse. Dev. Cell 1, 423–434.

Mulholland, D.J., Dedhar, S., Coetzee, G.A., Nelson, C.C. 2005. Interaction of nuclear receptors with the Wnt/beta-catenin/Tcf signaling axis: Wnt you like to know? Endocr. Rev. 26, 898–915.

Munemitsu, S., Albert, I., Souza, B., Rubinfeld, B., Polakis, P. 1995. Regulation of intracellular beta-catenin levels by the adenomatous polyposis coli (APC) tumor-suppressor protein. Proc. Natl. Acad. Sci. USA 92, 3046–3050.

Myers, L.C., Kornberg, R.D. 2000. Mediator of transcriptional regulation. Annu. Rev. Biochem. 69, 729–749.

Naishiro, Y., Yamada, T., Idogawa, M., Honda, K., Takada, M., Kondo, T., Imai, K., Hirohashi, S. 2005. Morphological and transcriptional responses of untransformed intestinal epithelial cells to an oncogenic beta-catenin protein. Oncogene 24, 3141–3153.

Nakaya, M.A., Biris, K., Tsukiyama, T., Jaime, S., Rawls, J.A., Yamaguchi, T.P. 2005. Wnt3a links left-right determination with segmentation and anteroposterior axis elongation. Development 132, 5425–5436.

Natarajan, L., Witwer, N.E., Eisenmann, D.M. 2001. The divergent *Caenorhabditis elegans* beta-catenin proteins BAR-1, WRM-1 and HMP-2 make distinct protein interactions but retain functional redundancy *in vivo*. Genetics 159, 159–172.

Nelson, W.J., Nusse, R. 2004. Convergence of Wnt, beta-catenin, and cadherin pathways. Science 303, 1483–1487.

Neufeld, K.L., Nix, D.A., Bogerd, H., Kang, Y., Beckerle, M.C., Cullen, B.R., White, R.L. 2000a. Adenomatous polyposis coli protein contains two nuclear export signals and shuttles between the nucleus and cytoplasm. Proc. Natl. Acad. Sci. USA 97, 12085–12090.

Neufeld, K.L., Zhang, F., Cullen, B.R., White, R.L. 2000b. APC-mediated downregulation of beta-catenin activity involves nuclear sequestration and nuclear export. EMBO Rep. 1, 519–523.

Niehrs, C., Kazanskaya, O., Wu, W., Glinka, A. 2001. Dickkopf1 and the Spemann-Mangold head organizer. Int. J. Dev. Biol. 45, 237–240.

Nieto, M.A. 2002. The snail superfamily of zinc-finger transcription factors. Nat. Rev. Mol. Cell Biol. 3, 155–166.

Nishita, M., Hashimoto, M.K., Ogata, S., Laurent, M.N., Ueno, N., Shibuya, H., Cho, K.W. 2000. Interaction between Wnt and TGF-beta signalling pathways during formation of Spemann's organizer. Nature 403, 781–785.

Noordermeer, J., Klingensmith, J., Perrimon, N., Nusse, R. 1994. Dishevelled and armadillo act in the wingless signalling pathway in *Drosophila*. Nature 367, 80–83.

Nusse, R., Varmus, H.E. 1992. Wnt genes. Cell 69, 1073–1087.

Ogryzko, V.V., Schiltz, R.L., Russanova, V., Howard, B.H., Nakatani, Y. 1996. The transcriptional coactivators p300 and CBP are histone acetyltransferases. Cell 87, 953–959.

Orsulic, S., Huber, O., Aberle, H., Arnold, S., Kemler, R. 1999. E-cadherin binding prevents beta-catenin nuclear localization and beta-catenin/LEF-1-mediated transactivation. J. Cell Sci. 112(Pt 8), 1237–1245.

Orsulic, S., Peifer, M. 1996. An *in vivo* structure-function study of armadillo, the beta-catenin homologue, reveals both separate and overlapping regions of the protein required for cell adhesion and for wingless signaling. J. Cell Biol. 134, 1283–1300.

Papkoff, J., Rubinfeld, B., Schryver, B., Polakis, P. 1996. Wnt-1 regulates free pools of catenins and stabilizes APC-catenin complexes. Mol. Cell. Biol. 16, 2128–2134.

Park, J.I., Kim, S.W., Lyons, J.P., Ji, H., Nguyen, T.T., Cho, K., Barton, M.C., Deroo, T., Vleminckx, K., Moon, R.T., McCrea, P.D. 2005. Kaiso/p120-catenin and TCF/beta-catenin complexes coordinately regulate canonical Wnt gene targets. Dev. Cell 8, 843–854.

Park, W.S., Oh, R.R., Park, J.Y., Kim, P.J., Shin, M.S., Lee, J.H., Kim, H.S., Lee, S.H., Kim, S.Y., Park, Y.G., An, W.G., Kim, H.S., et al. 2001. Nuclear localization of beta-catenin is an important prognostic factor in hepatoblastoma. J. Pathol. 193, 483–490.

Parker, D.S., Jemison, J., Cadigan, K.M. 2002. Pygopus, a nuclear PHD-finger protein required for Wingless signaling in *Drosophila*. Development 129, 2565–2576.

Payre, F., Vincent, A., Carreno, S. 1999. ovo/svb integrates Wingless and DER pathways to control epidermis differentiation. Nature 400, 271–275.

Peifer, M., Wieschaus, E. 1990. The segment polarity gene armadillo encodes a functionally modular protein that is the *Drosophila* homolog of human plakoglobin. Cell 63, 1167–1176.

Peifer, M., Rauskolb, C., Williams, M., Riggleman, B., Wieschaus, E. 1991. The segment polarity gene armadillo interacts with the wingless signaling pathway in both embryonic and adult pattern formation. Development 111, 1029–1043.

Peifer, M., Sweeton, D., Casey, M., Wieschaus, E. 1994. Wingless signal and Zeste-white 3 kinase trigger opposing changes in the intracellular distribution of Armadillo. Development 120, 369–380.

Piepenburg, O., Vorbruggen, G., Jackle, H. 2000. *Drosophila* segment borders result from unilateral repression of hedgehog activity by wingless signaling. Mol. Cell 6, 203–209.

Pinto, D., Clevers, H. 2005. Wnt, stem cells and cancer in the intestine. Biol. Cell 97, 185–196.

Pinto, D., Gregorieff, A., Begthel, H., Clevers, H. 2003. Canonical Wnt signals are essential for homeostasis of the intestinal epithelium. Genes Dev. 17, 1709–1713.

Polakis, P. 1997. The adenomatous polyposis coli (APC) tumor suppressor. Biochim. Biophys. Acta 1332, F127–F147.

Polakis, P. 2000. Wnt signaling and cancer. Genes Dev. 14, 1837–1851.

Polakis, P. 2002. Casein kinase 1: A wnt'er of disconnect. Curr. Biol. 12, R499.

Popperl, H., Schmidt, C., Wilson, V., Hume, C.R., Dodd, J., Krumlauf, R., Beddington, R.S. 1997. Misexpression of Cwnt8C in the mouse induces an ectopic embryonic axis and causes a truncation of the anterior neuroectoderm. Development 124, 2997–3005.

Poy, F., Lepourcelet, M., Shivdasani, R.A., Eck, M.J. 2001. Structure of a human Tcf4-beta-catenin complex. Nat. Struct. Biol. 8, 1053–1057.

Prieve, M.G., Waterman, M.L. 1999. Nuclear localization and formation of beta-catenin-lymphoid enhancer factor 1 complexes are not sufficient for activation of gene expression. Mol. Cell. Biol. 19, 4503–4515.

Ren, B., Robert, F., Wyrick, J.J., Aparicio, O., Jennings, E.G., Simon, I., Zeitlinger, J., Schreiber, J., Hannett, N., Kanin, E., Volkert, T.L., Wilson, C.J., et al. 2000. Genome-wide location and function of DNA binding proteins. Science 290, 2306–2309.

Reya, T., Clevers, H. 2005. Wnt signalling in stem cells and cancer. Nature 434, 843–850.

Reya, T., O'Riordan, M., Okamura, R., Devaney, E., Willert, K., Nusse, R., Grosschedl, R. 2000. Wnt signaling regulates B lymphocyte proliferation through a LEF-1 dependent mechanism. Immunity 13, 15–24.

Riese, J., Yu, X., Munnerlyn, A., Eresh, S., Hsu, S.C., Grosschedl, R., Bienz, M. 1997. LEF-1, a nuclear factor coordinating signaling inputs from wingless and decapentaplegic. Cell 88, 777–787.

Riggleman, B., Wieschaus, E., Schedl, P. 1989. Molecular analysis of the armadillo locus: Uniformly distributed transcripts and a protein with novel internal repeats are associated with a *Drosophila* segment polarity gene. Genes Dev. 3, 96–113.

Rocheleau, C.E., Downs, W.D., Lin, R., Wittmann, C., Bei, Y., Cha, Y.H., Ali, M., Priess, J.R., Mello, C.C. 1997. Wnt signaling and an APC-related gene specify endoderm in early *C. elegans* embryos. Cell 90, 707–716.

Rocheleau, C.E., Yasuda, J., Shin, T.H., Lin, R., Sawa, H., Okano, H., Priess, J.R., Davis, R.J., Mello, C.C. 1999. WRM-1 activates the LIT-1 protein kinase to transduce anterior/posterior polarity signals in *C. elegans*. Cell 97, 717–726.

Roose, J., Huls, G., van Beest, M., Moerer, P., van der Horn, K., Goldschmeding, R., Logtenberg, T., Clevers, H. 1999. Synergy between tumor suppressor APC and the beta-catenin-Tcf4 target Tcf1. Science 285, 1923–1926.

Roose, J., Molenaar, M., Peterson, J., Hurenkamp, J., Brantjes, H., Moerer, P., van de Wetering, M., Destree, O., Clevers, H. 1998. The *Xenopus* Wnt effector XTcf-3 interacts with Groucho-related transcriptional repressors. Nature 395, 608–612.

Rosin-Arbesfeld, R., Cliffe, A., Brabletz, T., Bienz, M. 2003. Nuclear export of the APC tumour suppressor controls beta-catenin function in transcription. EMBO J. 22, 1101–1113.

Rosin-Arbesfeld, R., Townsley, F., Bienz, M. 2000. The APC tumour suppressor has a nuclear export function. Nature 406, 1009–1012.

Rottbauer, W., Saurin, A.J., Lickert, H., Shen, X., Burns, C.G., Wo, Z.G., Kemler, R., Kingston, R., Wu, C., Fishman, M. 2002. Reptin and pontin antagonistically regulate heart growth in zebrafish embryos. Cell 111, 661–672.

Sachdev, S., Bruhn, L., Sieber, H., Pichler, A., Melchior, F., Grosschedl, R. 2001. PIASy, a nuclear matrix-associated SUMO E3 ligase, represses LEF1 activity by sequestration into nuclear bodies. Genes Dev. 15, 3088–3103.

Sakamoto, I., Kishida, S., Fukui, A., Kishida, M., Yamamoto, H., Hino, S., Michiue, T., Takada, S., Asashima, M., Kikuchi, A. 2000. A novel beta-catenin-binding protein inhibits beta-catenin-dependent Tcf activation and axis formation. J. Biol. Chem. 275, 32871–32878.

Sansom, O.J., Reed, K.R., Hayes, A.J., Ireland, H., Brinkmann, H., Newton, I.P., Batlle, E., Simon-Assmann, P., Clevers, H., Nathke, I.S., Clarke, A.R., Winton, D.J., et al. 2004. Loss of Apc *in vivo* immediately perturbs Wnt signaling, differentiation, and migration. Genes Dev. 18, 1385–1390.

Schohl, A., Fagotto, F. 2003. A role for maternal beta-catenin in early mesoderm induction in *Xenopus*. EMBO J. 22, 3303–3313.

Schweizer, L., Nellen, D., Basler, K. 2003. Requirement for Pangolin/dTCF in *Drosophila* Wingless signaling. Proc. Natl. Acad. Sci. USA 100, 5846–5851.

Shetty, P., Lo, M.C., Robertson, S.M., Lin, R. 2005. C. elegans TCF protein, POP-1, converts from repressor to activator as a result of Wnt-induced lowering of nuclear levels. Dev. Biol. 285, 584–592.

Shin, T.H., Yasuda, J., Rocheleau, C.E., Lin, R., Soto, M., Bei, Y., Davis, R.J., Mello, C.C. 1999. MOM-4, a MAP kinase kinase kinase-related protein, activates WRM-1/LIT-1 kinase to transduce anterior/posterior polarity signals in *C. elegans*. Mol. Cell 4, 275–280.

Shulewitz, M., Soloviev, I., Wu, T., Koeppen, H., Polakis, P., Sakanaka, C. 2006. Repressor roles for TCF-4 and Sfrp1 in Wnt signaling in breast cancer. Oncogene in press.

Siegfried, K.R., Kimble, J. 2002. POP-1 controls axis formation during early gonadogenesis in *C. elegans*. Development 129, 443–453.

Siegfried, E., Wilder, E.L., Perrimon, N. 1994. Components of wingless signalling in *Drosophila*. Nature 367, 76–80.

Siegfried, K.R., Kidd, A.R., 3rd, Chesney, M.A., Kimble, J. 2004. The sys-1 and sys-3 genes cooperate with Wnt signaling to establish the proximal-distal axis of the *Caenorhabditis elegans* gonad. Genetics 166, 171–186.

Sierra, J., Yoshida, T., Joazeiro, C.A., Jones, K.A. 2006. The APC tumor suppressor counteracts beta-catenin activation and H3K4 methylation at Wnt target genes. Genes Dev. 20, 586–600.

Sinner, D., Rankin, S., Lee, M., Zorn, A.M. 2004. Sox17 and beta-catenin cooperate to regulate the transcription of endodermal genes. Development 131, 3069–3080.

Smith, W.C., Harland, R.M. 1991. Injected Xwnt-8 RNA acts early in *Xenopus* embryos to promote formation of a vegetal dorsalizing center. Cell 67, 753–765.

Sokol, S., Christian, J.L., Moon, R.T., Melton, D.A. 1991. Injected Wnt RNA induces a complete body axis in *Xenopus* embryos. Cell 67, 741–752.

Song, L.N., Herrell, R., Byers, S., Shah, S., Wilson, E.M., Gelmann, E.P. 2003. Beta-catenin binds to the activation function 2 region of the androgen receptor and modulates the effects of the N-terminal domain and TIF2 on ligand-dependent transcription. Mol. Cell. Biol. 23, 1674–1687.

Spink, K.E., Polakis, P., Weis, W.I. 2000. Structural basis of the Axin-adenomatous polyposis coli interaction. EMBO J. 19, 2270–2279.

Stadeli, R., Basler, K. 2005. Dissecting nuclear Wingless signalling: Recruitment of the transcriptional co-activator Pygopus by a chain of adaptor proteins. Mech. Dev. 122, 1171–1182.

Standley, H.J., Destree, O., Kofron, M., Wylie, C., Heasman, J. 2006. Maternal XTcf1 and XTcf4 have distinct roles in regulating Wnt target genes. Dev. Biol. 289, 318–328.

Suh, E.K., Gumbiner, B.M. 2003. Translocation of beta-catenin into the nucleus independent of interactions with FG-rich nucleoporins. Exp. Cell Res. 290, 447–456.

Sun, Y., Kolligs, F.T., Hottiger, M.O., Mosavin, R., Fearon, E.R., Nabel, G.J. 2000. Regulation of beta-catenin transformation by the p300 transcriptional coactivator. Proc. Natl. Acad. Sci. USA 97, 12613–12618.

Tago, K., Nakamura, T., Nishita, M., Hyodo, J., Nagai, S., Murata, Y., Adachi, S., Ohwada, S., Morishita, Y., Shibuya, H., Akiyama, T. 2000. Inhibition of Wnt signaling by ICAT, a novel beta-catenin-interacting protein. Genes Dev. 14, 1741–1749.

Takada, S., Stark, K.L., Shea, M.J., Vassileva, G., McMahon, J.A., McMahon, A.P. 1994. Wnt-3a regulates somite and tailbud formation in the mouse embryo. Genes Dev. 8, 174–189.

Takash, W., Canizares, J., Bonneaud, N., Poulat, F., Mattei, M.G., Jay, P., Berta, P. 2001. SOX7 transcription factor: Sequence, chromosomal localisation, expression, transactivation and interference with Wnt signalling. Nucleic Acids Res. 29, 4274–4283.

Takemaru, K.I., Moon, R.T. 2000. The transcriptional coactivator CBP interacts with beta-catenin to activate gene expression. J. Cell Biol. 149, 249–254.

Takemaru, K., Yamaguchi, S., Lee, Y.S., Zhang, Y., Carthew, R.W., Moon, R.T. 2003. Chibby, a nuclear beta-catenin-associated antagonist of the Wnt/Wingless pathway. Nature 422, 905–909.

Tao, Q., Yokota, C., Puck, H., Kofron, M., Birsoy, B., Yan, D., Asashima, M., Wylie, C.C., Lin, X., Heasman, J. 2005. Maternal wnt11 activates the canonical wnt signaling pathway required for axis formation in *Xenopus* embryos. Cell 120, 857–871.

Tetsu, O., McCormick, F. 1999. Beta-catenin regulates expression of cyclin D1 in colon carcinoma cells. Nature 398, 422–426.

Thompson, B., Townsley, F., Rosin-Arbesfeld, R., Musisi, H., Bienz, M. 2002. A new nuclear component of the Wnt signalling pathway. Nat. Cell Biol. 4, 367–373.

Thompson, B.J. 2004. A complex of Armadillo, Legless, and Pygopus coactivates dTCF to activate wingless target genes. Curr. Biol. 14, 458–466.

Thorpe, C.J., Moon, R.T. 2004. Nemo-like kinase is an essential co-activator of Wnt signaling during early zebrafish development. Development 131, 2899–2909.

Thorpe, C.J., Schlesinger, A., Carter, J.C., Bowerman, B. 1997. Wnt signaling polarizes an early *C. elegans* blastomere to distinguish endoderm from mesoderm. Cell 90, 695–705.

Tolwinski, N.S., Wieschaus, E. 2001. Armadillo nuclear import is regulated by cytoplasmic anchor Axin and nuclear anchor dTCF/Pan. Development 128, 2107–2117.

Tolwinski, N.S., Wieschaus, E. 2004. Rethinking WNT signaling. Trends Genet. 20, 177–181.

Tolwinski, N.S., Wehrli, M., Rives, A., Erdeniz, N., DiNardo, S., Wieschaus, E. 2003. Wg/Wnt signal can be transmitted through arrow/LRP5,6 and Axin independently of Zw3/Gsk3beta activity. Dev. Cell 4, 407–418.

Townsley, F.M., Cliffe, A., Bienz, M. 2004a. Pygopus and Legless target Armadillo/beta-catenin to the nucleus to enable its transcriptional co-activator function. Nat. Cell Biol. 6, 626–633.

Townsley, F.M., Thompson, B., Bienz, M. 2004b. Pygopus residues required for its binding to Legless are critical for transcription and development. J. Biol. Chem. 279, 5177–5183.

Treisman, J.E., Luk, A., Rubin, G.M., Heberlein, U. 1997. Eyelid antagonizes wingless signaling during *Drosophila* development and has homology to the Bright family of DNA-binding proteins. Genes Dev. 11, 1949–1962.

Tsuji, S., Hashimoto, C. 2005. Choice of either beta-catenin or Groucho/TLE as a co-factor for Xtcf-3 determines dorsal-ventral cell fate of diencephalon during *Xenopus* development. Dev. Genes Evol. 215, 275–284.

Tutter, A.V., Fryer, C.J., Jones, K.A. 2001. Chromatin-specific regulation of LEF-1-beta-catenin transcription activation and inhibition *in vitro*. Genes Dev. 15, 3342–3354.

Valenta, T., Lukas, J., Korinek, V. 2003. HMG box transcription factor TCF-4's interaction with CtBP1 controls the expression of the Wnt target Axin2/Conductin in human embryonic kidney cells. Nucleic Acids Res. 31, 2369–2380.

van de Wetering, M., Castrop, J., Korinek, V., Clevers, H. 1996. Extensive alternative splicing and dual promoter usage generate Tcf-1 protein isoforms with differential transcription control properties. Mol. Cell. Biol. 16, 745–752.

van de Wetering, M., Cavallo, R., Dooijes, D., van Beest, M., van Es, J., Loureiro, J., Ypma, A., Hursh, D., Jones, T., Bejsovec, A., Peifer, M., Mortin, M., et al. 1997. Armadillo coactivates transcription driven by the product of the *Drosophila* segment polarity gene dTCF. Cell 88, 789–799.

van de Wetering, M., Oosterwegel, M., Dooijes, D., Clevers, H. 1991. Identification and cloning of TCF-1, a T lymphocyte-specific transcription factor containing a sequence-specific HMG box. EMBO J. 10, 123–132.

van de Wetering, M., Sancho, E., Verweij, C., de Lau, W., Oving, I., Hurlstone, A., van der Horn, K., Batlle, E., Coudreuse, D., Haramis, A.-P., Tjon-Pon-Fong, M., Moerer, P. 2002. The beta-catenin/TCF-4 complex imposes a crypt progenitor phenotype on colorectal cancer cells. Cell 111, 241–250.

van Es, J.H., Jay, P., Gregorieff, A., van Gijn, M.E., Jonkheer, S., Hatzis, P., Thiele, A., van den Born, M., Begthel, H., Brabletz, T., Taketo, M.M., Clevers, H., et al. 2005. Wnt signalling induces maturation of Paneth cells in intestinal crypts. Nat. Cell Biol. 7, 381–386.

van Genderen, C., Okamura, R.M., Farinas, I., Quo, R.G., Parslow, T.G., Bruhn, L., Grosschedl, R. 1994. Development of several organs that require inductive epithelial-mesenchymal interactions is impaired in LEF-1-deficient mice. Genes Dev. 8, 2691–2703.

Verbeek, S., Izon, D., Hofhuis, F., Robanus-Maandag, E., te Riele, H., van de Wetering, M., Oosterwegel, M., Wilson, A., MacDonald, H.R., Clevers, H. 1995. An HMG-box-containing T-cell factor required for thymocyte differentiation. Nature 374, 70–74.

Vleminckx, K., Kemler, R., Hecht, A. 1999. The C-terminal transactivation domain of beta-catenin is necessary and sufficient for signaling by the LEF-1/beta-catenin complex in *Xenopus* laevis. Mech. Dev. 81, 65–74.

Waltzer, L., Bienz, M. 1998. *Drosophila* CBP represses the transcription factor TCF to antagonize Wingless signalling. Nature 395, 521–525.

Wei, C.L., Wu, Q., Vega, V.B., Chiu, K.P., Ng, P., Zhang, T., Shahab, A., Yong, H.C., Fu, Y., Weng, Z., Liu, J., Zhao, X.D., et al. 2006. A global map of p53 transcription-factor binding sites in the human genome. Cell 124, 207–219.

Weiske, J., Huber, O. 2005. The histidine triad protein Hint1 interacts with Pontin and Reptin and inhibits TCF-beta-catenin-mediated transcription. J. Cell Sci. 118, 3117–3129.

Wiechens, N., Heinle, K., Englmeier, L., Schohl, A., Fagotto, F. 2004. Nucleo-cytoplasmic shuttling of Axin, a negative regulator of the Wnt-beta-catenin Pathway. J. Biol. Chem. 279, 5263–5267.

Willert, K., Shibamoto, S., Nusse, R. 1999. Wnt-induced dephosphorylation of axin releases beta-catenin from the axin complex. Genes Dev. 13, 1768–1773.

Willis, T.G., Zalcberg, I.R., Coignet, L.J., Wlodarska, I., Stul, M., Jadayel, D.M., Bastard, C., Treleaven, J.G., Catovsky, D., Silva, M.L., Dyer, M.J. 1998. Molecular cloning of translocation t(1;14)(q21;q32) defines a novel gene (BCL9) at chromosome 1q21. Blood 91, 1873–1881.

Wolf, D., Rodova, M., Miska, E.A., Calvet, J.P., Kouzarides, T. 2002. Acetylation of beta-catenin by CREB-binding protein (CBP). J. Biol. Chem. 277, 25562–25567.

Xing, Y., Clements, W.K., Le Trong, I., Hinds, T.R., Stenkamp, R., Kimelman, D., Xu, W. 2004. Crystal structure of a beta-catenin/APC complex reveals a critical role for APC phosphorylation in APC function. Mol. Cell 15, 523–533.

Yamaguchi, T.P., Takada, S., Yoshikawa, Y., Wu, N., McMahon, A.P. 1999. T (Brachyury) is a direct target of Wnt3a during paraxial mesoderm specification. Genes Dev. 13, 3185–3190.

Yamamoto, H., Ihara, M., Matsuura, Y., Kikuchi, A. 2003. Sumoylation is involved in beta-catenin-dependent activation of Tcf-4. EMBO J. 22, 2047–2059.

Yamamoto, H., Kishida, S., Kishida, M., Ikeda, S., Takada, S., Kikuchi, A. 1999. Phosphorylation of axin, a Wnt signal negative regulator, by glycogen synthase kinase-3beta regulates its stability. J. Biol. Chem. 274, 10681–10684.

Yang, C.K., Kim, J.H., Li, H., Stallcup, M.R. 2006. Differential use of functional domains by coiled-coil coactivator in its synergistic coactivator function with beta-catenin or GRIP1. J. Biol. Chem. 281, 3389–3397.

Yang, J., Tan, C., Darken, R.S., Wilson, P.A., Klein, P.S. 2002. Beta-catenin/Tcf-regulated transcription prior to the midblastula transition. Development 129, 5743–5752.

Yang, X., van Beest, M., Clevers, H., Jones, T., Hursh, D.A., Mortin, M.A. 2000. Decapentaplegic is a direct target of dTcf repression in the *Drosophila* visceral mesoderm. Development 127, 3695–3702.

Yoda, A., Kouike, H., Okano, H., Sawa, H. 2005. Components of the transcriptional Mediator complex are required for asymmetric cell division in *C. elegans*. Development 132, 1885–1893.

Yokoya, F., Imamoto, N., Tachibana, T., Yoneda, Y. 1999. Beta-catenin can be transported into the nucleus in a Ran-unassisted manner. Mol. Biol. Cell 10, 1119–1131.

Yook, J.I., Li, X.Y., Ota, I., Fearon, E.R., Weiss, S.J. 2005. Wnt-dependent regulation of the E-cadherin repressor snail. J. Biol. Chem. 280, 11740–11748.

Yost, C., Torres, M., Miller, J.R., Huang, E., Kimelman, D., Moon, R.T. 1996. The axis-inducing activity, stability, and subcellular distribution of beta-catenin is regulated in *Xenopus* embryos by glycogen synthase kinase 3. Genes Dev. 10, 1443–1454.

Zeng, L., Fagotto, F., Zhang, T., Hsu, W., Vasicek, T.J., Perry, W.L., 3rd, Lee, J.J., Tilghman, S.M., Gumbiner, B.M., Costantini, F. 1997. The mouse Fused locus encodes Axin, an inhibitor of the Wnt signaling pathway that regulates embryonic axis formation. Cell 90, 181–192.

Zeng, Y.A., Verheyen, E.M. 2004. Nemo is an inducible antagonist of Wingless signaling during *Drosophila* wing development. Development 131, 2911–2920.

Zhang, H., Emmons, S.W. 2000. A *C. elegans* mediator protein confers regulatory selectivity on lineage-specific expression of a transcription factor gene. Genes Dev. 14, 2161–2172.

Zhou, P., Byrne, C., Jacobs, J., Fuchs, E. 1995. Lymphoid enhancer factor 1 directs hair follicle patterning and epithelial cell fate. Genes Dev. 9, 700–713.

Zhou, B.P., Deng, J., Xia, W., Xu, J., Li, Y.M., Gunduz, M., Hung, M.C. 2004. Dual regulation of Snail by GSK-3beta-mediated phosphorylation in control of epithelial-mesenchymal transition. Nat. Cell Biol. 6, 931–940.

Ziegler, S., Rohrs, S., Tickenbrock, L., Moroy, T., Klein-Hitpass, L., Vetter, I.R., Muller, O. 2005. Novel target genes of the Wnt pathway and statistical insights into Wnt target promoter regulation. FEBS J. 272, 1600–1615.

Zorn, A.M., Barish, G.D., Williams, B.O., Lavender, P., Klymkowsky, M.W., Varmus, H.E. 1999. Regulation of Wnt signaling by Sox proteins: XSox17 alpha/beta and XSox3 physically interact with beta-catenin. Mol. Cell 4, 487–498.

Wnt signaling and the establishment of cell polarity

Gretchen L. Dollar and Sergei Y. Sokol

Department of Molecular, Cell, and Developmental Biology,
Mount Sinai School of Medicine, New York, New York

Contents

Advances in Developmental Biology
Volume 17 ISSN 1574-3349
DOI: 10.1016/S1574-3349(06)17002-7

Wnt signaling involves multiple mediators and molecular targets functions to regulate cell division, polarity, shape, motility, and fate. In addition to the best-studied regulation of gene expression by the Wnt/β-catenin or canonical pathway, Wnt proteins signal via Frizzled and Disheveled to regulate epithelial cell polarity, asymmetric cell divisions, and cell movements by noncanonical pathways. These include the planar cell polarity pathway in the fly wing and a similar pathway that controls convergent extension movements in vertebrates and mammalian cochlear cell polarization. Although historically cell polarity decisions have been associated exclusively with the noncanonical pathways, there is an emerging role for β-catenin-dependent Wnt signaling in cell polarization. This chapter discusses multiple targets of Wnt pathways implicated in cell polarity regulation during embryonic development and the relationship between canonical and noncanonical Wnt pathways.

1. Wnt-signaling pathways

The Wnt pathway is a complex signaling network that involves multiple signaling mediators and molecular targets. In the β-catenin-dependent pathway, commonly referred to as the "canonical" Wnt pathway, signal transduction is activated by the binding of Wnt ligands to the seven transmembrane Frizzled (Fz) receptor and the coreceptor LRP5/6 (Vinson et al., 1989; Tamai et al., 2000, 2004). This results in the activation of Disheveled (Dsh), a cytoplasmic protein containing a conserved DIX, DEP, and PDZ domain, and a basic domain of unknown function (Klingensmith et al., 1994; Pan et al., 2004). Dsh acts through an unknown mechanism to inhibit the ability of Axin, APC, and GSK-3 to target β-catenin for destruction. Stabilized β-catenin accumulates in the cytoplasm and the nucleus where, in combination with TCF-transcription factors, it activates target gene transcription (Logan and Nusse, 2004).

Wnt pathways that are not β-catenin dependent are referred to as "noncanonical" pathways. These include the calcium pathway, which will not be considered in this chapter, and the planar cell polarity (PCP) pathway. PCP signaling acts through Fz and Dsh, where it is thought to diverge from the canonical pathway. Other "core" PCP components include the seven transmembrane atypical cadherin Flamingo/Starry night (Fmi) (Chae et al., 1999; Usui et al., 1999), Strabismus/Van Gogh (Stbm/Vangl1/2), a four transmembrane protein with a C-terminal PDZ-domain-binding motif (Wolff and Rubin, 1998), the LIM domain protein Prickle (Pk) (Gubb et al., 1999), and the ankyrin-repeat protein Diego (Dgo) (Feiguin et al., 2001). These proteins act together in a feedback loop to spatially restrict Fz signaling, resulting in asymmetry in the plane of the tissue. PCP signaling has also been implicated in the control of convergent extension (CE) movements that occur during gastrulation and neurulation in vertebrate embryos (Wallingford et al., 2002) and in the polarization of mammalian cochlear cells (Montcouquiol et al., 2003).

The relationship between canonical and noncanonical Wnt pathways is not well understood. On one hand, there is no evidence for the involvement of Wingless (Wg), a *Drosophila* Wnt ligand, in fly PCP signaling, and a good argument can be made that PCP can be established in *Drosophila* without direct ligand involvement (Ma et al., 2003; Simon, 2004). However, a Wnt signal appears to be required for both CE and inner ear polarity in vertebrate embryos (Heisenberg et al., 2000; Dabdoub et al., 2003; Dabdoub and Kelley, 2005). Although historically cell polarity decisions have been associated exclusively with the noncanonical pathways, there is an emerging role for β-catenin-dependent Wnt signaling in cell polarization (Kimura-Yoshida et al., 2005; Price et al., 2006). In addition, data suggests that the role for Fz and Dsh is not restricted to planar cell polarity but may regulate apical–basal (A/B) polarity as well (Dollar et al., 2005). This chapter will discuss multiple molecular targets of Wnt pathways implicated in cell polarity regulation during embryonic development.

2. Apical–basal and Planar Cell Polarity

All epithelial cells have distinct A/B polarity. That is, some proteins are exclusively localized to the top or the bottom of the cell. In epithelial cells, these domains are separated by cell junctions, which serve to prevent the mixing of proteins from the two regions and maintain a rigid epithelial structure. Neuronal precursors also have A/B polarity, which is required for neuronal cell fate choice through asymmetric cell division (Roegiers and Jan, 2004). Migrating cells are dynamically polarized to create a leading edge required for cell movement, and many A/B determinants contribute to this polarization (Plant et al., 2003). A/B determinants include an apical complex consisting of atypical protein kinase C (aPKC) and the PDZ-containing proteins Par3 [Bazooka (Baz) in *Drosophila*] and Par6. The Par6/aPKC complex can bind to Lethal giant larvae (Lgl), a basolateral determinant, resulting in phosphorylation and inactivation of Lgl on the apical side. Lgl is then directed to the basolateral surface, where it functions in complex with Discs Large (Dlg) and Scribble (Scrib) to specify the basolateral domain, presumably by modulating vesicular transport (Musch et al., 2002; Betschinger et al., 2003; Plant et al., 2003; Yamanaka et al., 2003; Hutterer et al., 2004; Gangar et al., 2005). Binding of Par3 and Lgl to the Par6/aPKC complex is mutually exclusive. Therefore, inactivation of Lgl at the apical surface promotes the formation of the Par3/Par6/aPKC complex. In epithelial cells, this regulation results in the mutual exclusion of these molecules and promotes the formation of cell junctions (Yamanaka et al., 2003). Another protein complex, consisting of the transmembrane protein Crumbs (Crb), the PDZ-containing protein Stardust (Std/Pals1), and Pals1-associated tight junction protein (Patj), is localized to the apical domain by Par3 and is required for A/B establishment (Bilder et al., 2003). Crb functions redundantly with Par3 to antagonize Lgl activity, and Crb is not required for apical identity in the absence of Lgl (Tanentzapf and Tepass, 2003). Loss of Lgl, Dlg, or Scrib results in an expansion of the apical domain at the expense of the basolateral domain, and loss of Par3, Par6, aPKC, Crb, or Std causes an expansion of the basolateral domain at the expense of the apical domain (Bilder et al., 2000, 2003; Tanentzapf and Tepass, 2003; Hutterer et al., 2004).

In *Drosophila* neuroblasts, some mechanisms of A/B polarity establishment are conserved (Bilder, 2001). For example, the Par3/Par6/aPKC complex is localized to an apical crescent, and phosphorylation of Lgl by Par6/aPKC inactivates it at the apical surface (Betschinger et al., 2003). However, Lgl is not required for the apical localization of Par3/Par6/aPKC in neuroblasts but is required for the localization of the cell fate determinant Numb, a Notch antagonist, to a basal crescent (Ohshiro et al., 2000; Peng et al., 2000). The Par3/Par6/aPKC complex is also required for the apical targeting of the proteins Inscuteable (Insc), Partner of Inscuteable (Pins), and Gαi, which orient the mitotic spindle perpendicular to the basal Numb crescent to result in asymmetric cell division. The basal daughter cell exclusively inherits Numb and differentiates into a neuron, and the apical daughter remains a neuroblast (Ohshiro et al., 2000; Peng et al., 2000; Fig. 1B).

PCP forms in an axis perpendicular to the original A/B axis within the plane of the tissue. PCP has been described in a number of epithelial tissues, such as the orientation of hairs in the *Drosophila* wing and notum and in the mouse epidermis. PCP is also seen in other cell types such as the *Drosophila* eye, sensory organ precursor (SOP) cells of the *Drosophila* peripheral nervous system, and the sensory hair cells of the mouse inner ear (Roegiers et al., 2001; Montcouquiol and Kelley, 2003; Montcouquiol et al., 2003; Strutt, 2003). While molecular and genetic interactions between the core PCP proteins are largely conserved in different cell types (Fig. 1C), the downstream targets of the pathway can be quite diverse. The cellular mechanisms that involve the PCP pathway components vary from the direct influence on the cytoskeletal organization in the wing and mitotic spindle orientation in sensory organ precursors to mediating gene transcription in specific photoreceptor cells in the eye (Bellaiche et al., 2001a; Strutt, 2002; Strutt et al., 2002). The establishment of PCP is usually considered to be independent of the establishment of the A/B polarity; however, evidence is accumulating that many key A/B-polarity determinants play a role in PCP signaling (see Section 3). Therefore, PCP may be derived from the preexistent A/B polarity in the same epithelial tissue.

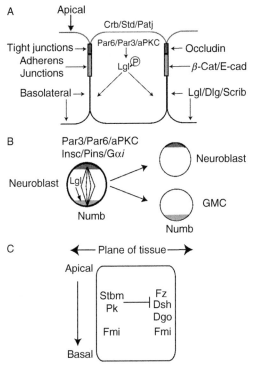

Fig. 1. Apical–basal and planar cell polarity. (A) Schematic diagram illustrating apical–basal (A/B) epithelial polarity and the subcellular localization of essential proteins. The apically localized Par6/Par3/aPKC and Crb/Std/Patj complexes antagonize the basolateral polarity determinants, including Lgl. Upon phosphorylation by aPKC, Lgl is targeted to the basolateral cortex, where it functions with Dlg and Scrib to establish the basolateral domain. β-Catenin and E-cadherin are localized to adherens junctions and Occludin to tight junctions. (B) An interaction of the Par3/Par6/aPKC complex and Lgl is also required to polarize *Drosophila* neuroblasts. In these cells, Lgl is necessary to form a basal crescent of Numb, a negative regulator of Notch. The Insc/Pins/Gαi complex, recruited apically by Par3, orients the mitotic spindle along the A/B axis. Upon division, the daughter cell that inherits Numb differentiates into a ganglion mother cell (GMC), while the other daughter cell remains a neuroblast. (C) Planar cell polarity (PCP) is generated perpendicular to the A/B axis. Core PCP components are localized asymmetrically in the plane of tissue, and their molecular interactions are highly conserved among various cell types. Fz, Dsh, and Dgo localize to one side of the cell, whereas Stbm and Pk localize to the opposite side of the cell where they inhibit Fz signaling. Fmi is localized to both lateral cortexes where it may promote PCP signaling across cell membranes. (See Color Insert.)

3. Regulation of apical–basal polarity by Wnt signaling

Although A/B polarity has been considered to be a prerequisite for PCP establishment, Wnt pathway components have not been previously implicated in A/B polarity decisions. However, several experiments have suggested a genetic link between A/B and PCP determinants. For example, PCP defects of the inner ear were described in mice homozygous mutant for the basolateral determinant *Scrib*, as well as *Vangl2/+;Scrib/+* double heterozygotes, demonstrating a genetic interaction between a PCP and an A/B polarity determinant (Bilder and Perrimon, 2000; Montcouquiol et al., 2003). Other evidence for the involvement of A/B determinants in PCP signaling comes from the report that aPKC and Par1 are required for CE movements in *Xenopus* embryos (Kusakabe and Nishida, 2004). Par1 was identified by virtue of its requirement for the first asymmetric division of the *Caenorhabditis elegans* embryo (Kemphues et al., 1988; Bowerman and Shelton, 1999) and has since been shown to be required for the basolateral identity of mammalian epithelial cells (Hurov et al., 2004; Suzuki et al., 2004). Par1 was first implicated in the positive regulation of the canonical Wnt pathway and was shown to phosphorylate Dsh, even though the functional consequence of this phosphorylation was not clear (Sun et al., 2001). More recently, different Par1 isoforms have been shown to act preferentially in canonical Wnt or PCP signaling. Moreover, the phosphorylation of Dsh by Par1 may affect Fz recruitment of Dsh to the membrane, which is thought to be important for noncanonical signaling (Ossipova et al., 2005). aPKC phosphorylation of Par1 is required for Par1 membrane localization and is essential for both proper A/B polarity and CE movements, indicating a conserved molecular mechanism (Hurov et al., 2004; Kusakabe and Nishida, 2004; Suzuki et al., 2004). Functional interactions between A/B and PCP determinants have also been described in *Drosophila* SOP cells. The anterior localization of Dlg is dependent on Stbm, and the posterior localization of Par3 is dependent on Fz activity (Bellaiche et al., 2001b, 2004).

Despite these indications, demonstration of direct interactions between A/B and PCP components has remained somewhat elusive. However, two papers have now shown such interactions (Djiane et al., 2005; Dollar et al., 2005). One of these studies has demonstrated the binding of Lgl to Dsh in a yeast two-hybrid system and *in vivo* in *Xenopus* embryos (Dollar et al., 2005), suggesting that A/B polarity may be subjected to direct regulation by the Wnt pathway. The region of Dsh that binds to Lgl contains the DIX and the conserved basic domain, regions not previously associated with PCP signaling but required for canonical signaling (Penton et al., 2002; Pan et al., 2004).

Consistent with other epithelial cell types, in the superficial ectoderm of *Xenopus* embryos Lgl and aPKC are localized to the basolateral and the apical domains, respectively (Chalmers et al., 2003; Dollar et al., 2005; Fig. 2A–C), suggesting a conserved mechanism of A/B establishment in this cell type. Moreover, overexpression of Lgl in these cells results in A/B polarity defects, likely due to a saturation of endogenous aPKC regulation, and a subsequent gain-of-function (GOF) activity at the apical surface. In support of this, the expression of a GFP-Lgl construct mutated for consensus aPKC

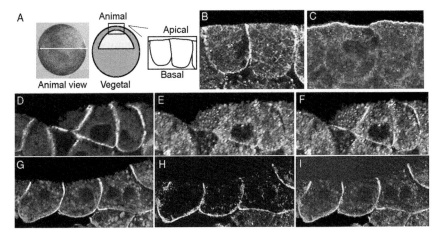

Fig. 2. Frizzled regulates Lgl localization. (A–C) A/B polarity of *Xenopus* superficial ectoderm. (A) Schematic of cross-sections are shown in this figure. (B) Lgl is localized to the basolateral cortex. (C) aPKC is localized to the apical cortex. This subcellular localization suggests a conserved mechanism for polarity determination. (D–F) The mislocalization of Lgl by Frizzled 8 in the ectoderm. (D) Dsh-GFP coexpressed with Fz8 is localized to the basolateral domain. This suggests an asymmetry of Fz signaling along the A/B axis. (E) Fz8 dissociates Lgl from the basolateral membrane, suggesting that it can influence A/B polarity. (F) Merged image of D and E. (G) Fz7 also recruits Dsh to the basolateral membrane but has no effect on Lgl localization (H). This indicates that the effect of Fz8 on Lgl localization is specific and does not result from Dsh recruitment to the membrane. (I) Merged image of G and H. (See Color Insert.)

phosphorylation sites is localized primarily to the apical surface and results in similar polarity defects at lower levels (Dollar et al., 2005).

This vertebrate system was utilized to investigate the role of Dsh for Lgl function and A/B polarity. Strikingly, depletion of Dsh caused a loss of Lgl GOF activity and delocalization of Lgl from the membrane. Moreover, depletion of Lgl or Dsh caused similar A/B polarity defects, including the loss of aPKC and ectopic Occludin localization at the apical surface. This suggests that Dsh plays a role in A/B polarity through the regulation of Lgl. It should be noted, however, that the basolateral localization of β-catenin or β-1-integrin was not significantly altered upon Dsh or Lgl depletion (Dollar et al., 2005). This could be due to a more stable or static localization of these proteins, the presence of other Dsh and Lgl isoforms in *Xenopus* embryos or other redundant mechanisms for A/B polarity determination in this cell type.

How Dsh regulates Lgl localization is not entirely clear. Dsh may regulate Lgl protein stability, as Dsh depletion leads to a marked decrease in GFP-Lgl, but not GFP coexpressed in the same embryos. Consistent with this, over-expression of Dsh causes an increase in endogenous Lgl protein levels (Dollar et al., 2005). Regardless of mechanism, the regulation of Lgl membrane localization by Dsh seems to be evolutionarily conserved in the *Drosophila* ovary: Lgl was delocalized from the membrane in 38% of *dsh* mutant follicle cell clones. *Dsh* mutant clones do not phenotypically resemble *lgl* mutant clones, as loss of Lgl but not Dsh leads to overproliferation of cells, the loss of cell adhesion, and a disorganized, multilayered epithelium (Abdelilah-Seyfried et al., 2003). Intriguingly, while membrane Lgl localization was lost in a percentage of *dsh* mutant clones, nuclear Lgl staining was still evident. The functional significance of nuclear Lgl is not known but could potentially explain the phenotypic differences between *lgl* and *dsh* mutant follicle cell clones. While wild-type Lgl localizes almost exclusively to the membrane in *Xenopus* animal caps, an Lgl construct deleted for the C-terminal domain that binds to Dsh shows high levels of nuclear localization (Dollar and Sokol, unpublished data). The mechanism of Lgl regulation by Dsh and its possible connection to nuclear localization will be interesting topics for future research and may reveal new functions for Lgl in the nucleus.

Overexpression of the Fz8 receptor also resulted in Lgl membrane delocalization, suggesting that signaling can influence Dsh regulation of Lgl, as opposed to a generic housekeeping role for Dsh in Lgl protein stability and/or localization. While Fz8 overexpression consistently results in a slight decrease of GFP-Lgl, it does not result in the same dramatic reduction in protein levels caused by the knockdown of Dsh. This suggests a model in which Fz8 may locally destabilize Lgl at the membrane, but the mechanism of this delocalization is speculative. Nonetheless, evidence suggests that the effect on Lgl localization is Fz receptor specific because Fz7 had no effect on Lgl localization or stability (Dollar et al., 2005; Fig. 2D–I), while both Fz7 and Fz8 specifically recruit Dsh to the basolateral domain (Fig. 2D, F, G, I). This raises the question of how various Fz receptors differentially affect downstream effectors, a subject that is not well investigated. In *Drosophila* wing discs, the asymmetric localization of different Fz receptors along the A/B axis affects their signaling specificity. Fz1, localized to the apical region, acts primarily in PCP signaling and Fz2, localized uniformly along the A/B axis, is primarily involved in canonical Wnt signaling (Wu et al., 2004). While basolateral recruitment of Dsh in *Xenopus* ectoderm implies a role for Fz receptors in A/B polarity, the subcellular localization does not explain the differential effect of Fz7 and Fz8 on Lgl localization. Future experiments to determine the mechanism of Fz receptor specificity in Lgl regulation may provide further insight into how different Fz receptors are utilized *in vivo* to affect signaling outcome.

While it is not known if Fz/Dsh regulation of Lgl involves PCP signaling, canonical signaling, or represents a new signaling pathway, this regulation potentially has many implications. Lgl was first identified as a tumor suppressor, and the inappropriate activation of the Wnt pathway has been implicated in many types of cancer (Michler et al., 1985; Bilder et al., 2000;

Bilder, 2004). Furthermore, loss of polarity is a hallmark of tumor progression (Logan and Nusse, 2004). Defining interactions between the Wnt pathway and polarity determinants should provide further understanding of how the uncontrolled proliferation of tumor cells is linked to the loss of cell polarity.

These results also suggest a potential role for Lgl in the establishment of PCP. Localized inhibition of Lgl on one side of a cell due to relatively higher Fz signaling would be expected to promote the subcellular localization of Par6/Par3/aPKC, since Lgl competes with Par3 for binding in this complex (Yamanaka et al., 2003). Consistent with this, Fz and Dsh colocalize with Par6/Par3/aPKC to the posterior lateral cortex of *Drosophila* SOP cells after PCP signaling (Bellaiche et al., 2004; Fig. 6B). Moreover, both depletion of Lgl and the expression of an Lgl construct mutant for aPKC phosphorylation sites result in CE defects in *Xenopus* embryos (Dollar and Sokol, unpublished data), suggesting a role in PCP signaling. The analysis of Dsh and Lgl membrane localization in both wild-type and mutant cells undergoing PCP signaling *in vivo* or in explant assays should give further insight into how Fz/Dsh signaling may control cell polarity through the regulation of A/B determinants.

4. PCP signaling in the *Drosophila* wing

The *Drosophila* wing is a relatively simple and well-studied epithelial tissue in which PCP signaling occurs, and it can be used as a paradigm to illustrate basic physical and genetic interactions of the core PCP proteins and their effects on signaling. These basic mechanisms will then be extended to discuss more complex systems with PCP signaling such as the *Drosophila* eye, mammalian cochlea, and vertebrate CE movements of gastrulation and neurulation.

In wing epithelial cells, core PCP proteins segregate to the proximal or distal lateral cortices and direct the accumulation of actin at the distal vertex of the cell, a prerequisite to the wing hair. This results in the wing hairs pointing uniformly in a proximal-to-distal orientation (Fig. 3A, B). Many of the PCP genes were identified through their mutant wing phenotype with characteristic whorls of wing hairs (Fig. 3A). Subsequent molecular analysis has revealed disruption of the planar asymmetry of PCP proteins in these mutants, resulting in a central position of actin accumulation and randomized hair orientation within the apical region. The PCP proteins show a dynamic subcellular distribution during development. Initially, all PCP proteins colocalize to the apical cortex in an Fmi- and Fz-dependent manner (Strutt, 2001; Das et al., 2002). As development proceeds, Fz, Dsh, and Dgo are localized to the distal cortex, Pk and Stbm are localized to the proximal cortex, and Fmi is localized to both lateral cortices (Axelrod, 2001; Strutt, 2001; Tree et al., 2002; Bastock et al., 2003; Fig. 3B). The localization of

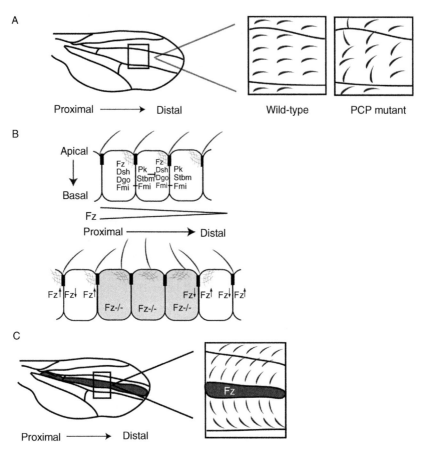

Fig. 3. PCP signaling during *Drosophila* wing development. (A) An illustration of *Drosophila* wing hair orientation. Hairs point uniformly in the proximal-to-distal direction in wild-type wings. PCP mutants exhibit a characteristic disruption of hair orientation resulting in a whorl pattern. (B) Schematic representation of individual wing hair cells illustrating PCP establishment. The core PCP components localize to the proximal and distal cortices, perpendicular to the A/B axis. Pk and Stbm are localized to the proximal side of the cell where they act to restrict Fz, Dsh, and Dgo to the distal side of the cell. Dgo is a positive regulator of Fz signaling. Fmi is localized to both lateral cortices where it likely promotes signaling across the membrane through homophilic interactions. The localized Fz activity results in the accumulation of the actin prehair at the distal vertex and a wing hair that points distally. Fz has a non-cell autonomous function (illustrated in the bottom panel). A *fz* mutant clone not only results in the random orientation of hair cells within the clone (dark cells) but also reorients hairs of wild-type cells on the distal end to point toward the clone. This is due to a relative change in Fz levels across cell membranes and is propagated several cell lengths away. (C) An assay for evaluating genetic interactions between PCP components in the wing. Expression of an *fz* transgene from the *dpp* promoter results in a proximal-to-distal stripe of *fz* expression. This results in hairs pointing away from the Fz expression domain, illustrating a non-cell autonomous role for Fz.

these proteins is mutually regulated, as loss of any one of the core proteins results in the delocalization of the others. The exception is apical Fmi localization, which appears to be controlled by components of both proximal and distal complexes (Klein and Mlodzik, 2005). These proteins are thought to act in an intercellular feedback loop to amplify preexisting small proximal-to-distal differences in Fz activity (Tree et al., 2002). This feedback loop reinforces higher Fz activity at the distal cortex of one cell relative to the proximal cortex of its neighbor, resulting in wing hairs pointing distally away from the highest Fz activity (Fig. 3A, B).

This model has been supported by genetic studies, which have provided further insight into how the opposing PCP protein complexes regulate each other. Overexpression of Fz from the *dpp* promoter (*dpp* > *fz*) results in an ectopic stripe of Fz expression along the proximal–distal axis and reorients surrounding hairs to point away from the Fz expression domain (Adler et al., 1997; Fig. 3C). Consistent with this observation, a mutant *fz* clone reverses the gradient of Fz activity at the distal clone boundary and reorients hairs in wild-type cells so that they point toward the mutant clone (Fig. 3B). These experiments illustrate a non-cell autonomous role for Fz (Strutt and Strutt, 2002; Fig. 3B, C).

The *dpp* promoter overexpression assay has been advantageous in establishing the genetic interactions of PCP genes. For example, ectopic expression of Dgo from the *dpp* promoter (*dpp* > *dgo*) phenocopies *dpp* > *fz*, while *dpp* > *pk* has the opposite effect, providing evidence that Dgo promotes and Pk antagonizes the Fz signal (Feiguin et al., 2001; Tree et al., 2002). Consistent with this, coexpressed Pk attenuates the ability of *dpp* > *dgo* to reorient hairs laterally. Stbm has no effect on *dpp* > *dgo* on its own but increases the effect of Pk on *dpp* > *dgo* when expressed together (Jenny et al., 2005).

Competition assays demonstrated that Pk and Dgo bind to Dsh in a mutually exclusive manner (Jenny et al., 2005), suggesting that Dgo may stimulate Fz/Dsh activity by binding Dsh and antagonizing the suppressive effect of Pk. Stbm is required to recruit Pk to the membrane, where they form a complex with Dsh to antagonize Fz signaling, although Stbm cannot antagonize Fz signaling on its own (Bastock et al., 2003; Jenny et al., 2005).

These physical and genetic interactions support a model in which Pk and Dgo influence PCP signaling through direct competition for Dsh binding (Jenny et al., 2005). Stbm may act to restrict Pk inhibition of Fz to the proximal side of the cell (Bastock et al., 2003; Jenny et al., 2005). It is not clear how Dgo binding to Dsh promotes Fz-PCP signaling. Dgo could act passively by sequestering Dsh from Pk. Alternatively, Dgo could play a more active role in promoting the ability of Dsh to signal. The interactions between the core PCP proteins ultimately result in localized Fz signaling at the distal tip of the cell, resulting in the localized activation of downstream effectors. In the wing, these include the small GTPase RhoA and Rho-associated kinase (dROK), a downstream effector of RhoA. dROK directly regulates the actin cytoskeleton, defining the precise location and orientation of the wing hair (Strutt et al., 1997; Fanto et al., 2000; Winter et al., 2001).

The atypical cadherins Fat (Ft) (Mahoney et al., 1991) and Dachsous (Ds) (Clark et al., 1995), and the type II transmembrane receptor four-jointed (Fj) have been proposed to represent long-range signals, which establish a bias in proximal-to-distal Fz activity in the wing (Villano and Katz, 1995; Ma et al., 2003). Fj is expressed in a distal-to-proximal gradient, whereas Ds is highest at the hinge and lower distally (Zeidler et al., 2000; Ma et al., 2003). However, while the gradients of these proteins are essential for PCP determination in the eye, they are not required in a gradient for wing PCP (Simon, 2004). Therefore, the signal required for the initial bias in Fz activity in the wing has yet to be identified.

5. PCP signaling in the *Drosophila* eye

PCP signaling in the eye imaginal disc is responsible for creating the highly ordered structure of the approximately 800 ommatidia that make up the adult *Drosophila* compound eye. Molecular and genetic interactions of the core PCP genes are largely conserved in the eye. However, PCP signaling in the eye results in gene transcription and cell fate specification (Strutt et al., 2002). Ommatidial clusters, each composed of eight photoreceptor cells (R1–R8), are formed behind the morphogenetic furrow that sweeps from the posterior to the anterior of the eye imaginal disc (Wolff and Ready, 1991). Photoreceptor cell fates are specified in a temporal (R8, followed by R2/R5, R3/R4, R2/R1, and R7) and spatial manner (Fig. 4A). The R3/R4 cell fate choice is determined spatially, with the cell closest to the dorsal/ventral (D/V) midline (equator) adopting the R3 fate and the neighboring precursor induced to become R4 (Mollereau and Domingos, 2005). Genetic mosaic analysis has shown that this fate is attributed to a higher level of Fz signaling in the future R3 cell (Zheng et al., 1995). The R3/R4 cell fate choice governs the direction of the subsequent 90° rotation of the ommatidial cluster and results in clusters of opposite chirality that are mirror images across the D/V boundary (Wolff and Ready, 1991; Mlodzik, 1999). Genetic and molecular interactions of the core PCP components are conserved from the wing to the eye. However, in the eye, PCP molecules act across the R3/R4 cell border, and PCP signaling primarily determines cell fate choice through gene transcription and subsequent Notch signaling (Strutt et al., 2002; Fig. 4B). The rotation of ommatidial clusters is secondary to the R3/R4 fate choice and requires Fz and EGF signaling (Strutt and Strutt, 2003). In *fz* and *dsh* mutants, the R3/R4 fate choice is perturbed, resulting in ommatidia with misplaced R3 and R4 cells, symmetrical R3/R4 cells, random direction of rotation, and over- or under-rotation (Zheng et al., 1995; Fig. 4A). Genetic mosaic analysis has determined that Fz, Dsh, and Dgo are required in the R3

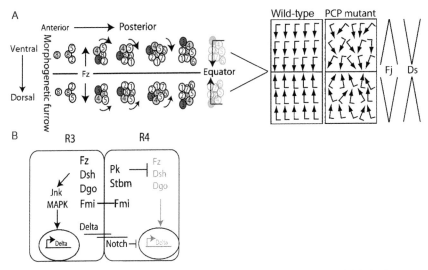

Fig. 4. PCP in the *Drosophila* eye. (A) Schematic representation of ommatidial development. Photoreceptor cell fates are specified posterior to the morphogenetic furrow in a temporal and spatial manner, as indicated. The R3/R4 cell fate choice is determined by a higher relative level of Fz signaling in the cell closest to the equator (the D/V boundary), which specifies the R3 fate. The relative location of the R3/R4 cells determines the direction of rotation of the ommatidial cluster, resulting in mirror image ommatidia across the equator. (B) Molecular interactions between the PCP genes are conserved across R3/R4 cell borders. Higher Fz signaling results in the downregulation of Notch signaling and R3 cell fate specification. (See Color Insert.)

cell (Zheng et al., 1995; Jenny et al., 2005), and Stbm and Pk are required in R4, where, as in the wing, they act to antagonize Fz signaling. Fmi is required in both R3 and R4, most likely to promote cell-to-cell communication through homophilic interactions (Usui et al., 1999; Feiguin et al., 2001; Das et al., 2002).

As in the wing, Fj and Ds are expressed in opposite gradients from the equator and the poles respectively (Fig. 4A). However, unlike in the wing, these proteins are required in a gradient for normal PCP establishment and act through Ft to bias higher Fz activity in the R3 cell (Zeidler et al., 2000; Yang et al., 2002; Simon, 2004).

Despite the genetic and molecular conservation of core PCP signaling in the eye, the downstream consequences are quite different. PCP specification of the R3/R4 cell fate has been shown to act through different downstream effectors than in the wing. Nevertheless, some conserved factors, such as the Rho family of GTPases, may be required for the ommatidial rotation, as well as involved redundantly in cell fate specification (Strutt et al., 1997; Fanto et al., 2000; Mlodzik, 2002). Factors shown to be involved in downstream gene regulation include the Msn/STE20-like kinase (Paricio et al., 1999) and the JNK/p38 MAPK cascade (Boutros et al., 1998; Weber et al., 2000).

These downstream effectors lead to expression of the Notch ligand *Delta* in R3 and subsequent Notch activation and *Delta* repression in R4 (Das et al., 2002; Strutt et al., 2002; Fig. 4B).

A study has demonstrated that A/B-polarity determinants can directly act upon PCP components to establish PCP in the eye (Djiane et al., 2005). Fz was shown to contain consensus aPKC phosphorylation sites, and to be phosphorylated by aPKC *in vitro*. Overexpression of a Fz1 construct that cannot be phosphorylated by aPKC had activity very similar to wild-type Fz1, whereas an aPKC phospho-mimetic Fz1 had very little GOF activity, despite its proper apical localization and ability to recruit Dsh to the membrane (Djiane et al., 2005). This suggests that aPKC phosphorylation may downregulate Fz activity.

The direct binding of A/B polarity determinants, including Par6, Par3, aPKC, Sdt, Patj, Scrib, Dlg, and Lgl, to Fz1 was analyzed in a yeast two-hybrid assay (Djiane et al., 2005; Fig. 1A). Of these, only Patj was found to bind directly to Fz, suggesting it may act as an adaptor molecule for binding to aPKC. Coimmunoprecipitation experiments in *Drosophila* S2 cell extracts confirmed Fz1 binding to Patj and additionally showed binding to Par3 and aPKC, indicating that Fz1 and Patj form a complex with aPKC and/or Par3 *in vivo*. Patj is the only member of the apical polarity complexes that does not have an A/B polarity mutant phenotype (Pielage et al., 2003). A constitutively active aPKC overexpressed in the eye resulted in characteristic PCP defects, and the expression of an Fz1 construct lacking Patj-binding sites caused a stronger GOF effect than wild-type Fz1, further illustrating the role of A/B determinants in PCP establishment.

Consistent with these findings, aPKC and Patj are downregulated in R3/R4 photoreceptor cells where Fz/PCP signaling is active (Djiane et al., 2005), and Par3 becomes enriched in the R3/R4 pair upon PCP establishment. Furthermore, removal of one copy of *baz/par3* suppresses the GOF phenotype of Fz1 but not the GOF phenotype of Dsh or Fmi, indicating that Par3 acts directly on Fz1. Together, these results suggest that Patj-mediated phosphorylation of Fz1 by aPKC downregulates Fz signaling and that Par3 can counteract this downregulation. Importantly, this study shows that A/B polarity is not just a prerequisite to PCP but that A/B components can directly act on the PCP proteins to coordinate polarization cues.

6. PCP signaling is conserved in mammalian epithelium

The best-characterized example of PCP in mammals is the uniform orientation of stereociliary bundles in sensory hair cells within the mouse cochlea. Each of these bundles consists of a group of actin-based stereocilia arranged in a crescent shape at the abneural edge of the cell, with the vertex oriented perpendicular to the A/B axis (Fig. 5A, B). This planar orientation of stereociliary bundles is crucial for proper auditory perception (Hudspeth and Jacobs, 1979). The orientation of stereociliary bundles in both inner and outer hair cells (IHCs and OHCs) is severely disrupted in *loop-tail* (*Lp*) mice (Fig. 5B), which are mutant for *Vangl2*, a mammalian homologue of *Stbm* (Kibar et al., 2001; Murdoch et al., 2001; Montcouquiol et al., 2003), suggesting that PCP regulation is conserved in mammals. This disruption is evident at the first sign of polarity, when the centrally located cilium, which will become the kinocilium at the vertex of the future bundle, moves to the abneural edge of developing hair cells. In *lp/lp* embryos, the cilium moves to the periphery, but the orientation is random (Montcouquiol et al., 2003)

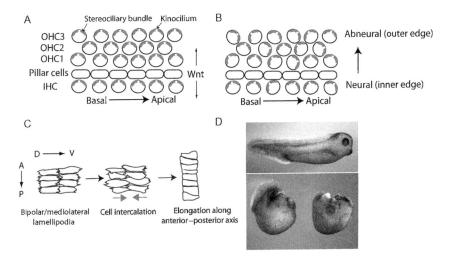

Fig. 5. Noncanonical Wnt signaling in vertebrates. (A) Schematic representation of a cross-sectional view of mechanosensory hair cells within the mouse inner ear, illustrating cellular organization and PCP of stereociliary bundles [inner hair cell (IHC), outer hair cell (OHC)]. Stereociliary bundles are composed of the kinocilium, which is derived from an initial central cilium, and 50–200 modified microvilli called stereocilia that are densely packed together to form a bundle. Stereocilia contain a dense core of filamentous actin surrounding the centrally located kinocilium. Stereociliary bundles are oriented perpendicular to the A/B axis, so that the kinocilium is at a 90° angle to the plane of tissue. Pillar cells, which separate the IHCs from the OHCs, have the highest level of Wnt ligand expression, suggesting a direct effect of Wnt ligands on planar polarity in vertebrates. Consistent with this, Fz3, Fz6, and Dsh-GFP are localized in a domain opposite to the kinocilium. (B) Schematic example of stereociliary bundle misorientation which occurs in PCP mutants. (C) Cell movements that occur during CE of vertebrate embryos. Cells polarize along the mediolateral axis, intercalate, and converge along the D/V axis, resulting in elongation along the A/P axis. (D) Morpholino knockdown of Dsh in *Xenopus* embryos causes CE defects. A severe case is shown that also exhibits an open neural tube (bottom panel). Top panel shows a control sibling embryo. (See Color Insert.)

(as illustrated in Fig. 5B). Other conserved genes from the PCP pathway have also been implicated in this process. Mice mutant for *Celsr1*, the mammalian homologue of *fmi*, have misoriented stereociliary bundles (Curtin et al., 2003), as do *dvl1−/−; dvl2−/−* and *fz3−/−; fz6−/−* double mutants (Wang et al., 2005, 2006). A conserved role for Fz in planar polarity signaling in mammals is supported by the observation that *fz6* mutant mice have hair whorls strikingly similar to the whorls in the *Drosophila* wing of PCP mutants (Guo et al., 2004).

Genetic and localization analysis of core PCP components in mammals provide further evidence for conservation of PCP signaling. *Dvl2* genetically interacts with *Vangl2*, as stereociliary bundles are misoriented in *dvl2−/−; Lp/+* double-mutant mice, but are normal in each individual mutation (Wang et al., 2005). Dvl2-GFP, Fz3, and Fz6 are localized asymmetrically in the plane of epithelium within the hair-cell opposite to the kinocilium (Wang et al., 2005, 2006). Furthermore, asymmetric localization of these proteins is disrupted in *lp/lp* mice (Wang et al., 2005, 2006). These data are consistent with the demonstrated interactions of PCP components in the *Drosophila* eye and wing, and strongly suggest a conserved mechanism for PCP establishment in mammalian epithelia.

The best evidence for an instructive Wnt gradient in PCP establishment comes from studies of the mammalian cochlea. Wnt7a is expressed asymmetrically in the cochlea, with the highest levels being present in cells located between the inner and outer hair cell rows when stereociliary bundles are being oriented (Dabdoub et al., 2003), consistent with a graded Wnt7a expression across IHCs and OHCs (Fig 5A). Furthermore, orientation is disrupted when cochlear explants were maintained in Wnt7a-cultured medium when compared to those in control media, suggesting that localized Wnt7a expression is important *in vivo* (Dabdoub and Kelley, 2005). Wnt7a mutant mice have normal stereociliary bundle orientation, but Wnt7a is likely to be redundant with other Wnts. Many Wnt ligands and Fz receptors are expressed in the cochlea, making loss-of-function analysis of these genes in this system technically problematic (Daudet et al., 2002). However, disruption of heparan sulfate proteoglycans (HSPGs), which are involved in Wnt/Fz interactions and are required for the formation of Wnt morphogen gradients, resulted in a disruption of stereociliary bundle formation, strongly suggesting that a Wnt gradient is required for polarization (Kjellen and Lindahl, 1991; Kispert et al., 1996; Baeg and Perrimon, 2000; Baeg et al., 2001; Dabdoub et al., 2003; Baeg et al., 2004). Consistent with this, treatment of cochlear explants with the Wnt inhibitors sFRP1 (Leyns et al., 1997; Xu et al., 1998; Uren et al., 2000; Jones and Jomary, 2002) or WIF1 (Hsieh et al., 1999) resulted in similar misoriented bundles as Wnt7a treatment (Dabdoub and Kelley, 2005). These Wnt inhibitors are expressed in the cochlea (Leimeister et al., 1998) and therefore may function to create a gradient of Wnt/PCP-signaling activity *in vivo* (Kawano and Kypta, 2003).

7. Noncanonical Wnt signaling and polarized cell movements

7.1. Vertebrate convergent extension

Morphogenetic movements that occur during gastrulation and neurulation are crucial for the establishment of the body axis during early vertebrate development. A major component of these movements is CE, a process of mediolateral cell intercalation in dorsal mesoderm and other tissues which results in tissue elongation. CE has been studied through overexpression and knockdown analysis in *Xenopus* and zebrafish embryos, as well as the analysis of knockout mice. Direct visualization of living *Xenopus* tissue explants undergoing CE has been particularly informative in the understanding of individual cell behavior (Wallingford et al., 2000). During CE, dorsal mesodermal cells become polarized, extend lamellipodia along the mediolateral axis, and form contacts with neighboring cells. The cells then intercalate mediolaterally, resulting in convergence (shortening) along the mediolateral axis and extension (elongation) of the tissue along the anterior–posterior (A/P) axis (Fig. 5C). Despite little apparent resemblance to the more static establishment of epithelial planar cell polarity, CE requires the same molecules that are involved in fly and mammalian PCP. This is perhaps not surprising, given that both cellular processes involve a coordinated polarization of cells within the tissue.

The Wnt pathway has long been implicated in the regulation of CE movements. Overexpression of Wnt ligands and Fz receptors interferes with proper axis elongation (Moon et al., 1993; Du et al., 1995; Deardorff et al., 1998; Itoh et al., 1998; Gradl et al., 1999; Djiane et al., 2000). Overexpression of core PCP factors, such as Stbm (Darken et al., 2002; Park and Moon, 2002) and Pk (Takeuchi et al., 2003), has been shown to cause similar CE defects, suggesting that Wnt/Fz/Dsh signals through the PCP pathway to control this process. The requirement for PCP signaling in *Xenopus* CE movements has been supported by loss-of-function studies. Dominant negative forms of Dsh (Sokol, 1996) and Wnt11 (Tada and Smith, 2000), and morpholino-mediated depletion of Pk (Carreira-Barbosa et al., 2003; Takeuchi et al., 2003; Veeman et al., 2003) and Stbm (Darken et al., 2002) interfere with CE and result in embryos with shortened body axes and in neural tube closure defects (Fig. 5D).

The regulation of CE by the PCP pathway is conserved in other vertebrates, as demonstrated by mutant analysis. *Silberblick* (*slb*)/*wnt11* mutant zebrafish exhibit insufficient migration of prechordal mesoderm cells, resulting in cyclopia and other midline defects (Heisenberg and Nusslein-Volhard, 1997). These defects are rescued by a truncated Dsh that does not signal through the canonical Wnt pathway (Heisenberg et al., 2000; Tada and Smith, 2000). Consistent with these effects, morpholino knockdown of Pk1 in zebrafish leads to defective CE, enhancement of *slb/wnt11* and *pipetail* (*ppt*)/*wnt5* phenotypes, and suppression of Wnt11 rescue of the *slb* phenotype (Carreira-Barbosa et al., 2003). Pk1 overexpression also inhibits CE movements and enhances the *slb* phenotype, and *pk1* genetically interacts with *trilobite* (*tri*), a mutant in the zebrafish *stbm* homologue (Carreira-Barbosa et al., 2003; Veeman et al., 2003).

Unlike in the cochlea, Wnt11 has been proposed to act as a permissive, rather than an instructive signal, as injection of Wnt11 RNA in one or two-cell zebrafish embryos rescues the *slb/wnt11* phenotype (Heisenberg et al., 2000). Nonetheless, these results demonstrate that Wnt ligands function to control CE in zebrafish.

7.2. Vertebrate neurulation

PCP-like signaling may be required during zebrafish neurulation for cell polarization and migration following cell division (Ciruna et al., 2006). After a neural plate cell divides, the basal daughter cell maintains contact with the basement membrane and subsequently reincorporates into the neuroepithelium at its original location. The apical daughter cell polarizes along the mediolateral axis and intercalates across the midline to incorporate into the opposite side of the developing neural tube (Concha and Adams, 1998; Geldmacher-Voss et al., 2003). GFP-Pk is asymmetrically localized to the anterior cortex of neural progenitors. This localization is lost during cell division and reestablished in both daughter cells. In maternal and zygotic trilobite (*tri*)/*Stbm* mutants, the asymmetric localization of GFP-Pk is lost, and cells fail to intercalate across the midline or reintegrate into the neuroepithelium. Remarkably, blocking cell division rescues the neurulation defects in 90% of embryos, suggesting that repolarization after cell division is a novel, previously unrecognized function of PCP signaling (Ciruna et al., 2006).

PCP molecular components are also required for neural tube closure in mammals, although the mechanism is not well studied. Mutant mice with cochlear stereociliary bundle orientation defects also have craniorachischisis, a perinatally lethal defect characterized by an open neural tube from the midbrain/hindbrain boundary to the end of the spinal cord. These include mice homozygous mutant for *Celsr* (*fmi*), *Vangl2*, and to lesser extent *fz3/6* and *dvll/2* double homozygous mutants, suggesting these genes may act redundantly (Kibar et al., 2001; Murdoch et al., 2001; Hamblet et al., 2002; Curtin et al., 2003; Wang et al., 2006). These data demonstrate that PCP proteins are involved in a number of diverse developmental and cellular processes in both *Drosophila* and vertebrate development.

7.3. Downstream mediators of Wnt-dependent cell movements

Although the same molecules are required for both epithelial PCP and CE movements, it is not fully established whether the underlying mechanisms are conserved. Available evidence suggests that at least some of the molecular interactions are similar. For example, as in the fly, Pk and Stbm affect the localization of each other in vertebrate systems. When expressed individually in *Xenopus* animal pole cells, Stbm-GFP is localized uniformly at the cell membrane and Pk-GFP is localized mostly to the cytoplasm. However, when they are coexpressed, Pk and Stbm colocalize to patchy regions around the cell membrane (Jenny et al., 2003). Two independent mutations in the *Vangl2* gene, both of which cause neural tube defects in mice, impair the binding of Vangl2 to Dsh, suggesting this interaction is necessary for Vangl2 function (Torban et al., 2004).

Some evidence suggests that PCP signaling regulates the mediolateral polarization of lamellipodia in cells undergoing CE. In cells expressing a dominant interfering Dsh construct, cellular projections still occur but never become polarized along the mediolateral axis (Wallingford et al., 2000). Lamellipodia formation depends on cytoskeletal components and down-stream effectors implicated in CE include direct regulators of the cytoskeleton such as Rho, Rac, and ROK (Habas et al., 2001, 2003; Kim and Han, 2005). Consistent with this, morpholino-mediated depletions of Rac1 and RhoA result in CE defects (Habas et al., 2001, 2003). Furthermore, RhoA activation requires the Dsh-binding protein Daam1 (Habas et al., 2001, 2003), and overexpression of ROK rescues the neural tube closure defect and the inhibition of dorsal explant CE caused by a dominant negative Rho construct (Kim and Han, 2005). These results demonstrate that some down-stream PCP components that may function to control the actin cytoskeleton in *Drosophila* wing and/or eye are also involved in CE movements.

A link between CE cell movements and the orientation of stereociliary bundles in mouse cochlear development has been described. Prior to orientation of the stereociliary bundles, a postmitotic sensory organ primordia undergoes cell intercalation to form a thinner and longer mature sensory organ in a process similar to CE (Chen et al., 2002; McKenzie et al., 2004; Wang et al., 2005). Mice deficient for PCP components, including *dvl1, dvl2* and *Lp*, show incomplete extension and have shorter and wider cochlear ducts. Furthermore, there is a strong correlation between the degree of extension of the primordial sense organ and the severity of stereociliary bundle misorientation (Wang et al., 2005), suggesting that the two processes are connected.

7.4. Neural crest migration

The PCP pathway may be involved in other types of cell movements as well. PCP signaling has been implicated in *Xenopus* neural crest cell migration, in which cells do not move as a coordinated tissue but individually follow each other in defined pathways (Nieto, 2001; De Calisto et al., 2005).

The requirement for PCP signaling in this process is based on the inhibition of neural crest migration by overexpression of a form of Dsh that is believed to disrupt PCP but not canonical Wnt signaling (Axelrod et al., 1998). The expression of Wnt11 adjacent to those cells that undergo migration is consistent with its postulated role in directing cell movements (De Calisto et al., 2005). Also, both Wnt11 and a dominant negative form of Wnt11 inhibited the migration of cells expressing *Slug*, a marker for neural crest cell specification, consistent with a specific role for Wnts in cell movements (De Calisto et al., 2005). Future studies of the role of Wnt signaling in cell migration will enhance our understanding of the regulation of cell movements by extracellular signals.

8. Wnt signaling and asymmetric cell division

Wnt signaling has been demonstrated to influence asymmetric cell division, which is an important mechanism for creating cellular diversity. For asymmetric division to occur, both asymmetric localization of cytoplasmic determinants and the specific orientation of the mitotic spindle are essential (Fig. 6A). Wnt pathway components have been shown to play a role in the asymmetric division of SOPs in the *Drosophila* peripheral nervous system and in the *C. elegans* early embryo.

8.1. Mitotic spindle orientation in Drosophila *SOPs*

In *Drosophila* neural precursors, asymmetric localization of cell fate determinants is coordinated with mitotic spindle orientation so that cell division results in determinants localized exclusively to one daughter cell (Figs. 1A and 6A, B). In both *Drosophila* neuroblasts and SOPs, cell fate determinants including Numb, a negative regulator of Notch, are asymmetrically localized in a cortical crescent perpendicular to the mitotic spindle. This results in the inheritance of Numb by one daughter cell, which determines the Notch-dependent cell fates of the two daughter cells (Frise et al., 1996; Guo et al., 1996).

Conserved Notch signaling regulates cell fate choice between equipotent precursors in a number of tissues throughout development (Schweisguth, 2004). Notch acts in a self-regulating feedback loop, in which an initial bias in Notch signaling in one cell is amplified by a transcriptional response in the receiving cell, resulting in the downregulation of Notch inhibitors and the upregulation of the Notch ligand Delta (Schweisguth, 2004; Fig. 4B). Numb downregulates Notch by targeting Notch or Sanpodo, a positive regulator of Notch, to the lysosome for degradation (Berdnik et al., 2002; O'Connor-Giles and Skeath, 2003; Hutterer and Knoblich, 2005). Therefore, inheritance of

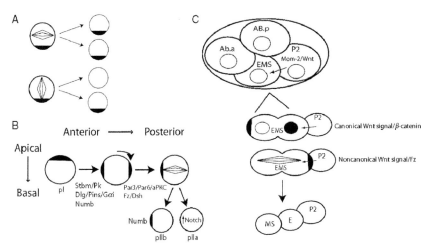

Fig. 6. Wnt signaling and asymmetric cell division. (A) The alignment of the mitotic spindle with respect to the asymmetric localization of cell fate determinants is crucial for the generation of different daughter cells. (B) *Drosophila* SOP division. Originally, A/B and PCP components are localized along the apical surface of the pI cell. After PCP signaling, Stbm/Pk and Dlg/Pins/Gαi complexes are localized to the anterior cortex, and Par3/Par6/aPKC and Fz/Dsh are localized to the posterior cortex, and the Notch inhibitor Numb forms an anterior crescent, which suppresses Notch in one of the daughter cells (pIIb). The proper mitotic spindle alignment depends on PCP components and results in two daughters of different fates. (C) Wnt signaling in EMS asymmetric cell division of the *C. elegans* four-cell embryo. Mom-2/Wnt signaling from the P2 cell to the dividing EMS cell results in an increase of nuclear Wrm-1/β-catenin in the cell proximal to P2, while Wrm-1 is cortical in the distal cell. In parallel, noncanonical Mom-2/Wnt signaling results in the localization of Fz-GFP closest to the location of Wnt signal, and orientation of the mitotic spindle parallel to the signal. The cell proximal to the P2 is specified to become the E cell, an endoderm precursor. The cell that did not receive high levels of Wnt signal becomes the mesodermal MS cell.

Numb by one of the two daughter cells during mitosis biases that daughter cell to become the signal-receiving cell.

Although the A/B axis of a *Drosophila* neuroblast is perpendicular to the plane of cell division (Fig. 1B), in the sensory organ precursor, the pI cell, the A/B axis is reoriented 90° and becomes parallel to the plane of the tissue. This directs the Numb crescent to the anterior lateral cortex and orients the mitotic spindle perpendicular to the original A/B axis resulting in cell division within the plane of the tissue. This division produces a posterior pIIa daughter cell and an anterior pIIb daughter cell, with pIIb exclusively inheriting Numb (Fig. 6B). This axis reorientation in the plane of tissue requires core PCP components which direct the formation of an anterior crescent of Dlg, Pins, and Gαi and localize the apical complex Par6/Par3/aPKC to the posterior cortex (Bellaiche et al., 2001b; Schaefer et al., 2001). In neuroblasts, the Par6/Par3/aPKC complex directs orientation of the mitotic spindle, but in SOPs mitotic spindle orientation is regulated by core

PCP components (Bellaiche et al., 2001a, 2004; Roegiers et al., 2001). The molecular interactions of PCP proteins are conserved in SOP cells. Stbm and Pk colocalize to the anterior cortex, while Fz and Dsh localize to the posterior cortex of the pI cell (Fig. 6B). This localization of core PCP components in SOPs is interdependent, as the PCP proteins are uniformly distributed at the apical surface in *fz, Stbm, pk,* or *dsh* mutants. The exception is Pk, which is cytoplasmic in *Stbm* mutants. In *fz, Stbm,* and *pk* mutant pupae, the Numb crescent is still localized asymmetrically but is located randomly along the A/P axis. The mitotic spindle still aligns perpendicular to this crescent, regardless of its position, suggesting that additional mechanisms exist for aligning the spindle with the cell fate determinants (Bellaiche et al., 2001a; Roegiers et al., 2001; Roegiers and Jan, 2004). Molecular interactions of Dlg and Pins couple PCP signaling with the orientation of the mitotic spindle in SOPs. Stbm acts through binding to Dlg to promote the localization of Pins to the anterior lateral cortex during prophase; in the majority of *Stbm* mutants, Pins is not localized to a crescent and was also found in the cytoplasm (Bellaiche et al., 2004). Conversely, Dsh acts antagonistically to Pins localization, as Pins spreads along the cortex in *dsh* mutants. However, Stbm and Dsh each act redundantly to other factors in localizing Pins to an asymmetric cortical crescent, as the Pins crescent is still formed during prometaphase in these mutants, albeit randomly along the A/B axis. Dsh is likely to act redundantly with the Par3/aPKC complex, as at prometaphase Pins is correctly localized in single *baz/par3* mutants but is mislocalized in the majority of *dsh; baz/par3* double mutants. On the other hand, the Pins crescent still forms in the majority of prometaphase *stbm; baz/par3* double mutants. This suggests that Stbm acts independently of Dsh to localize Pins and redundantly with an as yet unknown factor (Bellaiche et al., 2004). The redundancy of molecular mechanisms involved to bias Notch signaling during asymmetric cell division indicates the vital importance of this mechanism of cell fate specification. Lgl is also required to bias Notch signaling in both neuroblasts and SOPs, as Numb is mislocalized in *lgl* mutant neuroblasts and SOPs. In *lgl* knockout mice, Numb is mislocalized in the neuroepithelial cells and neural progenitors divide abnormally (Klezovitch et al., 2004), indicating that this mechanism for cell fate specification is conserved in mammals.

Since Dsh is capable of regulating Lgl it is tempting to speculate that Wnt signaling is involved in Lgl-dependent asymmetric cell division. Accumulating evidence suggests a direct role for Wnt signaling in Notch-dependent cell fate specification. As discussed previously, Fz antagonizes Notch signaling in the *Drosophila* R3 photoreceptor through gene transcription (Strutt et al., 2002). Direct interactions between Dsh and the Notch intracellular domain (ICD) (Axelrod et al., 1996) have been proposed to be required for R3/R4 cell fate specification (Strutt et al., 2002). Upon ligand binding, the ICD is cleaved from the membrane and translocates to the nucleus to activate gene transcription (Schweisguth, 2004). Notch has also been shown to directly associate with and positively regulate β-catenin, resulting in increased signaling (Hayward et al., 2005). Consistent with this, Notch enhances the transcriptional activity of LEF-1/TCF (Ross and Kadesch, 2001). Conversely, ablation of Notch 1 in skin leads to enhanced β-catenin signaling (Nicolas et al., 2003), suggesting interactions between these pathways are cell-type specific. Finally, GSK-3β can phosphorylate the Notch ICD and modulate Notch signaling and stability (Foltz et al., 2002). These interactions suggest a complex relationship between the Notch and the Wnt pathways. Further analysis of the role of these two pathways in asymmetric cell division is an important goal of future research.

8.2. Wnt signaling and asymmetric cell division of the C. elegans embryo

The first mitotic cell division of the *C. elegans* embryo is asymmetric, resulting in a larger anterior blastomere (AB) and a smaller posterior cell (P1) (Bowerman and Shelton, 1999). This asymmetry is dependent on the *partitioning defective* (Par) genes, in which mutations result in a symmetric first cell division (Kemphues et al., 1988). Many Par genes have subsequently been implicated in A/B polarity determination in a variety of cell types (Bilder et al., 2003; Suzuki et al., 2004). In contrast, Wnt signaling is not required in the *C. elegans* embryo until the four-cell stage, in which the posterior blastomere P2 signals to the anterior EMS cell, resulting in its polarization and asymmetric division (Fig. 6C). The anterior EMS daughter, MS, generates mesoderm and the posterior daughter, E, generates the entire endoderm of the mature animal. This P2 to EMS signaling is instructive for endoderm development, as the relative position of the P2 to the EMS cell determines which region of the EMS cell will take on this fate (Thorpe et al., 1997; Goldstein et al., 2006).

The Wnt pathway was implicated in this asymmetric division when a subset of the *mom* (*mo*re *m*esoderm) mutations were shown to encode members of the Wnt pathway, including *mom-2/wnt* and *mom-5/fz*. Consistent with this finding, the *wrm-1/β-catenin* mutation caused a *mom*-like phenotype (Rocheleau et al., 1997; Thorpe et al., 1997). Moreover, a mutation in *pop-1/TCF* caused the opposite phenotype of more endoderm (Lin et al., 1995, 1998). Pop-1/TCF inhibits E cell fate in the absence of a Wnt signal by directly repressing endodermal genes (Calvo et al., 2001). The Wnt pathway acts to lower Pop-1/TCF levels in the E cell, thereby allowing endoderm induction (Bei et al., 2002). E cell specification requires the Wnt components Apr-1/APC, Sgg-1/GSK-3, and Kin-19/PP2C, as RNAi-mediated inactivation of these genes causes *mom*-like defects (Fire et al., 1998).

For E cell fate specification, Wnt signaling directs the nuclear localization of Wrm-1/β-catenin. Initially, Wrm-1-GFP is weakly localized in the nucleus of all cells of the four-cell embryo, including those not exposed to Wnt signals (Takeshita and Sawa, 2005). During mitosis, Wrm-1-GFP is found at the posterior cortex of the EMS, and Wrm-1 nuclear levels become higher in the E cell and lost from the MS cell (Nakamura et al., 2005). This suggests a mechanism by which Wnt signaling releases Wrm-1 from the cortex to signal in the nucleus and specifies E cell fate in the proximal cell (Nakamura et al., 2005). This is a conserved mechanism throughout *C. elegans* development, as Wnt-dependent Wrm-1 cortical release and subsequent asymmetric cell division has been observed in other cell types undergoing Wnt-dependent asymmetric cell division later in development (Takeshita and Sawa, 2005).

Mom-2/Wnt signaling from the P2 to EMS cell is also required for the 90° rotation of the EMS spindle, which reorients from the left–right axis to align along the A/P axis just prior to EMS division (Herman, 2002). Mutations in *mom-2/wnt*, *mom-5/fz*, and sgg-1/GSK-3 RNAi cause EMS spindle orientation defects, whereas wrm-1 RNAi or mutations in *pop-1/TCF* do not (Schlesinger et al., 1999). This suggests that these pathways diverge at GSK-3 and that EMS spindle orientation is controlled via a noncanonical Wnt pathway.

Wnt ligands may also play a positional role in EMS spindle orientation. Blastomere isolation experiments have shown that placement of the P2 cell in relation to the EMS cell causes the reorientation of the mitotic spindle in the EMS cell so that it aligns with the P2 cell, and this is dependent on the relative position of the *mom-2/Wnt* signal (Goldstein et al., 2006). In addition, these experiments showed that the direction of a Wnt signal can determine the polarized localization of the LIN-17/Fz:GFP receptor (Goldstein et al., 2006; Fig. 6C). These results provide the first direct evidence for the significance of the position of a Wnt signal for the localization of an Fz receptor and the direct polarization of adjacent cells.

9. A new model system for PCP

The denticles of the *Drosophila* embryonic ectoderm have been described as a new PCP system (Price et al., 2006). As with other PCP models, many mechanisms are conserved, but there are notable differences. Denticles are actin-containing hair-like projections that are present in six rows on each segment of the ventral *Drosophila* embryonic epidermis. As cells narrow along the A/P axis and extend along the dorsal/ventral axis during dorsal closure, cells that will make denticles are already evident, as they are more narrow along the A/P axis relative to cells that will make naked cuticle. Initially, these cells accumulate cortical actin uniformly at the apical cortex, but subsequently the actin cytoskeleton becomes condensed along the posterior margin of each cell and denticles elongate posteriorly (Dickinson and Thatcher, 1997).

In *wg* and *dsh* mutant embryos, actin still condenses to a single spot per cell, but the localization is random along the A/P axis, often forming in the center of the apical domain, much like actin prehair accumulation in *fz* and *dsh* mutant wing cells (Wong and Adler, 1993; Price et al., 2006). Despite the similarities between cuticle denticles and wing hairs, many differences are apparent. For example, denticle polarity was shown to depend on the canonical Wg and Hedgehog signaling pathways, both of which are required for the establishment of embryonic segment polarity (Hidalgo, 1991; Price et al., 2006). This requirement is likely to be indirect through their role in transcriptional regulation, but this question remains to be investigated.

Similar to other tissues in which PCP occurs Fmi, Fz-GFP, and Dsh-GFP initially accumulate uniformly around the cortex, and later become polarized to the A/P cell margins. However, mutations in core PCP components, such as *Stbm*, *fmi*, a PCP-specific *dsh* mutation, and zygotic *fz1* and *fz2* mutants, show polarity defects only in denticle rows 1 and 2, and *pk* mutants have very few defects (Price et al., 2006). One explanation proposed for this observation is that these proteins are thought to reinforce long-range polarity cues in adjacent cells. The relatively short distance of required signaling across each denticle may abrogate the need for these genes, with defects only occurring in denticles furthest from the signal (Price et al., 2006). It remains to be seen if long-range cues, such as Fj, Ft, and Ds, play a role in denticle polarity. Nonetheless, further study of PCP in this tissue is likely to bring insight into new mechanisms of polarity specification by the Wnt pathway and other cell-signaling pathways.

10. Future directions

The appreciation and understanding of a role for Wnt signaling in the regulation of cell polarity have advanced greatly in recent years. However, these advances demonstrate the diverse adaptation of Wnt components to control a wide variety of cellular processes, and many recent findings raise as many questions as they answer. For example, the finding that Wnt pathway components can regulate A/B polarity raises the question of how these pathways interact in processes involving loss of polarity, including epithelial to mesenchymal transitions in normal development and during tumor progression. In addition, the evidence that A/B-polarity determinants play more than a passive role in the establishment of PCP, and the demonstration that these molecules are also involved in CE movements, suggest a higher degree of cross talk between polarity pathways than was previously recognized.

One way to further elucidate mechanisms of Wnt signaling that establish cell polarity is in the further identification of downstream effectors, and the study of how these downstream effectors are coordinated to result in differential outcomes for individual cell types. This would include interactions between the PCP signaling and β-catenin-dependent Wnt pathways, and how the activation or repression of one can influence the other. How the

Wnt pathway interacts with other signaling pathways, such as the Notch pathway, in asymmetric cell division and cell fate choice is another important area for future research. While recent studies in this area have advanced our understanding of these processes, they also illustrate the complexity of these signaling interactions. Insights into the mechanisms of Wnt-signaling pathways in different tissues, both the conservation and diversity, will provide important information on how cell polarity and movements are regulated during development, and how we may prevent these mechanisms from going awry both during development and in later incidences of cancer.

References

Abdelilah-Seyfried, S., Cox, D.N., Jan, Y.N. 2003. Bazooka is a permissive factor for the invasive behavior of discs large tumor cells in *Drosophila* ovarian follicular epithelia. Development 130(9), 1927–1935.

Adler, P.N., Krasnow, R.E., Liu, J. 1997. Tissue polarity points from cells that have higher Frizzled levels towards cells that have lower Frizzled levels. Curr. Biol. 7(12), 940–949.

Axelrod, J.D. 2001. Unipolar membrane association of Dishevelled mediates Frizzled planar cell polarity signaling. Genes Dev. 15(10), 1182–1187.

Axelrod, J.D., Matsuno, K., Artavanis-Tsakonas, S., Perrimon, N. 1996. Interaction between Wingless and Notch signaling pathways mediated by dishevelled. Science 271(5257), 1826–1832.

Axelrod, J.D., Miller, J.R., Shulman, J.M., Moon, R.T., Perrimon, N. 1998. Differential recruitment of Dishevelled provides signaling specificity in the planar cell polarity and Wingless signaling pathways. Genes Dev. 12(16), 2610–2622.

Baeg, G.H., Perrimon, N. 2000. Functional binding of secreted molecules to heparan sulfate proteoglycans in *Drosophila*. Curr. Opin. Cell Biol. 12(5), 575–580.

Baeg, G.H., Lin, X., Khare, N., Baumgartner, S., Perrimon, N. 2001. Heparan sulfate proteoglycans are critical for the organization of the extracellular distribution of Wingless. Development 128(1), 87–94.

Baeg, G.H., Selva, E.M., Goodman, R.M., Dasgupta, R., Perrimon, N. 2004. The Wingless morphogen gradient is established by the cooperative action of Frizzled and heparan sulfate proteoglycan receptors. Dev. Biol. 276(1), 89–100.

Bastock, R., Strutt, H., Strutt, D. 2003. Strabismus is asymmetrically localised and binds to Prickle and Dishevelled during *Drosophila* planar polarity patterning. Development 130(13), 3007–3014.

Bei, Y., Hogan, J., Berkowitz, L.A., Soto, M., Rocheleau, C.E., Pang, K.M., Collins, J., Mello, C.C. 2002. SRC-1 and Wnt signaling act together to specify endoderm and to control cleavage orientation in early *C. elegans* embryos. Dev. Cell 3(1), 113–125.

Bellaiche, Y., Gho, M., Kaltschmidt, J.A., Brand, A.H., Schweisguth, F. 2001a. Frizzled regulates localization of cell-fate determinants and mitotic spindle rotation during asymmetric cell division. Nat. Cell Biol. 3(1), 50–57.

Bellaiche, Y., Radovic, A., Woods, D.F., Hough, C.D., Parmentier, M.L., O'Kane, C.J., Bryant, P.J., Schweisguth, F. 2001b. The partner of inscuteable/discs-large complex is required to establish planar polarity during asymmetric cell division in *Drosophila*. Cell 106(3), 355–366.

Bellaiche, Y., Beaudoin-Massiani, O., Stuttem, I., Schweisguth, F. 2004. The planar cell polarity protein Strabismus promotes Pins anterior localization during asymmetric division of sensory organ precursor cells in *Drosophila*. Development 131(2), 469–478.

Berdnik, D., Torok, T., Gonzalez-Gaitan, M., Knoblich, J.A. 2002. The endocytic protein alpha-Adaptin is required for numb-mediated asymmetric cell division in *Drosophila*. Dev. Cell 3(2), 221–231.

Betschinger, J., Mechtler, K., Knoblich, J.A. 2003. The Par complex directs asymmetric cell division by phosphorylating the cytoskeletal protein Lgl. Nature 422(6929), 326–330.

Bilder, D. 2001. Cell polarity: Squaring the circle. Curr. Biol. 11(4), R132–R135.

Bilder, D. 2004. Epithelial polarity and proliferation control: Links from the *Drosophila* neoplastic tumor suppressors. Genes Dev. 18(16), 1909–1925.

Bilder, D., Perrimon, N. 2000. Localization of apical epithelial determinants by the basolateral PDZ protein Scribble. Nature 403(6770), 676–680.

Bilder, D., Li, M., Perrimon, N. 2000. Cooperative regulation of cell polarity and growth by *Drosophila* tumor suppressors. Science 289(5476), 113–116.

Bilder, D., Schober, M., Perrimon, N. 2003. Integrated activity of PDZ protein complexes regulates epithelial polarity. Nat. Cell Biol. 5(1), 53–58.

Boutros, M., Paricio, N., Strutt, D.I., Mlodzik, M. 1998. Dishevelled activates JNK and discriminates between JNK pathways in planar polarity and wingless signaling. Cell 94(1), 109–118.

Bowerman, B., Shelton, C.A. 1999. Cell polarity in the early *Caenorhabditis elegans* embryo. Curr. Opin. Genet. Dev. 9(4), 390–395.

Calvo, D., Victor, M., Gay, F., Sui, G., Luke, M.P., Dufourcq, P., Wen, G., Maduro, M., Rothman, J., Shi, Y. 2001. A POP-1 repressor complex restricts inappropriate cell type-specific gene transcription during *Caenorhabditis elegans* embryogenesis. EMBO J. 20(24), 7197–7208.

Carreira-Barbosa, F., Concha, M.L., Takeuchi, M., Ueno, N., Wilson, S.W., Tada, M. 2003. Prickle 1 regulates cell movements during gastrulation and neuronal migration in zebrafish. Development 130(17), 4037–4046.

Chae, J., Kim, M.J., Goo, J.H., Collier, S., Gubb, D., Charlton, J., Adler, P.N., Park, W.J. 1999. The *Drosophila* tissue polarity gene starry night encodes a member of the protocadherin family. Development 126(23), 5421–5429.

Chalmers, A.D., Strauss, B., Papalopulu, N. 2003. Oriented cell divisions asymmetrically segregate aPKC and generate cell fate diversity in the early *Xenopus* embryo. Development 130(12), 2657–2668.

Chen, P., Johnson, J.E., Zoghbi, H.Y., Segil, N. 2002. The role of Math1 in inner ear development: Uncoupling the establishment of the sensory primordium from hair cell fate determination. Development 129(10), 2495–2505.

Ciruna, B., Jenny, A., Lee, D., Mlodzik, M., Schier, A.F. 2006. Planar cell polarity signalling couples cell division and morphogenesis during neurulation. Nature 439(7073), 220–224.

Clark, H.F., Brentrup, D., Schneitz, K., Bieber, A., Goodman, C., Noll, M. 1995. Dachsous encodes a member of the cadherin superfamily that controls imaginal disc morphogenesis in *Drosophila*. Genes Dev. 9(12), 1530–1542.

Concha, M.L., Adams, R.J. 1998. Oriented cell divisions and cellular morphogenesis in the zebrafish gastrula and neurula: A time-lapse analysis. Development 125(6), 983–994.

Curtin, J.A., Quint, E., Tsipouri, V., Arkell, R.M., Cattanach, B., Copp, A.J., Henderson, D.J., Spurr, N., Stanier, P., Fisher, E.M., Nolan, P.M., Steel, K.P., et al. 2003. Mutation of Celsr1 disrupts planar polarity of inner ear hair cells and causes severe neural tube defects in the mouse. Curr. Biol. 13(13), 1129–1133.

Dabdoub, A., Kelley, M.W. 2005. Planar cell polarity and a potential role for a Wnt morphogen gradient in stereociliary bundle orientation in the mammalian inner ear. J. Neurobiol. 64(4), 446–457.

Dabdoub, A., Donohue, M.J., Brennan, A., Wolf, V., Montcouquiol, M., Sassoon, D.A., Hseih, J.C., Rubin, J.S., Salinas, P.C., Kelley, M.W. 2003. Wnt signaling mediates reorientation of outer hair cell stereociliary bundles in the mammalian cochlea. Development 130(11), 2375–2384.

Darken, R.S., Scola, A.M., Rakeman, A.S., Das, G., Mlodzik, M., Wilson, P.A. 2002. The planar polarity gene strabismus regulates convergent extension movements in *Xenopus*. EMBO J. 21(5), 976–985.

Das, G., Reynolds-Kenneally, J., Mlodzik, M. 2002. The atypical cadherin Flamingo links Frizzled and Notch signaling in planar polarity establishment in the *Drosophila* eye. Dev. Cell 2(5), 655–666.

Daudet, N., Ripoll, C., Moles, J.P., Rebillard, G. 2002. Expression of members of Wnt and Frizzled gene families in the postnatal rat cochlea. Brain Res. Mol. Brain Res. 105(1–2), 98–107.

De Calisto, J., Araya, C., Marchant, L., Riaz, C.F., Mayor, R. 2005. Essential role of non-canonical Wnt signalling in neural crest migration. Development 132(11), 2587–2597.

Deardorff, M.A., Tan, C., Conrad, L.J., Klein, P.S. 1998. Frizzled-8 is expressed in the Spemann organizer and plays a role in early morphogenesis. Development 125(14), 2687–2700.

Dickinson, W.J., Thatcher, J.W. 1997. Morphogenesis of denticles and hairs in *Drosophila* embryos: Involvement of actin-associated proteins that also affect adult structures. Cell Motil. Cytoskeleton 38(1), 9–21.

Djiane, A., Riou, J., Umbhauer, M., Boucaut, J., Shi, D. 2000. Role of frizzled 7 in the regulation of convergent extension movements during gastrulation in *Xenopus laevis*. Development 127(14), 3091–3100.

Djiane, A., Yogev, S., Mlodzik, M. 2005. The apical determinants aPKC and dPatj regulate Frizzled-dependent planar cell polarity in the *Drosophila* eye. Cell 121(4), 621–631.

Dollar, G.L., Weber, U., Mlodzik, M., Sokol, S.Y. 2005. Regulation of Lethal giant larvae by Dishevelled. Nature 437(7063), 1376–1380.

Du, S.J., Purcell, S.M., Christian, J.L., McGrew, L.L., Moon, R.T. 1995. Identification of distinct classes and functional domains of Wnts through expression of wild-type and chimeric proteins in *Xenopus* embryos. Mol. Cell. Biol. 15(5), 2625–2634.

Fanto, M., Weber, U., Strutt, D.I., Mlodzik, M. 2000. Nuclear signaling by Rac and Rho GTPases is required in the establishment of epithelial planar polarity in the *Drosophila* eye. Curr. Biol. 10(16), 979–988.

Feiguin, F., Hannus, M., Mlodzik, M., Eaton, S. 2001. The ankyrin repeat protein Diego mediates Frizzled-dependent planar polarization. Dev. Cell 1(1), 93–101.

Fire, A., Xu, S., Montgomery, M.K., Kostas, S.A., Driver, S.E., Mello, C.C. 1998. Potent and specific genetic interference by double-stranded RNA in *Caenorhabditis elegans*. Nature 391(6669), 806–811.

Foltz, D.R., Santiago, M.C., Berechid, B.E., Nye, J.S. 2002. Glycogen synthase kinase-3beta modulates notch signaling and stability. Curr. Biol. 12(12), 1006–1011.

Frise, E., Knoblich, J.A., Younger-Shepherd, S., Jan, L.Y., Jan, Y.N. 1996. The *Drosophila* Numb protein inhibits signaling of the Notch receptor during cell-cell interaction in sensory organ lineage. Proc. Natl. Acad. Sci. USA 93(21), 11925–11932.

Gangar, A., Rossi, G., Andreeva, A., Hales, R., Brennwald, P. 2005. Structurally conserved interaction of Lgl family with SNAREs is critical to their cellular function. Curr. Biol. 15(12), 1136–1142.

Geldmacher-Voss, B., Reugels, A.M., Pauls, S., Campos-Ortega, J.A. 2003. A 90-degree rotation of the mitotic spindle changes the orientation of mitoses of zebrafish neuroepithelial cells. Development 130(16), 3767–3780.

Goldstein, B., Takeshita, H., Mizumoto, K., Sawa, H. 2006. Wnt signals can function as positional cues in establishing cell polarity. Dev. Cell 10(3), 391–396.

Gradl, D., Kuhl, M., Wedlich, D. 1999. The Wnt/Wg signal transducer beta-catenin controls fibronectin expression. Mol. Cell. Biol. 19(8), 5576–5587.

Gubb, D., Green, C., Huen, D., Coulson, D., Johnson, G., Tree, D., Collier, S., Roote, J. 1999. The balance between isoforms of the prickle LIM domain protein is critical for planar polarity in *Drosophila* imaginal discs. Genes Dev. 13(17), 2315–2327.

Guo, M., Jan, L.Y., Jan, Y.N. 1996. Control of daughter cell fates during asymmetric division: Interaction of Numb and Notch. Neuron 17(1), 27–41.

Guo, N., Hawkins, C., Nathans, J. 2004. Frizzled6 controls hair patterning in mice. Proc. Natl. Acad. Sci. USA 101(25), 9277–9281.

Habas, R., Dawid, I.B., He, X. 2003. Coactivation of Rac and Rho by Wnt/Frizzled signaling is required for vertebrate gastrulation. Genes Dev. 17(2), 295–309.

Habas, R., Kato, Y., He, X. 2001. Wnt/Frizzled activation of Rho regulates vertebrate gastrulation and requires a novel Formin homology protein Daam1. Cell 107(7), 843–854.

Hamblet, N.S., Lijam, N., Ruiz-Lozano, P., Wang, J., Yang, Y., Luo, Z., Mei, L., Chien, K.R., Sussman, D.J., Wynshaw-Boris, A. 2002. Dishevelled 2 is essential for cardiac outflow tract Development, somite segmentation and neural tube closure. Development 129(24), 5827–5838.

Hayward, P., Brennan, K., Sanders, P., Balayo, T., DasGupta, R., Perrimon, N., Martinez Arias, A. 2005. Notch modulates Wnt signalling by associating with Armadillo/beta-catenin and regulating its transcriptional activity. Development 132(8), 1819–1830.

Heisenberg, C.P., Nusslein-Volhard, C. 1997. The function of silberblick in the positioning of the eye anlage in the zebrafish embryo. Dev. Biol. 184(1), 85–94.

Heisenberg, C.P., Tada, M., Rauch, G.J., Saude, L., Concha, M.L., Geisler, R., Stemple, D.L., Smith, J.C., Wilson, S.W. 2000. Silberblick/Wnt11 mediates convergent extension movements during zebrafish gastrulation. Nature 405(6782), 76–81.

Herman, M.A. 2002. Control of cell polarity by noncanonical Wnt signaling in *C. elegans*. Semin. Cell Dev. Biol. 13(3), 233–241.

Hidalgo, A. 1991. Interactions between segment polarity genes and the generation of the segmental pattern in *Drosophila*. Mech. Dev. 35(2), 77–87.

Hsieh, J.C., Kodjabachian, L., Rebbert, M.L., Rattner, A., Smallwood, P.M., Samos, C.H., Nusse, R., Dawid, I.B., Nathans, J. 1999. A new secreted protein that binds to Wnt proteins and inhibits their activities. Nature 398(6726), 431–436.

Hudspeth, A.J., Jacobs, R. 1979. Stereocilia mediate transduction in vertebrate hair cells (auditory system/cilium/vestibular system). Proc. Natl. Acad. Sci. USA 76(3), 1506–1509.

Hurov, J.B., Watkins, J.L., Piwnica-Worms, H. 2004. Atypical PKC phosphorylates PAR-1 kinases to regulate localization and activity. Curr. Biol. 14(8), 736–741.

Hutterer, A., Knoblich, J.A. 2005. Numb and alpha-Adaptin regulate Sanpodo endocytosis to specify cell fate in *Drosophila* external sensory organs. EMBO Rep. 6(9), 836–842.

Hutterer, A., Betschinger, J., Petronczki, M., Knoblich, J.A. 2004. Sequential roles of Cdc42, Par-6, aPKC, and Lgl in the establishment of epithelial polarity during *Drosophila* embryogenesis. Dev. Cell 6(6), 845–854.

Itoh, K., Jacob, J., Sokol, Y.S. 1998. A role for Xenopus Frizzled 8 in dorsal development. Mech. Dev. 74(1–2), 145–157.

Jenny, A., Darken, R.S., Wilson, P.A., Mlodzik, M. 2003. Prickle and Strabismus form a functional complex to generate a correct axis during planar cell polarity signaling. EMBO J. 22(17), 4409–4420.

Jenny, A., Reynolds-Kenneally, J., Das, G., Burnett, M., Mlodzik, M. 2005. Diego and Prickle regulate Frizzled planar cell polarity signalling by competing for Dishevelled binding. Nat. Cell Biol. 7(7), 691–697.

Jones, S.E., Jomary, C. 2002. Secreted Frizzled-related proteins: Searching for relationships and patterns. Bioessays 24(9), 811–820.

Kawano, Y., Kypta, R. 2003. Secreted antagonists of the Wnt signalling pathway. J. Cell Sci. 116(Pt. 13), 2627–2634.

Kemphues, K.J., Priess, J.R., Morton, D.G., Cheng, N.S. 1988. Identification of genes required for cytoplasmic localization in early *C. elegans* embryos. Cell 52(3), 311–320.

Kibar, Z., Vogan, K.J., Groulx, N., Justice, M.J., Underhill, D.A., Gros, P. 2001. Ltap, a mammalian homolog of *Drosophila* Strabismus/Van Gogh, is altered in the mouse neural tube mutant Loop-tail. Nat. Genet. 28(3), 251–255.

Kim, G.H., Han, J.K. 2005. JNK and ROKalpha function in the noncanonical Wnt/RhoA signaling pathway to regulate Xenopus convergent extension movements. Dev. Dyn. 232(4), 958–968.

Kimura-Yoshida, C., Nakano, H., Okamura, D., Nakao, K., Yonemura, S., Belo, J.A., Aizawa, S., Matsui, Y., Matsuo, I. 2005. Canonical Wnt signaling and its antagonist regulate anterior-posterior axis polarization by guiding cell migration in mouse visceral endoderm. Dev. Cell 9(5), 639–650.

Kispert, A., Vainio, S., Shen, L., Rowitch, D.H., McMahon, A.P. 1996. Proteoglycans are required for maintenance of Wnt-11 expression in the ureter tips. Development 122(11), 3627–3637.

Kjellen, L., Lindahl, U. 1991. Proteoglycans: Structures and interactions. Annu. Rev. Biochem. 60, 443–475.

Klein, T.J., Mlodzik, M. 2005. Planar cell polarization: An emerging model points in the right direction. Annu. Rev. Cell Dev. Biol. 21, 155–176.

Klezovitch, O., Fernandez, T.E., Tapscott, S.J., Vasioukhin, V. 2004. Loss of cell polarity causes severe brain dysplasia in Lgl1 knockout mice. Genes Dev. 18(5), 559–571.

Klingensmith, J., Nusse, R., Perrimon, N. 1994. The *Drosophila* segment polarity gene dishevelled encodes a novel protein required for response to the wingless signal. Genes Dev. 8(1), 118–130.

Kusakabe, M., Nishida, E. 2004. The polarity-inducing kinase Par-1 controls *Xenopus* gastrulation in cooperation with 14-3-3 and aPKC. EMBO J. 23(21), 4190–4201.

Leimeister, C., Bach, A., Gessler, M. 1998. Developmental expression patterns of mouse sFRP genes encoding members of the secreted frizzled related protein family. Mech. Dev. 75(1–2), 29–42.

Leyns, L., Bouwmeester, T., Kim, S.H., Piccolo, S., De Robertis, E.M. 1997. Frzb-1 is a secreted antagonist of Wnt signaling expressed in the Spemann organizer. Cell 88(6), 747–756.

Lin, R., Hill, R.J., Priess, J.R. 1998. POP-1 and anterior-posterior fate decisions in *C. elegans* embryos. Cell 92(2), 229–239.

Lin, R., Thompson, S., Priess, J.R. 1995. pop-1 encodes an HMG box protein required for the specification of a mesoderm precursor in early *C. elegans* embryos. Cell 83(4), 599–609.

Logan, C.Y., Nusse, R. 2004. The Wnt signaling pathway in development and disease. Annu. Rev. Cell Dev. Biol. 20, 781–810.

Ma, D., Yang, C.H., McNeill, H., Simon, M.A., Axelrod, J.D. 2003. Fidelity in planar cell polarity signalling. Nature 421(6922), 543–547.

Mahoney, P.A., Weber, U., Onofrechuk, P., Biessmann, H., Bryant, P.J., Goodman, C.S. 1991. The fat tumor suppressor gene in *Drosophila* encodes a novel member of the cadherin gene superfamily. Cell 67(5), 853–868.

McKenzie, E., Krupin, A., Kelley, M.W. 2004. Cellular growth and rearrangement during the development of the mammalian organ of Corti. Dev. Dyn. 229(4), 802–812.

Mechler, B.M., McGinnis, W., Gehring, W.J. 1985. Molecular cloning of lethal(2)giant larvae, a recessive oncogene of *Drosophila melanogaster*. EMBO J. 4(6), 1551–1557.

Mlodzik, M. 1999. Planar polarity in the *Drosophila* eye: A multifaceted view of signaling specificity and cross-talk. EMBO J. 18(24), 6873–6879.

Mlodzik, M. 2002. Planar cell polarization: Do the same mechanisms regulate *Drosophila* tissue polarity and vertebrate gastrulation? Trends Genet. 18(11), 564–571.

Mollereau, B., Domingos, P.M. 2005. Photoreceptor differentiation in *Drosophila*: From immature neurons to functional photoreceptors. Dev. Dyn. 232(3), 585–592.

Montcouquiol, M., Kelley, M.W. 2003. Planar and vertical signals control cellular differentiation and patterning in the mammalian cochlea. J. Neurosci. 23(28), 9469–9478.

Montcouquiol, M., Rachel, R.A., Lanford, P.J., Copeland, N.G., Jenkins, N.A., Kelley, M.W. 2003. Identification of Vangl2 and Scrb1 as planar polarity genes in mammals. Nature 423 (6936), 173–177.

Moon, R.T., Campbell, R.M., Christian, J.L., McGrew, L.L., Shih, J., Fraser, S. 1993. Xwnt-5A: A maternal Wnt that affects morphogenetic movements after overexpression in embryos of *Xenopus laevis*. Development 119(1), 97–111.

Murdoch, J.N., Doudney, K., Paternotte, C., Copp, A.J., Stanier, P. 2001. Severe neural tube defects in the loop-tail mouse result from mutation of Lpp1, a novel gene involved in floor plate specification. Hum. Mol. Genet. 10(22), 2593–2601.

Musch, A., Cohen, D., Yeaman, C., Nelson, W.J., Rodriguez-Boulan, E., Brennwald, P.J. 2002. Mammalian homolog of *Drosophila* tumor suppressor lethal (2) giant larvae interacts with basolateral exocytic machinery in Madin-Darby canine kidney cells. Mol. Biol. Cell 13(1), 158–168.

Nakamura, K., Kim, S., Ishidate, T., Bei, Y., Pang, K., Shirayama, M., Trzepacz, C., Brownell, D.R., Mello, C.C. 2005. Wnt signaling drives WRM-1/beta-catenin asymmetries in early *C. elegans* embryos. Genes Dev. 19(15), 1749–1754.

Nicolas, M., Wolfer, A., Raj, K., Kummer, J.A., Mill, P., van Noort, M., Hui, C.C., Clevers, H., Dotto, G.P., Radtke, F. 2003. Notch1 functions as a tumor suppressor in mouse skin. Nat. Genet. 33(3), 416–421.

Nieto, M.A. 2001. The early steps of neural crest development. Mech. Dev. 105(1–2), 27–35.

O'Connor-Giles, K.M., Skeath, J.B. 2003. Numb inhibits membrane localization of Sanpodo, a four-pass transmembrane protein, to promote asymmetric divisions in *Drosophila*. Dev. Cell 5(2), 231–243.

Ohshiro, T., Yagami, T., Zhang, C., Matsuzaki, F. 2000. Role of cortical tumour-suppressor proteins in asymmetric division of *Drosophila* neuroblast. Nature 408(6812), 593–596.

Ossipova, O., Dhawan, S., Sokol, S., Green, J.B. 2005. Distinct PAR-1 proteins function in different branches of Wnt signaling during vertebrate development. Dev. Cell 8(6), 829–841.

Pan, W.J., Pang, S.Z., Huang, T., Guo, H.Y., Wu, D., Li, L. 2004. Characterization of function of three domains in dishevelled-1: DEP domain is responsible for membrane translocation of dishevelled-1. Cell Res. 14(4), 324–330.

Paricio, N., Feiguin, F., Boutros, M., Eaton, S., Mlodzik, M. 1999. The *Drosophila* STE20-like kinase misshapen is required downstream of the Frizzled receptor in planar polarity signaling. EMBO J. 18(17), 4669–4678.

Park, M., Moon, R.T. 2002. The planar cell-polarity gene stbm regulates cell behaviour and cell fate in vertebrate embryos. Nat. Cell Biol. 4(1), 20–25.

Peng, C.Y., Manning, L., Albertson, R., Doe, C.Q. 2000. The tumour-suppressor genes lgl and dlg regulate basal protein targeting in *Drosophila* neuroblasts. Nature 408(6812), 596–600.

Penton, A., Wodarz, A., Nusse, R. 2002. A mutational analysis of dishevelled in *Drosophila* defines novel domains in the dishevelled protein as well as novel suppressing alleles of axin. Genetics 161(2), 747–762.

Pielage, J., Stork, T., Bunse, I., Klambt, C. 2003. The *Drosophila* cell survival gene discs lost encodes a cytoplasmic Codanin-1-like protein, not a homolog of tight junction PDZ protein Patj. Dev. Cell 5(6), 841–851.

Plant, P.J., Fawcett, J.P., Lin, D.C., Holdorf, A.D., Binns, K., Kulkarni, S., Pawson, T. 2003. A polarity complex of mPar-6 and atypical PKC binds, phosphorylates and regulates mammalian Lgl. Nat. Cell Biol. 5(4), 301–308.

Price, M.H., Roberts, D.M., McCartney, B.M., Jezuit, E., Peifer, M. 2006. Cytoskeletal dynamics and cell signaling during planar polarity establishment in the *Drosophila* embryonic denticle. J. Cell Sci. 119(Pt. 3), 403–415.

Rocheleau, C.E., Downs, W.D., Lin, R., Wittmann, C., Bei, Y., Cha, Y.H., Ali, M., Priess, J.R., Mello, C.C. 1997. Wnt signaling and an APC-related gene specify endoderm in early *C. elegans* embryos. Cell 90(4), 707–716.

Roegiers, F., Jan, Y.N. 2004. Asymmetric cell division. Curr. Opin. Cell Biol. 16(2), 195–205.

Roegiers, F., Younger-Shepherd, S., Jan, L.Y., Jan, Y.N. 2001. Two types of asymmetric divisions in the *Drosophila* sensory organ precursor cell lineage. Nat. Cell Biol. 3(1), 58–67.

Ross, D.A., Kadesch, T. 2001. The notch intracellular domain can function as a coactivator for LEF-1. Mol. Cell. Biol. 21(22), 7537–7544.

Schaefer, M., Petronczki, M., Dorner, D., Forte, M., Knoblich, J.A. 2001. Heterotrimeric G proteins direct two modes of asymmetric cell division in the *Drosophila* nervous system. Cell 107(2), 183–194.

Schlesinger, A., Shelton, C.A., Maloof, J.N., Meneghini, M., Bowerman, B. 1999. Wnt pathway components orient a mitotic spindle in the early *Caenorhabditis elegans* embryo without requiring gene transcription in the responding cell. Genes Dev. 13(15), 2028–2038.

Schweisguth, F. 2004. Notch signaling activity. Curr. Biol. 14(3), R129–R138.

Simon, M.A. 2004. Planar cell polarity in the *Drosophila* eye is directed by graded Four-jointed and Dachsous expression. Development 131(24), 6175–6184.

Sokol, S.Y. 1996. Analysis of Dishevelled signalling pathways during *Xenopus* development. Curr. Biol. 6(11), 1456–1467.

Strutt, D. 2003. Frizzled signalling and cell polarisation in *Drosophila* and vertebrates. Development 130(19), 4501–4513.

Strutt, D., Johnson, R., Cooper, K., Bray, S. 2002. Asymmetric localization of frizzled and the determination of notch-dependent cell fate in the *Drosophila* eye. Curr. Biol. 12(10), 813–824.

Strutt, D.I. 2001. Asymmetric localization of frizzled and the establishment of cell polarity in the *Drosophila* wing. Mol. Cell. 7(2), 367–375.

Strutt, D.I. 2002. The asymmetric subcellular localisation of components of the planar polarity pathway. Semin. Cell Dev. Biol. 13(3), 225–231.

Strutt, D.I., Weber, U., Mlodzik, M. 1997. The role of RhoA in tissue polarity and Frizzled signalling. Nature 387(6630), 292–295.

Strutt, H., Strutt, D. 2002. Nonautonomous planar polarity patterning in *Drosophila*: Dishevelled-independent functions of frizzled. Dev. Cell 3(6), 851–863.

Strutt, H., Strutt, D. 2003. EGF signaling and ommatidial rotation in the *Drosophila* eye. Curr. Biol. 13(16), 1451–1457.

Sun, T.Q., Lu, B., Feng, J.J., Reinhard, C., Jan, Y.N., Fantl, W.J., Williams, L.T. 2001. PAR-1 is a Dishevelled-associated kinase and a positive regulator of Wnt signalling. Nat. Cell Biol. 3(7), 628–636.

Suzuki, A., Hirata, M., Kamimura, K., Maniwa, R., Yamanaka, T., Mizuno, K., Kishikawa, M., Hirose, H., Amano, Y., Izumi, N., Miwa, Y., Ohno, S. 2004. aPKC acts upstream of PAR-1b in both the establishment and maintenance of mammalian epithelial polarity. Curr. Biol. 14(16), 1425–1435.

Tada, M., Smith, J.C. 2000. Xwnt11 is a target of *Xenopus* Brachyury: Regulation of gastrulation movements via Dishevelled, but not through the canonical Wnt pathway. Development 127(10), 2227–2238.

Takeshita, H., Sawa, H. 2005. Asymmetric cortical and nuclear localizations of WRM-1/beta-catenin during asymmetric cell division in *C. elegans*. Genes Dev. 19(15), 1743–1748.

Takeuchi, M., Nakabayashi, J., Sakaguchi, T., Yamamoto, T.S., Takahashi, H., Takeda, H., Ueno, N. 2003. The prickle-related gene in vertebrates is essential for gastrulation cell movements. Curr. Biol. 13(8), 674–679.

Tamai, K., Semenov, M., Kato, Y., Spokony, R., Liu, C., Katsuyama, Y., Hess, F., Saint-Jeannet, J.P., He, X. 2000. LDL-receptor-related proteins in Wnt signal transduction. Nature 407(6803), 530–535.

Tamai, K., Zeng, X., Liu, C., Zhang, X., Harada, Y., Chang, Z., He, X. 2004. A mechanism for Wnt coreceptor activation. Mol Cell. 13(1), 149–156.

Tanentzapf, G., Tepass, U. 2003. Interactions between the crumbs, lethal giant larvae and bazooka pathways in epithelial polarization. Nat. Cell Biol. 5(1), 46–52.

Thorpe, C.J., Schlesinger, A., Carter, J.C., Bowerman, B. 1997. Wnt signaling polarizes an early *C. elegans* blastomere to distinguish endoderm from mesoderm. Cell 90(4), 695–705.

Torban, E., Kor, C., Gros, P. 2004. Van Gogh-like2 (Strabismus) and its role in planar cell polarity and convergent extension in vertebrates. Trends Genet. 20(11), 570–577.

Tree, D.R., Shulman, J.M., Rousset, R., Scott, M.P., Gubb, D., Axelrod, J.D. 2002. Prickle mediates feedback amplification to generate asymmetric planar cell polarity signaling. Cell 109(3), 371–381.

Uren, A., Reichsman, F., Anest, V., Taylor, W.G., Muraiso, K., Bottaro, D.P., Cumberledge, S., Rubin, J.S. 2000. Secreted frizzled-related protein-1 binds directly to Wingless and is a biphasic modulator of Wnt signaling. J. Biol. Chem. 275(6), 4374–4382.

Usui, T., Shima, Y., Shimada, Y., Hirano, S., Burgess, R.W., Schwarz, T.L., Takeichi, M., Uemura, T. 1999. Flamingo, a seven-pass transmembrane cadherin, regulates planar cell polarity under the control of Frizzled. Cell 98(5), 585–595.

Veeman, M.T., Slusarski, D.C., Kaykas, A., Louie, S.H., Moon, R.T. 2003. Zebrafish prickle, a modulator of noncanonical Wnt/Fz signaling, regulates gastrulation movements. Curr. Biol. 13(8), 680–685.

Villano, J.L., Katz, F.N. 1995. Four-jointed is required for intermediate growth in the proximal-distal axis in *Drosophila*. Development 121(9), 2767–2777.

Vinson, C.R., Conover, S., Adler, P.N. 1989. A *Drosophila* tissue polarity locus encodes a protein containing seven potential transmembrane domains. Nature 338(6212), 263–264.

Wallingford, J.B., Rowning, B.A., Vogeli, K.M., Rothbacher, U., Fraser, S.E., Harland, R.M. 2000. Dishevelled controls cell polarity during *Xenopus* gastrulation. Nature 405(6782), 81–85.

Wallingford, J.B., Fraser, S.E., Harland, R.M. 2002. Convergent extension: The molecular control of polarized cell movement during embryonic development. Dev. Cell 2(6), 695–706.

Wang, J., Mark, S., Zhang, X., Qian, D., Yoo, S.J., Radde-Gallwitz, K., Zhang, Y., Lin, X., Collazo, A., Wynshaw-Boris, A., Chen, P. 2005. Regulation of polarized extension and planar cell polarity in the cochlea by the vertebrate PCP pathway. Nat. Genet. 37(9), 980–985.

Wang, Y., Guo, N., Nathans, J. 2006. The role of Frizzled3 and Frizzled6 in neural tube closure and in the planar polarity of inner-ear sensory hair cells. J. Neurosci. 26(8), 2147–2156.

Weber, U., Paricio, N., Mlodzik, M. 2000. Jun mediates Frizzled-induced R3/R4 cell fate distinction and planar polarity determination in the *Drosophila* eye. Development 127(16), 3619–3629.

Winter, C.G., Wang, B., Ballew, A., Royou, A., Karess, R., Axelrod, J.D., Luo, L. 2001. *Drosophila* Rho-associated kinase (Drok) links Frizzled-mediated planar cell polarity signaling to the actin cytoskeleton. Cell 105(1), 81–91.

Wolff, T., Ready, D.F. 1991. The beginning of pattern formation in the *Drosophila* compound eye: The morphogenetic furrow and the second mitotic wave. Development 113(3), 841–850.

Wolff, T., Rubin, G.M. 1998. Strabismus, a novel gene that regulates tissue polarity and cell fate decisions in *Drosophila*. Development 125(6), 1149–1159.

Wong, L.L., Adler, P.N. 1993. Tissue polarity genes of *Drosophila* regulate the subcellular location for prehair initiation in pupal wing cells. J. Cell Biol. 123(1), 209–221.

Wu, J., Klein, T.J., Mlodzik, M. 2004. Subcellular localization of frizzled receptors, mediated by their cytoplasmic tails, regulates signaling pathway specificity. PLoS Biol. 2(7), E158.

Xu, Q., D'Amore, P.A., Sokol, S.Y. 1998. Functional and biochemical interactions of Wnts with FrzA, a secreted Wnt antagonist. Development 125(23), 4767–4776.

Yamanaka, T., Horikoshi, Y., Sugiyama, Y., Ishiyama, C., Suzuki, A., Hirose, T., Iwamatsu, A., Shinohara, A., Ohno, S. 2003. Mammalian Lgl forms a protein complex with PAR-6 and aPKC independently of PAR-3 to regulate epithelial cell polarity. Curr. Biol. 13(9), 734–743.

Yang, C.H., Axelrod, J.D., Simon, M.A. 2002. Regulation of Frizzled by fat-like cadherins during planar polarity signaling in the *Drosophila* compound eye. Cell 108(5), 675–688.

Zeidler, M.P., Perrimon, N., Strutt, D.I. 2000. Multiple roles for four-jointed in planar polarity and limb patterning. Dev. Biol. 228(2), 181–196.

Zheng, L., Zhang, J., Carthew, R.W. 1995. Frizzled regulates mirror-symmetric pattern formation in the *Drosophila* eye. Development 121(9), 3045–3055.

Wnt signaling in *C. elegans*: New insights into the regulation of POP-1/TCF-mediated activation and repression

Hendrik C. Korswagen

Hubrecht Laboratory, Netherlands Institute for Developmental Biology and Center for Biomedical Genetics, 3584 CT, Utrecht, The Netherlands

Contents

The Wnt-signaling pathway is one of the key regulators of metazoan development. In the nematode *Caenorhabditis elegans*, Wnt signaling controls developmental processes such as cell fate specification, cell migration, and cell polarity. *C. elegans* utilizes conserved Wnt/β-catenin as well as novel signaling pathways to mediate Wnt signaling. In this chapter, I will give an

Advances in Developmental Biology
Volume 17 ISSN 1574-3349
DOI: 10.1016/S1574-3349(06)17003-9

overview of the different Wnt-signaling mechanisms in *C. elegans*. I will focus on interesting new results showing how transcriptional activation by the TCF/LEF-1 transcription factor POP-1 is regulated.

1. Introduction

The complex cell fate determinations and morphogenetic movements that generate the metazoan body plan are controlled by only a few families of signaling molecules, which are used repeatedly throughout development. One of these is the Wnt family of secreted lipid-modified glycoproteins. Wnt signaling controls cell fate specification and proliferation by regulating the expression of specific target genes and also controls cell polarity and migration by directly modulating the cytoskeleton (Cadigan and Nusse, 1997; Logan and Nusse, 2004). Studies in *Drosophila* and vertebrates have shown that these effects of Wnt are transduced by several, evolutionarily conserved signaling pathways (Huelsken and Behrens, 2002; Bejsovec, 2005): a canonical Wnt pathway that controls target gene expression and a non-canonical Wnt pathway [also termed the planar cell polarity (PCP) pathway] that interacts with Jun N-terminal kinase (JNK) and members of the Rho family of small GTPases to control cytoskeletal dynamics. In addition, Wnt signaling can trigger the release of intracellular calcium, leading to the activation of calcium/calmodulin-dependent protein kinase II (CamKII) and protein kinase C (PKC) (Kuhl et al., 2000).

Although these latter pathways are still poorly understood, a detailed picture has emerged of the canonical Wnt pathway (Huelsken and Behrens, 2002; Logan and Nusse, 2004; Cadigan and Liu, 2006). Central to this pathway are the effector β-catenin and a destruction complex that controls its stability. In the absence of Wnt signaling, this complex, which consists of the scaffolding protein Axin, the tumor suppressor gene product APC, and the protein kinases casein kinase Iα and GSK3β, targets β-catenin for degradation by the proteasome. Binding of Wnt to its receptor Frizzled and coreceptor low-density-lipoprotein receptor-related protein-5 or 6 (LRP5/6) blocks destruction complex function, allowing stabilization of β-catenin. It is not yet clear exactly how binding of Wnt to its receptors inhibits β-catenin degradation, but the evidence suggests that it involves two mechanisms that may function in parallel (Cadigan and Liu, 2006; Price, 2006). Binding of Wnt to its receptors leads to phosphorylation of conserved motifs within the intracellular domain of LRP5/6, which induces binding of Axin. The recruitment of Axin to the membrane may in turn lead to its degradation. Binding of Wnt to its receptors also activates the cytoplasmic protein Disheveled, which interacts with Axin and thereby directly inhibits destruction complex function. Following the inhibition of the destruction complex, the stabilized β-catenin accumulates in the nucleus and interacts with transcription factors of the

TCF/LEF-1 family (Cadigan and Nusse, 1997). In the absence of Wnt signaling, these sequence-specific transcription factors function as transcriptional repressors by interacting with Groucho corepressors. Binding of β-catenin displaces Groucho. Furthermore, the strong N- and C-terminal transactivation domains of β-catenin transform TCF/LEF-1 into an active bipartite transcription factor. The conversion of a transcriptional repressor into an activator leads to highly specific expression of Wnt target genes.

It is clear from the above that the intracellular level of free β-catenin needs to be tightly regulated in the absence of Wnt signaling. Mutations that disrupt destruction complex function, such as loss-of-function mutations in APC or Axin, are frequently found in cancer (Polakis, 2000; Moon et al., 2004). A tissue that is particularly sensitive to mutations in the Wnt pathway is the human colon. Wnt signaling maintains the undifferentiated characteristics of the intestinal-stem cells, and constitutive activation of the pathway leads to unchecked proliferation.

The canonical Wnt pathway has been highly conserved throughout evolution. It has been shown that Wnts and downstream pathway components, such as Disheveled, β-catenin, and TCF/LEF-1, are present in primitive animals such as *Hydra* (Hobmayer et al., 2000) and the sea anemone *Nematostella* (Kusserow et al., 2005), which clearly illustrates the central role of Wnt signaling in metazoan development. It is therefore not surprising that Wnt signaling also plays an important role in the development of the nematode *Caenorhabditis elegans*, a model organism that is widely used in developmental biology. Advantages of *C. elegans* are its relative simplicity (an adult hermaphrodite consists of only 959 somatic cells) and the extensive tool box that is available for genetic manipulation. The first Wnt pathway to be described in detail was the MOM-2 Wnt pathway in the early embryo (Rocheleau et al., 1997; Thorpe et al., 2000). An unexpected finding was that conserved β-catenin-destruction complex components, such as GSK3β and APC, function in an opposite manner to their vertebrate counterparts in the MOM-2/Wnt pathway, suggesting that Wnt signaling is fundamentally different in *C. elegans* (Han, 1997). This paradox was later solved by the discovery that *C. elegans* also has a canonical Wnt/β-catenin pathway.

2. Wnt signaling in *C. elegans*: Pathway components

Most of the important canonical Wnt pathway components are present in *C. elegans*, but there are some notable exceptions and differences (Korswagen, 2002; Eisenmann, 2005). Wnt inhibitors, such as Cerberus, Dickkopf, WIF, and secreted Frizzled-like proteins, are absent. Furthermore, a clear LRP5/6 homologue has not been found, suggesting that Wnt may signal without the aid of a coreceptor in the worm. However, several of the *C. elegans* Wnt-pathway components show considerable sequence divergence with their *Drosophila* and vertebrate counterparts, indicating that some of the missing pathway components may have been overlooked in sequence-based homology searches. Examples of such divergent proteins are the APC-related protein APR-1 (Rocheleau et al., 1997), the Axin homologue PRY-1 (Korswagen et al., 2002), and the three β-catenin-like proteins BAR-1, WRM-1, and HMP-2. The sequence divergence of the three β-catenin homologues is remarkable. The *C. elegans* β-catenins show only 20–30% overall amino acid sequence identity with *Drosophila* Armadillo and vertebrate β-catenin, whereas the β-catenin of the simple animal *Hydra* shows over 60% sequence identity (Hobmayer et al., 2000). Vertebrate and *Drosophila* β-catenin has a dual function in cellular adhesion and Wnt signaling. It interacts with classical cadherins and α-catenin to anchor the actin cytoskeleton to adherens junctions as well as with TCF/LEF-1 transcription factors to mediate Wnt signaling. These functions in cell adhesion and Wnt signaling are divided over separate proteins in *C. elegans* (Korswagen et al., 2000; Natarajan et al., 2001). Two β-catenins, BAR-1 and WRM-1, function specifically in Wnt signaling. BAR-1 is the only β-catenin that physically interacts with the single TCF/LEF-1 like transcription factor POP-1 and with PRY-1/Axin, and as discussed below, it is part of a canonical Wnt pathway. In contrast, WRM-1 interacts with the mitogen-activated protein kinase (MAPK) LIT-1/Nemo-like kinase (NLK) in a novel Wnt pathway to downregulate nuclear levels of POP-1/TCF (Rocheleau et al., 1999). The third β-catenin, HMP-2, functions specifically as a structural component of adherens junctions. HMP-2 is the only β-catenin that binds the classical cadherin HMR-1 and the α-catenin HMP-1 (Korswagen et al., 2000; Natarajan et al., 2001) and colocalizes with these proteins in adherens junctions (Costa et al., 1998). This functional separation explains the extensive sequence divergence, but it is not clear why *C. elegans* has evolved such dedicated β-catenins. Perhaps this is a mechanism to prevent cross talk between β-catenin in adherens junctions and the cytoplasmic signaling pool. Furthermore, the functional diversification of signaling β-catenins may enable a more complex regulation of Wnt target gene expression. *C. elegans* also has several potential homologues of downstream components of the noncanonical planar cell polarity pathway. However, it is not known if a planar cell polarity pathway has a function during *C. elegans* development.

3. A canonical Wnt pathway regulates Hox gene expression

A conserved canonical Wnt pathway controls several aspects of *C. elegans* larval development. This pathway utilizes a specific β-catenin, BAR-1, and interestingly, the main target genes appear to be Hox genes (Korswagen, 2002; Eisenmann, 2005). The best-studied examples of canonical-Wnt signaling are the Wnt-dependent migration of the Q neuroblasts and fate specification within the vulva lineage.

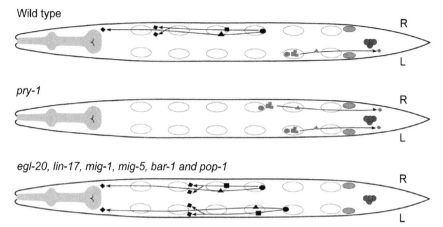

Fig. 1. The migration of the Q daughter cells is regulated by EGL-20/Wnt signaling. A dorsal view is shown. The complete Q cell lineages and their migrations are represented on both sides. Cells are in green and red when *mab-5* expression is activated or absent, respectively. *egl-20*/Wnt expressing cells are in blue. The gray circles represent the seam cells V1 to V6. In wild type (top panel), the Q daughter cells on the left side (QL) activate *mab-5* expression and migrate toward the posterior, whereas the Q daughter cells on the right side (R) migrate in the default anterior direction. Ectopic activation of *mab-5* expression in the QR.d of *pry-1*/Axin mutants results in posterior migration (middle panel). Loss of function of positive regulators of the EGL-20 pathway, such as *egl-20*/Wnt, *lin-17*/Fz, *mig-1*/Fz, *mig-5*/Dsh, *bar-1*/β-catenin, or *pop-1*/Tcf (bottom panel), results in anterior migration of the QL.d. (See Color Insert.)

The QL and the QR neuroblasts are initially present at similar anteroposterior positions on the left and right side of the animal, respectively (Fig. 1). During the first stage of larval development, both Q cells generate three neurons and two cells that undergo apoptosis (Sulston and Horvitz, 1977). Despite this similarity in lineage, the migration of the Q neuroblasts and their daughter cells is opposite between the two sides. On the left side the cells migrate toward the posterior, whereas on the right side migration is toward the anterior. The difference in migration direction between the two sides is controlled by the left/right asymmetric expression of the Hox gene *mab-5* (Salser and Kenyon, 1992). *mab-5* expression is restricted to the left side and directs the migration of the QL daughter cells toward the posterior. The QR daughter cells migrate in the default anterior direction. The expression of *mab-5* in QL is controlled by the Wnt EGL-20 (Harris et al., 1996; Maloof et al., 1999). In *egl-20* loss-of-function mutants, *mab-5* is not expressed in QL, and as a result, the QL daughter cells migrate toward the anterior. Overexpression of EGL-20 induces the opposite phenotype: ectopic expression of *mab-5* in QR and posterior migration of the QR daughter cells (Whangbo and Kenyon, 1999). EGL-20 is produced by a group of epidermal and muscle cells in the tail and forms a concentration gradient along the anteroposterior body axis (Whangbo and Kenyon, 1999; Coudreuse et al., 2006; Pan et al., 2006). The QL neuroblast is initially present at a more posterior position than the QR neuroblast and is therefore exposed to a higher concentration of EGL-20. The specific activation of EGL-20 signaling in QL is not the result of this difference in position, as the same effect is seen when the gradient is reversed by expressing EGL-20 in the pharynx. Instead, a difference in sensitivity to EGL-20 between the two Q cells determines that the EGL-20 pathway is only activated in QL (Whangbo and Kenyon, 1999). In addition to EGL-20, expression of *mab-5* requires the Frizzleds LIN-17 and MIG-1, the Disheveled MIG-5, BAR-1/β-catenin, and POP-1/TCF (Harris et al., 1996; Maloof et al., 1999; Korswagen et al., 2000; Herman, 2001; Pan et al., 2006). As discussed above, BAR-1 physically interacts with POP-1. This interaction is required for *mab-5* expression in QL, as overexpression of a dominant negative, N-terminally truncated POP-1 protein that lacks the BAR-1 interaction domain inhibits the expression of *mab-5* (Korswagen et al., 2000). This shows that BAR-1 and POP-1 form an active bipartite transcription factor that is required for Wnt target gene expression. The expression of *mab-5* is negatively regulated by the highly divergent, but functional Axin homologue PRY-1 (Korswagen et al., 2002). PRY-1 functions genetically upstream of BAR-1 and POP-1 (Maloof et al., 1999; Korswagen et al., 2002) and physically interacts with BAR-1, GSK-3, the APC-related protein APR-1, and MIG-5/Dsh, demonstrating that a conserved destruction complex negatively regulates BAR-1.

Canonical-Wnt signaling also plays an important role in the generation of the vulva. At the end of the first stage of larval development, 6 of the 12 ventral hypodermal Pn.p cells express the Hox gene *lin-39* and adopt the vulva precursor cell (VPC) fate. The remaining Pn.p cells fuse with the hypodermal syncytium. Later, inductive signaling from the somatic gonad and subsequent lateral signaling causes the VPCs to adopt one of three alternative vulval fates (Sundaram and Han, 1996; Wang and Sternberg, 2001). A LIN-3/EGF signal from the anchor cell in the somatic gonad activates a conserved receptor tyrosine kinase/Ras pathway in P6.p and induces the primary (1°) fate. Next, lateral signaling mediated by LIN-12/Notch specifies the secondary (2°) fate in P5.p and P7.p. Together, P5.p, P6.p, and P7.p generate the 22 cells of the vulva. The remaining VPCs adopt the tertiary (3°) fate and fuse with the hypodermal syncytium. In addition to inductive or lateral signaling, canonical Wnt/β-catenin signaling is also required for VPCs to select the appropriate fate. In *bar-1*/β-catenin mutants, both the generation and specification of the VPCs is defective (Eisenmann et al., 1998): VPCs frequently fuse with the hypodermal syncytium and P5.p to P7.p often adopt the 3° fate instead of the induced 1° or 2° fate. Overactivation of the pathway, such as in *pry-1*/Axin mutants or in transgenic animals overexpressing a constitutively active N-terminally truncated BAR-1 protein, leads to multiple VPCs adopting the vulva fate (Gleason et al., 2002). A likely Wnt target gene in the VPCs is the Hox gene *lin-39* (Eisenmann et al., 1998). In *bar-1* mutants, expression of *lin-39* is often lost in the VPCs. Furthermore, *bar-1* promoter-directed expression of *lin-39* can partially rescue the vulva phenotype of *bar-1* mutants. *lin-39* is also a target of the Ras pathway (Maloof and Kenyon, 1998), and Ras and Wnt signaling may cooperate in controlling the expression of this Hox gene (Gleason et al., 2002).

4. Wnt/MAPK signaling

Wnt/MAPK signaling specifies several asymmetric cell divisions during *C. elegans* development. Wnt/MAPK signaling is fundamentally different from the canonical, planar cell polarity, and calcium-signaling pathways in *Drosophila* and vertebrates. Unlike other pathways, Wnt/MAPK signaling in the worm utilizes the divergent β-catenin WRM-1 and LIT-1/NLK to downregulate nuclear levels of POP-1/TCF.

5. P2/EMS signaling: Spindle orientation and endoderm induction

At the four-cell stage, a signal from the posterior blastomere P2 polarizes the neighboring EMS cell to produce daughter cells with different fates (Fig. 2): the anterior daughter MS will form muscle cells, whereas the posterior daughter E will form the intestine. If contact between P2 and EMS is disrupted, both daughter cells adopt the MS fate (Goldstein, 1992). Blastomere isolation and recombination experiments have shown that P2 needs to contact EMS before it divides (Goldstein, 1993, 1995a). Signaling by P2 induces rotation of the EMS nucleus and centrosomes and orients the mitotic spindle perpendicular to the plane of contact between the two cells. This rotation polarizes EMS along the anteroposterior axis and is closely linked to the specification of endoderm fate in the posterior daughter (Goldstein, 1995b). An important determinant of MS and E fate is POP-1/TCF (Lin et al., 1995). As a result of the P2-derived polarizing signal, MS and E show a striking difference in nuclear levels of POP-1, with high levels of POP-1 in MS and low levels in E. Disruption of this POP-1 asymmetry abolishes the difference in fate between MS and E. Thus, when P2 signaling is prevented, nuclear POP-1 levels are equal in MS and E and both cells adopt the MS fate (Rocheleau et al., 1997; Thorpe et al., 1997). Mutation of *pop-1* causes the opposite phenotype, with both MS and E adopting the intestinal fate (Lin et al., 1995). The high level of POP-1 in MS ensures that endoderm fate is repressed; POP-1 interacts with UNC-37/Groucho and the histone deacetylase HDA-1 to repress transcription of the endoderm inducing gene *end-1* (Calvo et al., 2001). It has been shown that POP-1 also functions as an activator of endoderm fate in the E cell (Maduro et al., 2005; Shetty et al., 2005). This function of POP-1 has long been overlooked because it is masked

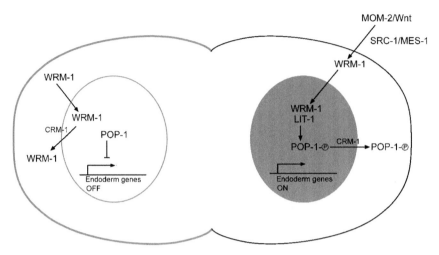

Fig. 2. A model of MOM-2/Wnt induced endoderm induction. At telophase of the EMS division, MOM-2/Wnt signaling, and parallel input from a MES-1/SRC-1 pathway induces accumulation of WRM-1 (indicated by the gray color) and LIT-1 in the posterior nucleus (E). LIT-1 phosphorylates POP-1, which results in PAR-5/14-3-3 and CRM-1-dependent nuclear export. The lower level of POP-1 in the posterior nucleus leads to activation of the endoderm determining gene *end-1*. At the anterior side, WRM-1 is retained at the cortex (gray color) and WRM-1 and LIT-1 are exported from the MS nucleus. The resulting high level of POP-1 represses *end-1*, enabling the cell to adopt the MS fate. The figure was adapted from (Nakamura et al., 2005).

by the parallel activation of *end-1* by the transcription factor SKN-1. Knock down of *pop-1* by RNAi not only results in derepression of *end-1* expression in MS but also in lower expression of *end-1* in E. High-level expression of *end-1* in E requires a TCF-binding site within the *end-1* promoter and the N-terminal β-catenin-binding domain of POP-1, suggesting that POP-1 interacts with a β-catenin to activate *end-1* transcription. Thus, the level of nuclear POP-1 may determine whether it functions as a repressor or an activator. The high level of POP-1 in MS turns POP-1 into a repressor. The low level of nuclear POP-1 in E may favor interaction of POP-1 with a coactivator that is present at a limiting concentration and may thereby activate target gene expression. This novel concept of POP-1/TCF regulation will be discussed further below.

The polarization of EMS by P2 is mediated by the Wnt MOM-2 (Rocheleau et al., 1997; Thorpe et al., 1997; Goldstein et al., 2006) and a parallel pathway consisting of the tyrosine kinases MES-1 and SRC-1 (Bei et al., 2002). MOM-2/Wnt is produced in P2, and its function depends on the Porcupine homologue MOM-1 and the uncloned gene *mom-3* (Rocheleau et al., 1997; Thorpe et al., 1997). In the EMS blastomere, polarization requires the Frizzled MOM-5 and GSK-3/GSK3β. Mutations in these genes affect both spindle rotation and endo-derm induction (Schlesinger et al., 1999). Further downstream components are the APC-related protein APR-1 and the divergent β-catenin WRM-1. Mutation of these genes disrupts endoderm induction but does not affect spindle rotation. Thus, the different effects of MOM-2/Wnt are mediated by pathways that branch at the level of GSK-3/GSK3β (Schlesinger et al., 1999). MOM-2/Wnt signaling induces endoderm fate by downregulating POP-1 in the nucleus of E. Although the role of GSK-3/GSK3β and APR-1 in this process is poorly understood, a detailed picture has emerged of the regulation of POP-1 by WRM-1/β-catenin (Rocheleau et al., 1999). In contrast to the β-catenin BAR-1, WRM-1 does not bind POP-1, but does physically interact with the NLK homologue LIT-1. In collaboration with the TGFβ activated kinase (TAK1) homologue MOM-4 (Meneghini et al., 1999; Shin et al., 1999), binding of WRM-1 activates LIT-1 kinase activity. LIT-1 in turn phosphorylates POP-1, which results in loss of POP-1 nuclear localization. Live imaging of a functional GFP::POP-1 fusion protein showed that MOM-2/Wnt signaling induces a dynamic nucleocytoplasmic redistribution of POP-1 that probably does not involve POP-1 degradation (Maduro et al., 2002). Nuclear localization of POP-1 is dependent on specific acetylation, a modification that enhances its nuclear retention, but this retention signal is overruled by LIT-1-mediated phosphorylation (Gay et al., 2003). Phosphorylation of POP-1 strongly enhances binding to the 14–3–3 protein PAR-5 and results in CRM-1-dependent nuclear export (Lo et al., 2004). LIT-1 is enriched in the nucleus of E and thus shows an asymmetry that is opposite to that of POP-1. Imaging of a functional WRM-1::GFP fusion protein revealed that the subcellular localization of WRM-1 is highly dynamic during the EMS division cycle (Nakamura et al., 2005). Early during EMS mitosis, WRM-1 is present at the cortex, but this cortical localization, which depends on MOM-5/ Frizzled, is lost at the site of contact with P2. During telophase, WRM-1 localizes at similar high levels in the newly formed MS and E nuclei. Later during the EMS division, the nuclear signal in MS becomes weaker due to CRM-1-dependent nuclear export, while high levels of WRM-1 are maintained in the nucleus of E. It is still unclear how MOM-2/Wnt signaling controls WRM-1 localization, but a likely possibility is that the WRM-1 released at the site of contact with P2 is somehow activated, allowing it to be retained in the nucleus of E. The high levels of WRM-1 and LIT-1 in E in turn activate nuclear export of POP-1 and thereby specify endoderm fate.

Asymmetric localization of POP-1 is not restricted to the EMS division. Many embryonic divisions along the anteroposterior axis display POP-1 asymmetry, suggesting that POP-1 is an important determinant of anteroposterior polarity throughout development (Lin et al., 1998). After the initial blastomere divisions in the early embryo, POP-1 asymmetry in the embryo may be induced independently of Wnt, as POP-1 asymmetry is still observed in *mom-1*/Porcupine mutants (Park and Priess, 2003). However, POP-1 asymmetry is fully dependent on MOM-5/Frizzled. MOM-5 is enriched at the posterior pole in many embryonic prophase cells that divide along the anteroposterior axis (Park et al., 2004). This is reminiscent of the asymmetric localization of

Frizzled in planar cell polarity signaling and suggests a parallel between these two Wnt pathways.

6. T cell polarity

Another asymmetric cell division that is polarized by Wnt/MAPK signaling is the division of the epidermal T cell during larval development (Sulston and Horvitz, 1977). The anterior daughter of this division gives rise to a hypodermal lineage, whereas the posterior daughter forms a sensory structure called the phasmid. The difference in fate of the T cell daughters is controlled by POP-1. A high level of nuclear POP-1 specifies the anterior hypodermal fate and a low level the posterior neural fate (Herman, 2001). POP-1 asymmetry is regulated by the Wnt LIN-44 and the Frizzled LIN-17. In *lin-44* mutants, POP-1 levels and the polarity of the T cell division are reversed (Herman et al., 1995). In *lin-17* mutants, on the other hand, POP-1 asymmetry is lost: both daughter cells have high POP-1 levels and adopt the anterior hypodermal fate (Sawa et al., 1996; Herman, 2001). T cell polarity also depends on WRM-1 and LIT-1 (Herman, 2001; Takeshita and Sawa, 2005). Live imaging of WRM-1 and LIT-1 localization during the division of T and V5.p (another epidermal cell of the hypodermis) revealed a dynamic pattern of subcellular localization. WRM-1 and LIT-1 are initially present at the anterior cortex of the cell but later redistribute to the posterior nucleus (Takeshita and Sawa, 2005). These observations imply that the downregulation of POP-1 levels in the posterior nuclei of the T and V5.p cells is mediated by a similar mechanism as in the EMS blastomere.

LIN-44 is produced by cells in the tip of the tail that are located just posterior of the T cell. It has been shown that this directional LIN-44 signal polarizes the T cell (Goldstein et al., 2006). When LIN-44 is expressed by cells located anterior to T, the polarity of the division is reversed. LIN-17 strongly accumulates at the side of the T cell that receives the highest level of LIN-44 (the posterior side in wild-type animals), which is the side that downregulates nuclear POP-1 levels and that will later adopt the neural fate. The accumulation of high levels of LIN-17 to one side of the cell may be important for activation of WRM-1 and LIT-1 signaling. Indeed, in *lin-44* mutants, LIN-17 is evenly distributed on the cell membrane of the T cell and both daughters fail to downregulate nuclear POP-1 levels.

7. Z1/4 polarity: Axis determination in the somatic gonad

The hermaphrodite gonadal primordium contains two somatic gonadal precursor cells, Z1 and Z4, which flank the two germ line precursor cells Z2 and Z3 (Kimble and Hirsh, 1979). The first division of Z1 and Z4 establishes the proximal–distal axis of the gonad. The distal daughter cells (Z1.a and Z4.p, respectively) adopt a leader cell fate (the distal tip cells) and are responsible for the outgrowth of the developing gonad, whereas one of the proximal daughter cells forms the anchor cell, which connects the gonad to the vulva. The asymmetric division of Z1 and Z4 is controlled by the Frizzled LIN-17 and WRM-1, LIT-1, and POP-1 (Siegfried and Kimble, 2002; Siegfried et al., 2004). POP-1 is present at high levels in the nuclei of the proximal daughter cells and at low levels in the distal daughter cells. Surprisingly, both high levels and complete absence of POP-1 induce the same phenotype, a symmetric division with both daughter cells adopting the proximal fate. The distal fate is only induced when POP-1 levels are low as a result of Wnt/MAPK signaling. The induction of distal fate also depends on the β-catenin interaction domain of POP-1 (Siegfried and Kimble, 2002). Analysis of mutants that fail to polarize the Z1 and Z4 divisions (called symmetrical sister or Sys mutants) resulted in the identification of a β-catenin-like protein that functions together with POP-1 in specifying the distal fate (Miskowski et al., 2001; Kidd et al., 2005). *sys-1* acts downstream of POP-1 and encodes a novel protein with three predicted Armadillo repeats, a sequence motif found in β-catenin and other proteins. SYS-1 binds to the β-catenin interaction domain of POP-1 and functions as a coactivator of POP-1-dependent transcription. A direct target gene of POP-1 and SYS-1 is *ceh-22*, which encodes a homeodomain containing protein similar to vertebrate Nkx2.5 (Lam et al., 2006).

The induction of distal fate in the Z1 and Z4 daughter requires low levels of POP-1. An attractive model is that SYS-1 is present in the nucleus at a limiting concentration (Kidd et al., 2005; Fig. 3). When POP-1 levels are downregulated by Wnt/MAPK signaling, the POP-1/SYS-1 ratio is altered to favor SYS-1-dependent activation of target genes. Several lines of evidence support this model. First, *sys-1* is haplo-insufficient, indicating that it is present at relatively low levels. Second, changing the POP-1/SYS-1 ratio by SYS-1 overexpression can induce distal fate when POP-1 levels are high (e.g., in the Z1 and Z4 proximal daughter cells or in the absence of WRM-1 or LIT-1). Finally, the model was directly tested by transfecting different ratios of POP-1 and SYS-1 in mammalian cells. When POP-1 was expressed in excess of SYS-1, transcription of a TCF reporter was repressed. However, when POP-1 and SYS-1 were expressed at equal levels or when excess SYS-1 was transfected, the reporter was activated. SYS-1-dependent activation of POP-1 may not be restricted to the somatic gonad. SYS-1 is also required for the induction of the posterior neural fate in the T cell lineage. Furthermore, the *sys-1* null phenotype is embryonic lethal and in many ways similar to the *pop-1* null phenotype (Kidd et al., 2005).

8. Multiple mechanisms regulate POP-1 transcriptional activity

It is clear from the above that Wnt signaling is surprisingly complex in *C. elegans*. In addition to a canonical Wnt/β-catenin pathway, there is an as yet novel Wnt/MAPK pathway that regulates nuclear levels of POP-1/TCF.

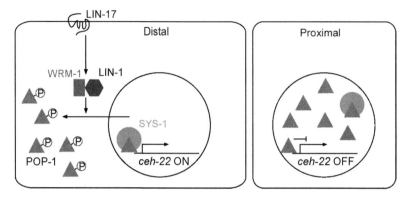

Fig. 3. A model of POP-1 and SYS-1-dependent target gene activation in the somatic gonad. The coactivator SYS-1 (green) is present at a limiting concentration in the proximal and distal daughter cells. In the proximal daughter cell, the low POP-1/SYS-1 ratio favors repression of distal specific target genes. In the distal daughter cell, the POP-1/SYS-1 ratio is increased by WRM-1 and LIT-1-dependent downregulation of nuclear POP-1 levels. As a result, POP-1 and SYS-1 activate the expression of the target gene *ceh-22*/Nkx2.5. The figure was adapted from (Kidd et al., 2005). (See Color Insert.)

Some aspects of this pathway may be conserved, as NLK has been shown to modulate TCF activity in vertebrates as well (Ishitani et al., 1999, 2003).

The TCF/LEF-1 transcription factor POP-1 plays a central role in canonical and Wnt/MAPK signaling and can function as a repressor as well as an activator. *C. elegans* has a surprisingly diverse set of β-catenin-like proteins that modulate POP-1 activity. The most familiar is BAR-1, a conventional β-catenin that is regulated by a conserved destruction complex and interacts with POP-1 to coactivate target gene expression. The β-catenin-like protein WRM-1 functions together with LIT-1/NLK to regulate nuclear levels of POP-1. The level of nuclear POP-1 determines whether it functions as a repressor or an activator. At high levels, POP-1 interacts with UNC-37/ Groucho to repress Wnt target gene expression (Calvo et al., 2001). At low levels, binding of POP-1 to the functional β-catenin SYS-1 (which is probably present at a limiting concentration) overcomes repression and turns POP-1 into an activator (Kidd et al., 2005). The regulation of nuclear levels of TCF/ LEF-1 is a novel concept in Wnt signaling. It will be interesting to see if similar mechanisms operate in other organisms.

In vertebrates, repression and activation of Wnt target genes are mediated by a diverse set of TCF/LEF-1 transcription factors (Liu et al., 2005). *C. elegans* has only a single TCF/LEF-1 transcription factor, POP-1 (Lin et al., 1995), but the repressor and activator functions of POP-1 are modulated by a set of functionally distinct β-catenin-like proteins. This suggests that vertebrates and nematodes may have found different solutions for a common problem: diversification and specialization at the level of TCF/ LEF-1, or at the level of β-catenin, to modulate the expression of a complex set of Wnt target genes.

References

Bei, Y., Hogan, J., Berkowitz, L.A., Soto, M., Rocheleau, C.E., Pang, K.M., Collins, J., Mello, C.C. 2002. SRC-1 and Wnt signaling act together to specify endoderm and to control cleavage orientation in early *C. elegans* embryos. Dev. Cell 3, 113–125.

Bejsovec, A. 2005. Wnt pathway activation: New relations and locations. Cell 120, 11–14.

Cadigan, K.M., Nusse, R. 1997. Wnt signaling: A common theme in animal development. Genes Dev. 11, 3286–3305.

Cadigan, K.M., Liu, Y.I. 2006. Wnt signaling: Complexity at the surface. J. Cell Sci. 119, 395–402.

Calvo, D., Victor, M., Gay, F., Sui, G., Luke, M.P., Dufourcq, P., Wen, G., Maduro, M., Rothman, J., Shi, Y. 2001. A POP-1 repressor complex restricts inappropriate cell type-specific gene transcription during *Caenorhabditis elegans* embryogenesis. EMBO J. 20, 7197–7208.

Costa, M., Raich, W., Agbunag, C., Leung, B., Hardin, J., Priess, J.R. 1998. A putative catenin-cadherin system mediates morphogenesis of the *Caenorhabditis elegans* embryo. J. Cell Biol. 141, 297–308.

Coudreuse, D.Y., Roel, G., Betist, M.C., Destree, O., Korswagen, H.C. 2006. Wnt gradient formation requires retromer function in Wnt producing cells. Science 312, 921–924.

Eisenmann, D.M. 2005. Wnt Signaling, Wormbook. http://www.wormbook.org, doi/10.1895/ wormbook.1.7.1.

Eisenmann, D.M., Maloof, J.N., Simske, J.S., Kenyon, C., Kim, S.K. 1998. The β-catenin homolog BAR-1 and LET-60 Ras coordinately regulate the Hox gene *lin-39* during *Caenorhabditis elegans* vulval development. Development 125, 3667–3680.

Gay, F., Calvo, D., Lo, M.C., Ceron, J., Maduro, M., Lin, R., Shi, Y. 2003. Acetylation regulates subcellular localization of the Wnt signaling nuclear effector POP-1. Genes Dev. 17, 717–722.

Gleason, J.E., Korswagen, H.C., Eisenmann, D.M. 2002. Activation of Wnt signaling bypasses the requirement for RTK/Ras signaling during *C. elegans* vulval induction. Genes Dev. 16, 1281–1290.

Goldstein, B. 1992. Induction of gut in *Caenorhabditis elegans* embryos. Nature 357, 255–257.

Goldstein, B. 1993. Establishment of gut fate in the E lineage of *C. elegans*: The roles of lineage-dependent mechanisms and cell interactions. Development 118, 1267–1277.

Goldstein, B. 1995a. An analysis of the response to gut induction in the *C. elegans* embryo. Development 121, 1227–1236.

Goldstein, B. 1995b. Cell contacts orient some cell division axes in the *Caenorhabditis elegans* embryo. J. Cell Biol. 129, 1071–1080.

Goldstein, B., Takeshita, H., Mizumoto, K., Sawa, H. 2006. Wnt signals can function as positional cues in establishing cell polarity. Dev. Cell 10, 391–396.

Han, M. 1997. Gut reaction to Wnt signaling in worms. Cell 90, 581–584.

Harris, J., Honigberg, L., Robinson, N., Kenyon, C. 1996. Neuronal cell migration in *C. elegans*: Regulation of Hox gene expression and cell position. Development 122, 3117–3131.

Herman, M. 2001. *C. elegans* POP-1/TCF functions in a canonical Wnt pathway that controls cell migration and in a non-canonical Wnt pathway that controls cell polarity. Development 128, 581–590.

Herman, M.A., Vassilieva, L.L., Horvitz, H.R., Shaw, J.E., Herman, R.K. 1995. The *C. elegans* gene *lin-44*, which controls the polarity of certain asymmetric cell divisions, encodes a Wnt protein and acts cell non-autonomously. Cell 83, 101–110.

Hobmayer, B., Rentzsch, F., Kuhn, K., Happel, C.M., von Laue, C.C., Snyder, P., Rothbacher, U., Holstein, T.W. 2000. WNT signalling molecules act in axis formation in the diploblastic metazoan *Hydra*. Nature 407, 186–189.

Huelsken, J., Behrens, J. 2002. The Wnt signalling pathway. J. Cell Sci. 115, 3977–3978.

Ishitani, T., Ninomiya-Tsuji, J., Nagai, S., Nishita, M., Meneghini, M., Barker, N., Waterman, M., Bowerman, B., Clevers, H., Shibuya, H., Matsumoto, K. 1999. The TAK1-NLK-MAPK-related pathway antagonizes signalling between β-catenin and transcription factor TCF. Nature 399, 798–802.

Ishitani, T., Ninomiya-Tsuji, J., Matsumoto, K. 2003. Regulation of lymphoid enhancer factor 1/T-cell factor by mitogen-activated protein kinase-related Nemo-like kinase-dependent phosphorylation in Wnt/β-catenin signaling. Mol. Cell. Biol. 23, 1379–1389.

Kidd, A.R., III, Miskowski, J.A., Siegfried, K.R., Sawa, H., Kimble, J. 2005. A β-catenin identified by functional rather than sequence criteria and its role in Wnt/MAPK signaling. Cell 121, 761–772.

Kimble, J., Hirsh, D. 1979. The postembryonic cell lineages of the hermaphrodite and male gonads in *Caenorhabditis elegans*. Dev. Biol. 70, 396–417.

Korswagen, H.C. 2002. Canonical and non-canonical Wnt signaling pathways in *Caenorhabditis elegans*: Variations on a common signaling theme. BioEssays 24, 801–810.

Korswagen, H.C., Herman, M.A., Clevers, H.C. 2000. Distinct β-catenins mediate adhesion and signalling functions in *C. elegans*. Nature 406, 527–532.

Korswagen, H.C., Coudreuse, D.Y., Betist, M.C., van de Water, S., Zivkovic, D., Clevers, H.C. 2002. The Axin-like protein PRY-1 is a negative regulator of a canonical Wnt pathway in *C. elegans*. Genes Dev. 16, 1291–1302.

Kuhl, M., Sheldahl, L.C., Malbon, C.C., Moon, R.T. 2000. Ca(2+)/calmodulin-dependent protein kinase II is stimulated by Wnt and Frizzled homologs and promotes ventral cell fates in *Xenopus*. J. Biol. Chem. 275, 12701–12711.

Kusserow, A., Pang, K., Sturm, C., Hrouda, M., Lentfer, J., Schmidt, H.A., Technau, U., von Haeseler, A., Hobmayer, B., Martindale, M.Q., Holstein, T.W. 2005. Unexpected complexity of the Wnt gene family in a sea anemone. Nature 433, 156–160.

Lam, N., Chesney, M.A., Kimble, J. 2006. Wnt signaling and CEH-22/tinman/Nkx2.5 specify a stem cell niche in *C. elegans*. Curr. Biol. 16, 287–295.

Lin, R., Thompson, S., Priess, J.R. 1995. *pop-1* encodes an HMG box protein required for the specification of a mesoderm precursor in early *C. elegans* embryos. Cell 83, 599–609.

Lin, R., Hill, R.J., Priess, J.R. 1998. POP-1 and anterior-posterior fate decisions in *C. elegans* embryos. Cell 92, 229–239.

Liu, F., van den Broek, O., Destree, O., Hoppler, S. 2005. Distinct roles for *Xenopus* Tcf/Lef genes in mediating specific responses to Wnt/β-catenin signalling in mesoderm development. Development 132, 5375–5385.

Lo, M.C., Gay, F., Odom, R., Shi, Y., Lin, R. 2004. Phosphorylation by the β-catenin/MAPK complex promotes 14-3-3-mediated nuclear export of TCF/POP-1 in signal-responsive cells in *C. elegans*. Cell 117, 95–106.

Logan, C.Y., Nusse, R. 2004. The Wnt signaling pathway in development and disease. Annu. Rev. Cell Dev. Biol. 20, 781–810.

Maduro, M.F., Lin, R., Rothman, J.H. 2002. Dynamics of a developmental switch: Recursive intracellular and intra-nuclear redistribution of *Caenorhabditis elegans* POP-1 parallels Wnt-inhibited transcriptional repression. Dev. Biol. 248, 128–142.

Maduro, M.F., Kasmir, J.J., Zhu, J., Rothman, J.H. 2005. The Wnt effector POP-1 and the PAL-1/Caudal homeoprotein collaborate with SKN-1 to activate *C. elegans* endoderm development. Dev. Biol. 285, 510–523.

Maloof, J.N., Kenyon, C. 1998. The Hox gene *lin-39* is required during *C. elegans* vulval induction to select the outcome of Ras signaling. Development 125, 181–190.

Maloof, J.N., Whangbo, J., Harris, J.M., Jongeward, G.D., Kenyon, C. 1999. A Wnt signaling pathway controls Hox gene expression and neuroblast migration in *C. elegans*. Development 126, 37–49.

Meneghini, M.D., Ishitani, T., Carter, J.C., Hisamoto, N., Ninomiya-Tsuji, J., Thorpe, C.J., Hamill, D.R., Matsumoto, K., Bowerman, B. 1999. MAP kinase and Wnt pathways converge to downregulate an HMG-domain repressor in *Caenorhabditis elegans*. Nature 399, 793–797.

Miskowski, J., Li, Y., Kimble, J. 2001. The *sys-1* gene and sexual dimorphism during gonadogenesis in *Caenorhabditis elegans*. Dev. Biol. 230, 61–73.

Moon, R.T., Kohn, A.D., De Ferrari, G.V., Kaykas, A. 2004. WNT and β-catenin signalling: Diseases and therapies. Nat. Rev. Genet. 5, 691–701.

Nakamura, K., Kim, S., Ishidate, T., Bei, Y., Pang, K., Shirayama, M., Trzepacz, C., Brownell, D.R., Mello, C.C. 2005. Wnt signaling drives WRM-1/β-catenin asymmetries in early *C. elegans* embryos. Genes Dev. 19, 1749–1754.

Natarajan, L., Witwer, N.E., Eisenmann, D.M. 2001. The divergent *Caenorhabditis elegans* β-catenin proteins BAR-1, WRM-1 and HMP-2 make distinct protein interactions but retain functional redundancy *in vivo*. Genetics 159, 159–172.

Pan, C.L., Howell, J.E., Clark, S.G., Hilliard, M., Cordes, S., Bargmann, C.I., Garriga, G. 2006. Multiple Wnts and Frizzled receptors regulate anteriorly directed cell and growth cone migrations in *Caenorhabditis elegans*. Dev. Cell 10, 367–377.

Park, F.D., Priess, J.R. 2003. Establishment of POP-1 asymmetry in early *C. elegans* embryos. Development 130, 3547–3556.

Park, F.D., Tenlen, J.R., Priess, J.R. 2004. *C. elegans* MOM-5/frizzled functions in MOM-2/Wnt-independent cell polarity and is localized asymmetrically prior to cell division. Curr. Biol. 14, 2252–2258.

Polakis, P. 2000. Wnt signaling and cancer. Genes Dev. 14, 1837–1851.

Price, M.A. 2006. CKI, there's more than one: Casein kinase I family members in Wnt and Hedgehog signaling. Genes Dev. 20, 399–410.

Rocheleau, C.E., Downs, W.D., Lin, R., Wittmann, C., Bei, Y., Cha, Y.H., Ali, M., Priess, J.R., Mello, C.C. 1997. Wnt signaling and an APC-related gene specify endoderm in early *C. elegans* embryos. Cell 90, 707–716.

Rocheleau, C.E., Yasuda, J., Shin, T.H., Lin, R., Sawa, H., Okano, H., Priess, J.R., Davis, R.J., Mello, C.C. 1999. WRM-1 activates the LIT-1 protein kinase to transduce anterior/posterior polarity signals in *C. elegans*. Cell 97, 717–726.

Salser, S.J., Kenyon, C. 1992. Activation of a *C. elegans* Antennapedia homologue in migrating cells controls their direction of migration. Nature 355, 255–258.

Sawa, H., Lobel, L., Horvitz, H.R. 1996. The *Caenorhabditis elegans* gene *lin-17*, which is required for certain asymmetric cell divisions, encodes a putative seven-transmembrane protein similar to the *Drosophila* frizzled protein. Genes Dev. 10, 2189–2197.

Schlesinger, A., Shelton, C.A., Maloof, J.N., Meneghini, M., Bowerman, B. 1999. Wnt pathway components orient a mitotic spindle in the early *Caenorhabditis elegans* embryo without requiring gene transcription in the responding cell. Genes Dev. 13, 2028–2038.

Shetty, P., Lo, M.C., Robertson, S.M., Lin, R. 2005. *C. elegans* TCF protein, POP-1, converts from repressor to activator as a result of Wnt-induced lowering of nuclear levels. Dev. Biol. 285, 584–592.

Shin, T.H., Yasuda, J., Rocheleau, C.E., Lin, R., Soto, M., Bei, Y., Davis, R.J., Mello, C.C. 1999. MOM-4, a MAP kinase kinase kinase-related protein, activates WRM-1/LIT-1 kinase to transduce anterior/posterior polarity signals in *C. elegans*. Mol. Cell 4, 275–280.

Siegfried, K.R., Kimble, J. 2002. POP-1 controls axis formation during early gonadogenesis in *C. elegans*. Development 129, 443–453.

Siegfried, K.R., Kidd, A.R., III, Chesney, M.A., Kimble, J. 2004. The *sys-1* and *sys-3* genes cooperate with Wnt signaling to establish the proximal-distal axis of the *Caenorhabditis elegans* gonad. Genetics 166, 171–186.

Sulston, J.E., Horvitz, H.R. 1977. Post-embryonic cell lineages of the nematode, *Caenorhabditis elegans*. Dev. Biol. 56, 110–156.

Sundaram, M., Han, M. 1996. Control and integration of cell signaling pathways during *C. elegans* vulval development. BioEssays 18, 473–480.

Takeshita, H., Sawa, H. 2005. Asymmetric cortical and nuclear localizations of WRM-1/β-catenin during asymmetric cell division in *C. elegans*. Genes Dev. 19, 1743–1748.

Thorpe, C.J., Schlesinger, A., Carter, J.C., Bowerman, B. 1997. Wnt signaling polarizes an early *C. elegans* blastomere to distinguish endoderm from mesoderm. Cell 90, 695–705.

Thorpe, C.J., Schlesinger, A., Bowerman, B. 2000. Wnt signalling in *Caenorhabditis elegans*: Regulating repressors and polarizing the cytoskeleton. Trends Cell Biol. 10, 10–17.

Wang, M., Sternberg, P.W. 2001. Pattern formation during *C. elegans* vulval induction. Curr. Top. Dev. Biol. 51, 189–220.

Whangbo, J., Kenyon, C. 1999. A Wnt signaling system that specifies two patterns of cell migration in *C. elegans*. Mol. Cell 4, 851–858.

The Wnt-signaling pathways in mammalian patterning and morphogenesis

Jianbo Wang, Leah Etheridge and
Anthony Wynshaw-Boris

*Departments of Pediatrics and Medicine, School of Medicine,
University of California, San Diego, California*

Contents

Advances in Developmental Biology
Volume 17 ISSN 1574-3349
DOI: 10.1016/S1574-3349(06)17004-0

Abbreviations

AER	Apical ectodermal ridge
AGM	Aorta–gonad–mesonephros
AME	Axial mesendoderm
APC	Adenomatous polyposis coli
AVE	Anterior visceral endoderm
β-gal	β-Galactosidase
BMP	Bone morphogenetic protein
CNC	Cardiac neural crest
Dkk	Dickkopf
FGF	Fibroblast growth factor
Fz	Frizzled
Gsk3β	Glycogen synthase kinase 3β
HSC	Hematopoietic stem cells
ICAT	Inhibitor of β-catenin and T cell factor
IM	Intermediate mesoderm
Lef	Lymphocyte enhancer factor
Lrp	Lipoprotein receptor related protein
mesd	Mesoderm development
MM	Metanephric mesenchyme
PCP	Planar cell polarity
SHH	Sonic hedgehog
Tcf	T cell transcription factor
UB	Ureteric bud
ZPA	Zone of polarizing activity

Together with the fibroblast growth factor (FGF), transforming growth factor (TGF)-β, and Hedgehog pathways, the Wnt pathway is one of the four major signaling pathways that are involved in almost every aspect of embryogenesis. Using the mouse as a model organism, research over the last decade has revealed numerous roles of the Wnt pathway in mammalian development. In this chapter, we focus on some of the recent advances in our understanding of how Wnt signaling, through collaboration with other pathways, regulates gastrulation and organogenesis, including neural patterning, heart development, hematopoiesis, and the formation of liver, limb, skin, kidney, lung, and gastrointestinal tract. We also discuss the important function of the novel noncanonical Wnt pathway in polarity establishment and morphogenesis, a process through which a tissue or embryo acquires its final shape. Interestingly, the canonical and noncanonical Wnt pathways share a number of critical components, raising the possibility that patterning and morphogenesis may be intimately linked.

1. Introduction

Genome sequencing has revealed that the mammalian genome encodes at least 19 Wnt ligands, 10 Frizzled (Fz) receptors, two coreceptors lipoprotein receptor related protein (Lrp5 and 6), and a large number of additional proteins related to the Wnt pathways. While other chapters discuss the detailed mechanisms of signal transduction, we will focus here on how these pathways impact on various aspects of mammalian development.

Depending on whether the pathways result in β-catenin-mediated gene transcription, we can divide the multiple Wnt pathways into canonical (β-catenin dependent) and noncanonical (β-catenin independent) Wnt pathways. These two pathways not only share some common components but also utilize distinct members. Through phenotypic characterization of genetic engineered, radiation-induced, or spontaneous mutations in many of the components in the mouse, we have now gained a basic understanding of both pathways in mammalian patterning and morphogenesis. In this chapter, we will first discuss the role of the canonical Wnt pathway in implantation, gastrulation, and organogenesis. We will then focus on the planar cell polarity (PCP) pathway, a primary branch of the noncanonical Wnt pathway, in mammalian polarity determination and morphogenesis.

2. The canonical Wnt pathway in mammalian patterning

As described in the previous chapters, in the canonical Wnt pathway, binding of Wnt to its receptor Fz and coreceptors Lrp5/6 increases the cytoplasmic level and nuclear accumulation of β-catenin. β-Catenin can then bind to transcription factors, such as T cell transcription factor/lymphocyte enhancer factor (Tcf/Lef), to activate target gene transcription. In this section, we will discuss the important developmental events regulated by this pathway and how an orchestrated interplay between the activation and inhibition of this pathway defines various patterning processes. We will also briefly discuss the integration of the canonical Wnt pathway with the other three major signaling pathways, FGF, Hedgehog, and TGF-β, in the global patterning during development.

2.1. Wnt signaling during gastrulation and implantation

Gastrulation is the single most important patterning event during which the initial symmetrical, one-layered epiblast gives rise to the three germ layers (ectoderm, mesoderm, and endoderm) with defined anterior–posterior (A/P) and dorsal–ventral (D/V) axes. Using a Wnt-reporter line where a *LacZ* transgene is placed behind multiple TCF/LEF-binding sites and a minimal promoter, Wnt-signaling activity can be conveniently visualized by β-galactosidase (β-gal) staining (Mohamed et al., 2004). Analysis of this Wnt-reporter line has clearly demonstrated that shortly before the onset of gastrulation and primitive streak formation at embryonic day 6.0 (E6.0), β-gal activity is detectable in an asymmetric pattern in a small group of epiblast cells near the embryonic–extraembryonic boundary. As gastrulation proceeds, β-gal staining is observed in the posterior part of the embryo and progresses distally, marking the forming primitive streak (Mohamed et al., 2004). Thus, the canonical Wnt signaling is active in the primitive streak from the beginning of gastrulation.

Analyses of several mouse mutants have provided solid evidence that the canonical Wnt pathway plays a critical role in primitive streak formation. First, the primitive streak fails to form in $Lrp5^{-/-};Lrp6^{-/-}$ double homozygotes, as well as a classical mouse mutant *mesoderm development* (*mesd*) that deletes the gene-encoding Mesd, a specific chaperone for Lrp5 and -6 trafficking to the cell membrane from the endoplasmic reticulum (Hsieh et al., 2003; Kelly et al., 2004). Consequently, gastrulation is completely disrupted and mesoderm and definitive endoderm do not form. The same defect is also observed in *Wnt3* mutants (Liu et al., 1999), indicating that Wnt3 is an indispensable Wnt ligand responsible for initiating gastrulation. Finally, because β-catenin plays a central role in the canonical Wnt pathway, disruption of this gene function is expected to completely abolish canonical Wnt signaling. The gastrulation defect is also recapitulated in *β-catenin*-null mutants (Huelsken et al., 2000; Huelsken and Birchmeier, 2001; Morkel et al., 2003). Collectively, these experiments indicate that Wnt/β-catenin signaling is an essential driving force of gastrulation in mammals.

In addition to the gastrulation defects, *β-catenin*-null mutants were also reported to display an abolishment of A-P axis specification as evidenced by the failure to orient distal visceral endoderm anteriorly to form the anterior visceral endoderm (AVE) (Huelsken et al., 2000; Huelsken and Birchmeier, 2001; Morkel et al., 2003). The A-P-patterning defect appears to be unique to *β-catenin* mutants, as the AVE is correctly translocated and specified in *Wnt3*, *Lrp5/6*, and *mesd* mutants despite the absence of the primitive streak (Liu et al., 1999; Hsieh et al., 2003; Kelly et al., 2004). Microarray and *in situ* analysis revealed that the expression of *Cripto*, the gene encoding a coreceptor for Nodal signaling that is essential for anterior translocation of the distal visceral endoderm, is missing in the epiblast of *β-catenin* mutants (Ding et al., 1998; Morkel et al., 2003). Further experiments with *β-catenin* gain-of-function mutants and reporter gene assays strongly suggest that *Cripto* may be a direct target of the Lef/Tcf/β-catenin complex (Morkel et al., 2003). In contrast, in *mesd* mutants, *Cripto* is expressed properly.

In light of the indispensable role of Lrp5/6 and its homologue Arrow in transmitting the Wnt signal in other model organisms, the phenotypic difference between *β-catenin* and *Lrp5/6* or *mesd* mutants is particularly intriguing. It was argued that a ligand-independent transcriptional activation by β-catenin is responsible for setting up the anterior organizer in mouse embryos (Morkel et al., 2003). However, this hypothesis cannot explain how broad *Cripto* expression in the epiblast can be activated by the extremely restricted pattern of β-catenin transactivation, as revealed by the β-gal staining in Wnt-reporter mice. Moreover, staining for nonphosphorylated β-catenin, accumulated as a consequence of Wnt-induced inactivation of glycogen synthase kinase 3β(GSK3β), also showed a highly restricted pattern resembling the β-gal staining (Mohamed et al., 2004). It is important to note that in addition to mediating Wnt signaling, β-catenin also has an essential function in the formation of the cadherins junction (Nelson and Nusse, 2004). Although removal of β-catenin does not appear to alter intercellular adhesion, presumably due to the compensatory substitution by plakoglobin (Huelsken et al., 2000; Lickert et al., 2002), it might interfere with morphogenetic movement that requires β-catenin-mediated dynamic regulation of cadherins junction formation.

Another discrepancy that needs to be resolved in the future is the role of the canonical Wnt signaling in the extraembryonic tissue during gastrulation. Analysis of the TCF/LEF-*LacZ* Wnt reporter revealed a transient Wnt activity in the extraembryonic visceral endoderm at E5.5, immediately before Wnt activity can be visualized in the adjacent epiblast (Mohamed et al., 2004). On the basis of this observation, it is tempting to speculate that Wnt signaling in the epiblast may be activated by a transient Wnt activity in the extraembryonic region. This hypothesis is also consistent with chimera analysis in *mesd* mutants where injected wild-type embryonic stem (ES) cells, which only contribute to the embryo proper, fail to rescue the gastrulation defects (Hsieh et al., 2003). Intriguingly, when wild-type ES cells are injected into *β-catenin* mutants, there is significant rescue of the gastrulation defects as determined by the formation of the headfold and the heart (Huelsken et al., 2000). One possible explanation for this difference in rescue is that there is a larger stock of β-catenin protein that persists from the maternal pool, which may be sufficient to support the initial Wnt activation in the extraembryonic visceral endoderm.

Exactly how canonical Wnt signaling promotes gastrulation and sustains mesoderm and definitive endoderm generation is not clear. On the basis of the fact that the T-box-containing transcription factor Brachyury (*T*) is directly activated by the canonical Wnt pathway (Yamaguchi et al., 1999b), the primary role of Wnt3 may be to induce mesodermal stem cell formation from pluripotent epiblast cells in the primitive streak (Liu et al., 1999; Yamaguchi, 2001). As *Wnt3* expression starts to diminish in the primitive streak by E7.5, the mesodermal stem cell induction/maintenance role is likely to be replaced by *Wnt3a*, whose expression become abundant at this stage (Liu et al., 1999; Yamaguchi, 2001). In contrast to *Wnt3* mutants, gastrulation and early A-P patterning are normal in *Wnt3a* mutants until neurulation, where *T* expression becomes diminished in the anterior primitive streak at the two-somite stage and significantly downregulated in the entire primitive streak at the six-somite stage. Consequently, mesodermally derived somites are not formed posterior to the forelimb (Yamaguchi et al., 1999b). A second function of the canonical Wnt pathway in early mesoderm patterning is to sustain the mesodermal fate and suppress the neural fate. In *Wnt3a* mutants, ectopic neural tissue is formed at the expense of paraxial mesoderm (Yoshikawa et al., 1997; Yamaguchi et al., 1999b).

For endoderm specification, β-catenin is critically required and appears to function in a cell-autonomous fashion. In chimeras where β-*catenin*-null ES cells are injected into wild-type embryos, the mutant cells can contribute to ectodermal and mesodermal lineages but rarely to endodermal lineage. Furthermore, specific ablation of β-*catenin* in the definitive endoderm precursors leads to a change of cell fate to cardiac mesoderm (Lickert et al., 2002). How β-catenin mediates the fate choice between endoderm and mesoderm is not clear.

Although a number of Wnts are expressed in preimplantation blastocysts (Kemp et al., 2005), there is no evidence for Wnt pathway activation, judged by β-gal staining in the Wnt-reporter mouse or immunostaining for stabilized, nonphosphorylated β-catenin as a result of Wnt-induced inactivation of GSK3β (Mohamed et al., 2004). Moreover, the fact that β-*catenin*, *mesd*, and *Lrp5/6* mutants can all implant successfully and do not show morphological defects until gastrulation starts suggest that the Wnt pathway is not required in the embryo *per se* for implantation. However, Wnts secreted by the blastocyst may trigger Wnt activation in the uterine epithelium, based on the observation that the Wnt pathway is specifically activated in the luminal epithelium adjacent to the blastocyst and that beads coated with exogenous Wnt7a can also trigger such activation. Furthermore, injection of a Wnt antagonist, secreted Fz-related protein (sFRP2), can significantly reduce implantation efficiency (Mohamed et al., 2005), suggesting activation of the Wnt pathway in the uterine epithelium may be essential for implantation to take place.

2.2. Wnt signaling in early brain and spinal cord development

Unlike mesodermal and endodermal cells, ectodermal cells are derived from the Wnt-free anterior epiblast that does not traverse through the primitive streak or the early organizer. Nevertheless, Wnt signaling is equally important in the patterning and differentiation of this germ layer. Upon neural tube closure, the ectoderm gives rise to the skin and the nervous system, including the brain and spinal cord.

Studies of brain patterning in the chick have led to a model in which Wnts produced from underlying paraxial mesoderm form an increasing gradient along the A-P axis and specify neural cells with initial rostral character into gradually more caudal character (Nordstrom et al., 2002). Whether the exact same mechanism governs mammalian-brain patterning remains to be tested. Nevertheless, a number of Wnts, such as Wnt3a, Wnt8b, and Wnt7b, are already expressed at E8.5 along the boundaries between the future telencephalon and diencephalon, prior to cephalic neural tube closure (Lee et al., 2000). Furthermore, by using another Wnt-reporter line called β-catenin-activated transgene driving expression of nuclear β-gal reporter (BAT-gal), similarly constructed by placing the *LacZ* transgene under the control of multiple LEF/TCF-binding sites, Wnt-signaling activity can already be visualized at this early stage as an increasing gradient from the telencephalon to the mesencephalon, with the anterior telencephalon devoid of any β-gal staining (Maretto et al., 2003), parallel to the increasing levels of Wnt gene expression observed along the A-P axis (Lee et al., 2000).

While the canonical Wnt signaling may be essential in specifying the caudal portion of the brain, it needs to be suppressed in the anterior region for proper forebrain patterning. Ectopic activation of Wnt signaling in the anterior region, through either overexpression of chicken *Wnt8c* or mutation of Wnt pathway repressor adenomatous polyposis coli (*Apc*), results in truncation of the anterior neural structures including the fore- and midbrain in the mouse (Popperl et al., 1997; Ishikawa et al., 2003). Mouse genetics have revealed that multiple mechanisms are involved in excluding Wnt signaling from the anterior neural plate.

First, extracellular inhibitors are secreted locally to prevent binding of Wnts to their receptors or coreceptors (Huelsken and Birchmeier, 2001; Mukhopadhyay et al., 2001; Yamaguchi, 2001; del Barco Barrantes et al., 2003; Robb and Tam, 2004). For example, two tissues that underlie the future forebrain, the AVE during early gastrulation and the axial mesendoderm (AME) derived from anterior primitive streak, express Dickkopf1 (Dkk1), a secreted Wnt antagonist that binds to Lrp5/6. Chimera analysis, though, indicated that Dkk1 expression in the AVE is not essential and the Dkk1 secreted from AME is sufficient for anterior head specification (Mukhopadhyay et al., 2001). Furthermore, BMP signaling also needs to be suppressed anteriorly for forebrain formation. Removing the secreted BMP antagonists Noggin and Chordin also leads to truncation of the forebrain (Bachiller et al., 2000), similar to the loss of function of *Dkk1*. Remarkably, anterior truncation is even observed in a portion of *Dkk1* and *noggin* double heterozygous mutants, providing compelling evidence that simultaneous inhibition of both Wnt and BMP signaling is required for forebrain specification (del Barco Barrantes et al., 2003). Unlike *Dkk1*, both *noggin* and *chordin* expressions are not in the AVE but limited to AME, again suggesting the essential role of anterior primitive streak derived AME in forebrain patterning.

Second, to further limit the amount of Wnt ligands in the anterior telencephalon, the transcription factor Six3 functions as a suppressor to directly inhibit *Wnt* gene expression in the anterior telencephalon. Loss of *Six3* leads to expression of *Wnt1* in the neural plate in a more rostral region and at an earlier stage and eventually, a severe truncation of the forebrain occurs (Lagutin et al., 2003). Finally, specification of anterior neural fate is also achieved through inhibition of *β*-catenin-mediated transactivation by inhibitor of *β*-catenin and T cell factor (ICAT), an 81 amino acid protein that disrupts the interaction between *β*-catenin and TCF. Gene targeting in the mouse has revealed that this gene is indispensable for forebrain development (Satoh et al., 2004).

In addition to blocking rostral neural fate, canonical Wnt signaling can also induce caudal neural fate, based on *in vitro* cell culture experiments. ES cells, when cultured over PA6 stromal cells, can differentiate into neurons. In this system, PA6-induced wild-type ES cells express only forebrain markers *Six3*, *BF-1*, and *Otx1*, but not hindbrain markers *Hoxb1* or spinal cord marker *Hoxc8*. Exposure of induced wild-type ES cells to Wnt3a, however, leads to suppression of the forebrain markers and activation of the hindbrain and spinal cord gene expression in a dosage-dependent fashion. The posteriorizing activity of Wnt3a can be effectively reversed by overexpression of ICAT. In contrast, PA6-induced *ICAT*-null ES cells express exclusively hindbrain and spinal cord markers and lack forebrain markers (Satoh et al., 2004).

While the above experiments strongly suggest that Wnts may function as classic morphogens in the global patterning of the mammalian brain along the A-P axis, definitive *in vivo* evidence is missing. Rather, various genetic manipulation experiments in the mouse have led to the argument that Wnt signaling may act as a mitogen to control the size of different compartments of the brain. For example, conditional *β-catenin* gain- and loss-of-function experiments have revealed that Wnt/*β-catenin* signaling is required in the entire central nervous system for expanding the progenitor cell population by simultaneously promoting cell proliferation and blocking apoptosis and differentiation (Zechner et al., 2003). Analyses of *Wnt3a* and *Lef1* mutant mouse embryos have reached a similar conclusion: in these mutants, hippocampal progenitor cells in the caudomedial cortical region are specified normally but show reduced proliferation subsequently, leading to a loss of hippocampus (Galceran et al., 2000; Lee et al., 2000). It is important to note that, however, neural cells may readily change their response to Wnts. A more recent study has shown that while Wnt signaling promotes proliferation of neural progenitor cells at midgestation stage, it induces neuronal differentiation once neurogenesis has initiated (Hirabayashi et al., 2004; Hirabayashi and Gotoh, 2005).

There is now clear evidence that canonical Wnt signaling also participates in D-V neural patterning. While Sonic hedgehog (SHH) signaling has long been known as the ventralizing morphogen, the signal(s) responsible for specifying dorsal neural fate has been elusive. In both the chick and the mouse, a number of *Wnt* genes are initially expressed in the lateral neuro-ectoderm and later in the dorsal midline of the brain and spinal cord as neurulation brings lateral neural plate to the dorsal region of a closed neural tube (Lee et al., 2000; Gunhaga et al., 2003). Studies of D-V patterning in the chick telencephalon have now demonstrated that canonical Wnts, acting in conjunction with FGF, can block ventral neural fate and induce dorsal neural fate. BMP signaling, in contrast, appears to be neither necessary nor sufficient for dorsal neural fate induction (Gunhaga et al., 2003). Consistent with these findings, conditional deletion of *β-catenin* in the mouse telenceph-alon results in reduction of dorsal neural markers and expansion of ventral neural markers. Conversely, activation of canonical Wnt signaling through expressing a stabilized *β-catenin* mutant suppresses the ventral telencephalic cell identity while expanding the dorsal identity (Backman et al., 2005). Wnts also play a similar role in patterning the dorsal neural tube in the spinal cord. Loss of *Wnt1* and *Wnt3a*, both of which are normally expressed in the roof plate, leads to severe reduction of the dorsal D1 and D2 neurons and an increase of the intermediate D3 neuron population. This defect appears to arise from an early cell fate shift in the progenitor population (Muroyama et al., 2002).

2.3. Wnt signaling in heart development

In contrast to the involvement of simultaneous inhibition of both BMP and Wnt signaling in head specification, studies in chick and *Xenopus* suggest that heart formation requires activity of BMP signaling coupled with inhibition of Wnt signaling (Huelsken and Birchmeier, 2001; Marvin et al., 2001; Schneider and Mercola, 2001; Yamaguchi, 2001). Consistent with these studies, over-expression of *Wnt8c* in the mouse completely inhibits or reduces the size of heart formation (Popperl et al., 1997). Chimera experiments further indicate that *β*-catenin, and consequently Wnt signaling, is not required in a cell autonomous fashion for the epiblast cells to contribute to cardiac mesoderm (Huelsken et al., 2000; Lickert et al., 2002). Conversely, conditional deletion of *β-catenin* in the definitive endoderm leads to ectopic formation of multiple hearts (Lickert et al., 2002).

While these studies all support a negative role of canonical Wnt signaling in mouse heart specification, it is intriguing that in mouse embryos carrying a hypomorphic mutation of *Apc*, the forebrain is truncated while cardiac mesoderm can be specified based on marker gene expression, albeit in an abnormal dorsal location (Ishikawa et al., 2003). In chimeric embryos, wild-type extraembryonic tissues can rescue the mislocalized cardiac mesoderm in the *Apc* hypomorphic mutants to form normal hearts but not forebrain truncation (Ishikawa et al., 2003). Further, *in vitro* studies have suggested that there may be a transient requirement for the canonical Wnt signaling during mammalian cardiomyogenic differentiation. *In vitro* studies using mouse P19CL6 cells, a pluripotent embryonal carcinoma cell line, have shown that *Wnt3a* and *Wnt8* are transiently upregulated shortly after cardiomyogenic stimulation, but then repressed following expression of *Nkx2.5*, an early cardiac-specific marker (Nakamura et al., 2003). An increase in canonical Wnt activity upon induction of cardiomyogenic differentiation was also observed using a TCF/LEF-dependent transcriptional assay. Finally, overexpression of a constitutively active form of GSK-3β blocked early differentiation, as shown by loss of cardiac-specific marker expression (Nakamura et al., 2003). Therefore, at least in this culture system, canonical Wnt signaling appears to be important for early cardiac myogenesis.

Canonical Wnt signals are also required for setting up correct left/right (L/R) asymmetry in the developing heart. A recent study has implicated Wnt3a-induced canonical Wnt signaling as a key regulator of the L-R asymmetry determination. This function is, at least in part, mediated through the activation of Notch signaling (Nakaya et al., 2005). The correct alignment of the aorta and pulmonary artery, with respect to the heart chambers, is essential for the formation of appropriate links between these structures. Consequently, severe cardiac abnormalities, such as transposition of great arteries (TGA), double outlet right ventricle (DORV), ventricular septal defects (VSD), and persistent truncus arteriosis (PTA) can arise following a misdirection of cardiac looping (Maclean and Dunwoodie, 2004). In wild-type embryos, hearts invariably loop to the right. In *Wnt3a*-null mice, however, heart looping becomes random. Furthermore, mice that are homozygous for a hypomorphic allele of *Wnt3a, vestigial tail* (*Wnt3a^{vt}*) and heterozygous for a null allele of *Delta-like 1* (*Wnt3a$^{vt/vt}$;Dll1$^{+/-}$*), display a number of cardiac morphological defects including PTA, TGA, and VSD (Nakaya et al., 2005). These L-R asymmetry defects in the heart demonstrate the importance of Wnt3a in cardiac laterality establishment. Further, they implicate a role for Notch signaling in the same regulatory pathway.

Wnt signals are also important for several morphogenetic events during later stages of heart development. In zebrafish canonical Wnt signals are required for endocardial cells to undergo the endothelial–mesenchymal transdifferentiation required for normal cardiac-valve formation (Hurlstone et al., 2003). Wnt signaling may also function in endocardial cushion formation in mammals, as canonical Wnt signaling was specifically identified in a subset of cells in the outflow tract and atrioventricular endocardial cushions in developing mouse hearts, using Wnt-reporter mice (Gitler et al., 2003). Several studies have also suggested a role for Wnt signals in regulating the integration of cardiac neural crest (CNC) cells during heart development, which contribute to the outflow tract cushion and are required for septation of the outflow tract. Attenuation of *Wnt1* expression in whole mouse embryo cultures by antisense oligonucleotide leads to conotruncal abnormalities, presumably due to a lack of neural crest cells in the developing hearts

(Augustine et al., 1993). Double mouse mutants that lack both *Wnt1* and *Wnt3a* show defects in neural crest cells (Ikeya et al., 1997) and *Dvl2*-deficient mice display severe outflow tract abnormalities and have defects in the CNC cell population (Hamblet et al., 2002). Conditional inactivation of the canonical Wnt repressor, *Apc* specifically in mouse neural crest also causes cardiac defects, including VSD and PTA, probably due to apoptosis in CNC cells (Hasegawa et al., 2002). These data suggest that a tightly controlled level of Wnt signaling is necessary for successful integration of CNC cells during mammalian heart morphogenesis.

2.4. Wnt signaling in hematopoiesis

Primitive hematopoiesis occurs in the extraembryonic yolk sac and leads to the production of nucleated erythroblasts, some megakaryocytes, and primitive macrophages (Baron, 2003). Embryonic hematopoiesis then follows, first in the aorta–gonad–mesonephros (AGM), and shortly after in the fetal liver. Here, definitive hematopoiesis occurs, during which a broad range of hematopoietic lineages are produced. Later in development, hematopoiesis shifts again to the bone marrow.

Recent studies indicate an important role for Wnt signaling in regulating hematopoiesis during mammalian development (Staal and Clevers, 2005). Expression of *Fz5*, *Wnt5a*, and *Wnt10b* has been found in both mouse embryonic yolk sac and fetal liver. Mice homozygous for a null mutation of *Fz5* die at E10.75 due to defects in yolk sac angiogenesis and reduced proliferation of endothelial cells in the yolk sac can be observed at earlier stages (Ishikawa et al., 2001). *Wnt10b* was also found to be expressed by murine fetal liver hematopoietic progenitors (Austin et al., 1997). The progenitor cell populations can be expanded greatly *in vitro* by addition of Wnt1-, Wnt5a-, or Wnt10b-conditioned media.

Human fetal bone marrow stromal cells have also been shown to express *Wnt2b*, *Wnt5a*, and *Wnt10b* and human CD34$^+$ hematopoietic progenitor cell number was also increased when cocultured with stromal lines transfected with these Wnts (Van Den Berg et al., 1998). In purified mouse bone marrow hematopoietic progenitor cells, the role of Wnt signaling in the maintenance and proliferation of hematopoietic stem cells (HSC) has also been more thoroughly characterized (Reya et al., 2003; Willert et al., 2003). Elevated Wnt signaling, through either retroviral infection of a stabilized *β-catenin* or treatment with purified Wnt3a protein, blocks differentiation and promotes self-renewal *in vitro* and the ability to reconstitute the blood cells *in vivo* when transplanted into lethally irradiated mice. Genes implicated in HSC self-renewal, such as *HoxB4* and *Notch1*, were shown to be upregulated upon activation of Wnt signaling in HSCs (Reya et al., 2003). Conversely, blocking Wnt signaling with overexpression of either Axin or a secreted form of the

Fz cysteine-rich domain, capable of competing with endogenous Fz receptor for Wnt binding, inhibits proliferation *in vitro* and the ability of HSC to reconstitute blood cells *in vivo*.

In addition to HSC maintenance, Wnt signaling is also important for the regulation of more committed hematopoietic cells. Normal thymocyte development requires canonical Wnt signaling, in particular at the pro-T cell stage (Staal and Clevers, 2003). The number of thymocytes in *Tcf1*-null mice is severely reduced and progenitor cells from the AGM, fetal liver, and fetal bone marrow of these mice fail to generate T cells in fetal thymic organ cultures (Schilham et al., 1998). Early thymocyte progenitor compartments, which are normally characterized by extensive proliferation, are severely affected in the *Tcf1*-null mutants, whereas the mature T cells that developed before thymopoiesis was completely arrested appear normal. Further, mice that lack both *Tcf1* and *Lef1* show an even more severe T cell phenotype, with a complete block in thymocyte development at the immature single-positive stage (Okamura et al., 1998). Moreover, *Wnt1* and *Wnt4* are expressed in mouse thymus and overexpression of these genes in fetal thymocytes increases cell number *in vitro* (Staal et al., 2001). Conversely, the number of immature thymocyte precursors is severely reduced in mice deficient for both *Wnt1* and *Wnt4* (Mulroy et al., 2002). Inhibition of Wnt signaling by overexpression of the extracellular Wnt-binding domain of Fz proteins in fetal liver lymphoid progenitors also leads to a failure of these cells to develop into T cells in fetal thymic organ cultures (Staal et al., 2001). In summary, these results highlight the importance of Wnt signaling in thymocyte production.

Wnt signaling has also been implicated in B cell development. Fetal liver B cells express *Lef1* and a reduced number of B220+ cells are found in the fetal liver and perinatal bone marrow of *Lef1*-null mice. Although differentiation of B cells in these mutants occurs normally, proliferation of pro-B cells is decreased, while apoptosis in this cell population is increased. Further, fetal liver pro-B cells in culture show increased levels of stabilization of β-catenin and enhanced proliferation, in response to Wnt3a. One of the Wnt receptors, *Fz9*, is also expressed in mouse B cells throughout their development, and expression levels are particularly high in early B cell populations (Ranheim et al., 2005). Mice lacking *Fz9* display a reduction in pro- and pre-B cells in the bone marrow, and these defects were found to be intrinsic to hematopoietic cells, as defective B cell development was observed following bone marrow implants from these mutants into a wild-type host, during competitive bone marrow reconstitution studies (Ranheim et al., 2005). Taken together these data indicate an essential role for Wnt signaling in the proliferation and survival of B cells.

2.5. *Wnt signaling in liver development*

Alterations in the regulation of canonical Wnt signaling have been implicated in 30–40% of human hepatocellular carcinoma and approximately 80%

of hepatoblastomas (Buendia, 2000). These alterations have been associated with activating mutations in β-catenin and inactivating mutations of Axin.

During normal liver development, β-catenin protein levels are regulated to peak at early stages (E10–E12), then decrease, and after E16 β-catenin levels are undetectable (Micsenyi et al., 2004). The importance of canonical Wnt signaling in liver development has been implicated by studies *in vitro* using antisense morpholino to β-*catenin* in embryonic mouse liver cultures (Monga et al., 2003). Subsequent reduced β-catenin protein levels caused a decrease in cell proliferation and an increase in apoptosis, suggesting a role for canonical Wnt signaling in regulating these processes. A further role for Wnt signaling in promoting the biliary differentiation of bipotential stem cells was also implied, as reduced β-catenin led to a decrease in CK19 expression. These results were supported by Wnt3a addition to these *in vivo* embryonic mouse liver cultures (Hussain et al., 2004). In controls, liver cultures show a loss in cell viability and growth in the absence of serum. However, in the presence of Wnt3a alone, under serum-free conditions, proliferation of CK-19-positive biliary cells and the formation of ductal structures were observed. Loss of c-kit expression with Wnt3a treatment also indicated a promotion of stem cell differentiation. Therefore, canonical Wnt signaling appears to enhance biliary differentiation by promoting stem cell specification as well as enhancing biliary proliferation and survival.

Mice that overexpress β-*catenin* in hepatocytes develop hepatomegaly due to increased cell proliferation. Surprisingly, the targets of the canonical pathway known to be involved in proliferation, such as *cyclin D1* and *c-myc*, were not overexpressed in these mutants (Cadoret et al., 2001). Suppression subtractive hybridization using RNA from the liver of β-*catenin* overexpression mice and control nontransgenic mice revealed that expression of genes involved in glutamine metabolism, including glutamine synthetase (GS), ornithine aminotransferase (OAT), and the glutamate transporter GLT-1, was upregulated in the transgenic mice livers (Cadoret et al., 2002). It was confirmed that increased expression of these genes is associated with β-*catenin* overexpression, rather than a secondary response to the increased cell proliferation, but whether these genes are direct or indirect targets remains to be established. Conditional deletion of the Wnt-signaling inhibitor *Apc* in mice livers also stimulated induction of GS, GLT-1, and OA (Colnot et al., 2004). As glutamine is critical for the expansion of proliferative cells, the authors have hypothesized that increased activities of these genes may offer a growth advantage to the cell by making it independent of a glutamine supply (Cadoret et al., 2002).

In another study using transgenic mice that overexpressed β-*catenin* specifically in the liver, extensive hepatocyte proliferation was also observed, as well as increased EGFR protein levels (Tan et al., 2005). A Tcf-binding site was identified in the EGFR promoter region and Wnt3a-induced activation of EGFR was confirmed using a reporter assay. Further, liver size was decreased in transgenic mice with inhibition of EGFR, suggesting that the upregulation of EGFR signaling by the Wnt pathway may be, at least in part, responsible for the observed proliferative effects in the liver.

2.6. Wnt signaling in limb development

Development of a mammalian limb requires a complex network of signals to coordinate the growth and patterning of skeletal-limb elements (Capdevila and Izpisua Belmonte, 2001). Wnt signaling is essential for several stages during this developmental process, including limb bud induction, outgrowth, and patterning, as well as late developmental phases, which include cartilage and bone formation (Church and Francis-West, 2002; Yang, 2003).

2.6.1. Limb induction and outgrowth

Limbs develop from primordial buds, undifferentiated mesodermal cells derived from the lateral plate, and somitic mesoderm, covered by ectoderm. Limb bud formation is mediated by a combination of FGF and Wnt signaling, and both *Fgf10*-null mouse embryos and those simultaneously lacking *Lef1* and *Tcf1*, transcription factors that mediate canonical Wnt signaling, develop severely retarded limb buds (Galceran et al., 1999; Sekine et al., 1999).

Prior to limb bud initiation, *Fgf10* expression in the chick limb mesenchyme is localized by Wnt signaling in the same tissue, stimulated by *Wnt-2b* in the forelimb and *Wnt-8c* in the hindlimb (Kawakami et al., 2001). However, in the mouse these Wnts were not found to be expressed at the limb induction stage (Agarwal et al., 2003). Initiation of *Fgf10* expression also appears normal in $Lef1^{-/-};Tcf1^{-/-}$ mouse embryos, although other Tcf family members, *Tcf3* and *Tcf4*, are still present, which may mediate a low level of canonical Wnt signaling (Agarwal et al., 2003). Therefore at present it is unknown whether in mammals limb bud initiation first requires an equivalent Wnt ligand or signaling in the early lateral plate mesoderm.

Both FGF and Wnt signaling probably act downstream of T-box transcription factors, specifically *Tbx5* in the presumptive forelimb area and *Tbx4* in the hindlimb area (Agarwal et al., 2003; Naiche and Papaioannou, 2003). $Tbx5^{-/-}$ and $Tbx4^{-/-}$ mice do not form forelimb or hindlimb buds, respectively (Agarwal et al., 2003; Naiche and Papaioannou, 2003), whereas $Fgf10^{-/-}$ and $Lef1^{-/-}Tcf1^{-/-}$ mouse embryos initiate limb bud formation but not outgrowth (Galceran et al., 1999; Sekine et al., 1999). In the chick flank, misexpression of *Tbx5* and *Tbx4* has also been shown to induce additional limb formation and induce both *Wnt* and *Fgf* expression (Takeuchi et al., 2003a).

The apical ectodermal ridge (AER) develops from a thickening of the ectoderm along the limb A-P axis. This specialized structure is required to maintain the outgrowth of the limb bud, and pattern the limb along the proximal–distal axis. Several studies have confirmed the functional importance of canonical Wnt signaling in AER initiation. The use of the BAT-gal and a similarly constructed TOPGAL Wnt-signaling reporter mouse, in which canonical Wnt signaling activates expression of *β*-gal, has demonstrated active Wnt signaling in the AER (Maretto et al., 2003; Topol et al., 2003). Defects in the formation of the AER have also been found in mouse embryos lacking *Lef1* and *Tcf1* (Galceran et al., 1999) or *LRP6* (Pinson et al., 2000) and upon conditional removal of *Wnt3* and *β-catenin* from the mouse limb ectoderm (Barrow et al., 2003; Soshnikova et al., 2003). Further, AER thickening and expression of AER markers were absent in mouse embryos when *β-catenin* was conditionally deleted only in the ventral limb ectoderm, indicating that Wnt-signaling activity in the dorsal ectoderm is not sufficient to sustain AER development (Barrow et al., 2003). Cumulatively, these studies have illustrated that canonical Wnt signaling is essential for initial AER formation in mice and identified *Wnt3* produced from the surface ectoderm as the signal responsible for activating Wnt signaling in this process.

When *β-catenin* is removed at a slightly later stage in limb development, after the induction of the ridge, the AER disappears, implicating canonical Wnt signaling with a continued role in the maintenance of the AER (Barrow et al., 2003). Activation of the Wnt pathway is able to sustain this structure by stimulating expression of the AER marker, *Fgf8* (Barrow et al., 2003; Soshnikova et al., 2003). *Fgf8* expression is absent in the limb ectoderm of *Lef1$^{-/-}$;Tcf1$^{-/-}$* mouse embryos (Galceran et al., 1999) and also reduced in *LRP6$^{-/-}$* mouse embryos (Pinson et al., 2000), indicating AER abnormalities occur with decreased canonical Wnt signaling. Conversely, mice lacking the canonical Wnt-signaling inhibitor, *Dkk1*, show a broadening of the AER along the D-V axis and expression of *Fgf8* is increased (Mukhopadhyay et al., 2001). The importance of *Fgf8* in the AER is evident as conditional disruption of this gene in mouse forelimb (Moon and Capecchi, 2000) and hindlimb (Lewandoski et al., 2000) produced smaller and abnormally shaped limb buds, and altered the expression of other AER regulatory genes. Loss of mesenchyme was also observed in the mutant forelimb buds, along with decreased *Fgf10* expression (Moon and Capecchi, 2000). Fgf8 in the ectoderm is thought to signal back to the mesenchyme to sustain expression of *Fgf10*, forming a positive feedback loop that is necessary to support the ongoing function of the AER (Xu et al., 1998). Therefore, canonical Wnt signaling stimulated by *Wnt3* in the ectoderm acts in both the initiation and maintenance of the AER, for further limb outgrowth, as well as proximal–distal patterning, to occur.

Recent studies have shown that conditional deletion of *β-catenin* specifically in mouse limb mesenchyme causes severe forelimb and hindlimb truncations (Hill et al., 2006). The limb buds of these mutants are growth retarded and expression of both *Fgf8* and *Fgf10* is reduced, indicating a specific role for mesenchymal Wnt signaling in AER maintenance, in addition to the one proposed for Wnt signaling in limb ectoderm.

Gap junctions are vital to the integrity of the AER as they mediate cell–cell communication, which is required for the cell-signaling processes involved in limb outgrowth. Wnt signaling may also function in the AER to modulate gap junction expression as ectopic expression of *Wnt1* in mouse limb mesenchyme causes alterations in expression of connexin 43 (Cx43), a gap junction component. In the *Wnt1*-expresing cells Cx43 is reduced, but in the surrounding mesenchyme it is increased. Whether these are direct effects has not been confirmed, although Cx43 has been identified as a target for Wnt signaling in PC12 and P19 cells (van der Heyden et al., 1998) and neonatal rat cardiomyocytes (Ai et al., 2000).

2.6.2. Limb patterning

As the limb buds begin to grow, coordinated-signaling mechanisms from specialized regions within the bud establish three axes. Patterning and elongation along the proximal–distal axis involve signaling from the AER, the A-P axis is determined by the zone of polarizing activity (ZPA), and the D-V axis is patterned by the limb bud ectoderm.

As discussed above, Wnt signaling is required to maintain the AER, which is directly responsible for proximal–distal patterning, and also participates in regulatory feedback signaling with the ZPA to indirectly affect A-P axis formation. *Fgf4* expression in the AER is required to maintain expression of *Shh* in the ZPA, which is the primary mediator of A-P patterning (Chiang et al., 1996). Shh is also able to modulate FGF4 levels in the AER, through activation of the BMP antagonist, *gremlin* (Zuniga et al., 1999) that prevents BMPs from inhibiting *Fgf4* expression, thereby creating a positive feedback loop that links the two signaling centers. The inhibitory action of BMPs on AER activity and growth was demonstrated using mouse limbs in culture (Niswander and Martin, 1993). Further, when the BMP antagonist Noggin was used to repress BMP signaling in the mouse limb *in vivo*, prolonged *Fgf8* expression was observed, indicating a delay in regression of the AER (Guha et al., 2002). In *gremlin* mutant mouse embryos, BMP signaling is increased in the limb bud, whereas *Fgf4* expression is undetectable in the AER and *Shh* expression is much reduced in the limb buds (Khokha et al., 2003), indicating that the Shh–Fgf feedback loop is disrupted by the increase in BMP signaling. Further, in the limb deformity (*ld*) mutant mouse, although *Shh* expression is initiated normally, *Fgf4* expression in the AER is blocked, which prevents further propagation of *Shh* (Haramis et al., 1995; Kuhlman and Niswander, 1997). This disruption of the Shh-Fgf4 feedback loop is due to a lack of *gremlin* expression (Zuniga et al., 1999), causing the AER to regresses prematurely (Zeller et al., 1989).

Expression of a dominant-active form of *β-catenin* in the mouse limb mesenchyme leads to a downregulation of AER markers *Fgf8*, *Fgf4*, *Fgf9*, and *Fgf17*, as well as mesenchymal *Fgf10*, indicating premature regression of the AER (Hill et al., 2006). *Shh* expression was also decreased, whereas *Bmp2*, *Bmp4*, *Bmp7*, the BMP target genes *Msx1* and *Msx2*, and the BMP antagonist, *gremlin* were all increased. The premature AER regression and truncated limbs in these mutant mice may therefore be due to an increase in BMP signaling and disruption of the feedback loop between the ZPA and the AER. The increase in *gremlin* may be a secondary effect, as expression of this gene can be activated by BMP signaling (Hill et al., 2006).

Wnt7a is also important for regulating *Shh* expression and consequently patterning along the limb A-P axis, as $Wnt7a^{-/-}$ mice display reduced expression of *Shh*, resulting in the absence of posterior limb digits (Parr and McMahon, 1995). In these mice the footpads are duplicated and develop on both the dorsal and ventral sides, indicating that Wnt signals are also required for dorsal patterning. Expression of *Wnt7a* is restricted to the dorsal limb ectoderm by *engrailed1* (*En-1*) expression in the ventral limb ectoderm, which represses *Wnt7a*. Loss of *En-1* function caused ectopic expression of *Wnt7a* in the ventral ectoderm, as well as the appearance of dorsal characteristics on the ventral side of limbs (Loomis et al., 1996; Cygan et al., 1997). This role of *Wnt7a* in D-V patterning is mediated through stimulation of *Lmx1b* expression in the underlying dorsal mesoderm (Cygan et al., 1997; Chen et al., 1998), and $Lmx1b^{-/-}$ mice show a duplication of ventral structures dorsally, similar to *Wnt7a* mutants (Chen et al., 1998). Mutations of *LMX1B* in humans cause the dominantly inherited skeletal malformation nail patella syndrome (NPS) (Dreyer et al., 1998).

Wnt7a is typically associated with stimulation of noncanonical Wnt pathways, and its role in dorsal patterning in chick has been proposed to be mediated via this mechanism (Kengaku et al., 1998). However, whether *Wnt7a* functions through canonical or noncanonical pathway in mammalian limb D-V patterning is still unclear. Reduced *Lmx1b* expression is found in the distal mesenchyme of forelimbs in mouse embryos that lack *Lrp6*, encoding one of the coreceptors for canonical Wnt signaling, and ventralized limbs develop in these mice, similar to those observed in $Wnt7a^{-/-}$ mice (Pinson et al., 2000; Adamska et al., 2005). *Wnt7a* has been proposed as the only ligand of Lrp6 in dorsal patterning, as the severity of the phenotype is not increased in $Wnt7a^{-/-};Lrp6^{+/-}$ mutants when compared to $Wnt7a^{-/-}$ single mutants (Pinson et al., 2000; Adamska et al., 2005). Further evidence to suggest that the canonical Wnt pathway is involved in D-V patterning comes from the observation that *Lmx1b* expression is reduced in the dorsal limb mesenchyme with specific deletion of *β-catenin* from limb mesenchyme and expanded into the ventral mesenchyme in embryos with expression of a dominant-active form of *β-catenin* (Hill et al., 2006). Taken together, these studies strongly suggest that *Wnt7a* regulates *Lmx1b* and specifies D-V patterning through canonical Wnt signaling.

2.6.3. Limb myogenesis

In addition to controlling the global patterning and outgrowth of the limb bud, Wnt signaling is also required at later stages of limb development to induce specific cellular differentiation and determine the morphology of individual limb structures including muscles, tendons, cartilage, and bone.

In response to signals from the lateral plate mesoderm, cells from the lateral portion of the somite dermomyotome migrate into the developing limb bud to provide the muscle precursors (Duprez, 2002; Francis-West et al., 2003). These then undergo specification and commence expression of *MyoD* and/ or *Myf5*, then coalesce to form the dorsal and ventral muscle masses. Following terminal differentiation these cells fuse into multinucleated slow or fast muscle fibers.

Signals from the surrounding tissues, such as the dorsal neural tube, notochord, and dorsal ectoderm, induce the expression of myogenic regulatory factors, *MyoD* and *Myf5*, in the myogenic precursors. *Wnt1* and *Wnt3a* are expressed in mouse dorsal neural tube and *Wnt7a* is expressed in the dorsal ectoderm (Parr et al., 1993), and these Wnt ligands play a role in skeletal myogenesis (Cossu and Borello, 1999). Using somite explants from mouse embryos, *Wnt1* was found to preferentially activate *Myf5* expression, whereas *Wnt7a* preferentially stimulated *MyoD* (Tajbakhsh et al., 1998), although whether these myogenic factors are differentially activated in the mouse limb *in vivo* remains unclear.

In mouse embryos that lack both *Wnt1* and *Wnt3a*, *Myf-5* expression is decreased. Markers, such as *Wnt6*, *Wnt11*, *Notch2*, and *noggin* are also absent, indicating that the medial compartment of the dermomyotome fail to form (Ikeya and Takada, 1998). Further, stable expression of either *Wnt3a* or activated *β-catenin* in P19 cells causes myogenesis of these cells (Petropoulos and Skerjanc, 2002). The Wnt antagonist, *Frzb1*, is expressed in presomitic mesoderm and newly formed somites. *In vitro* Frzb1 inhibits myogenesis in mouse presomitic mesoderm explants shown by a reduction in the number of *MyoD* and *Myf-5* expressing cells (Borello et al., 1999). *In vivo*, transplancental delivery of cells transiently expressing *Frzb1* to mouse embryos also appeared to inhibit skeletal myogenesis (Borello et al., 1999), although specific myogenic effects are difficult to analyze due to the overall growth retardation observed in these embryos (Duprez, 2002). The requirement of Wnt signaling for skeletal muscle formation has also become apparent from studies of the roles of *Wnt7a* in dorsal patterning. Via regulation of *Lmx1b*, Wnt7a is able to specify the dorsal muscle pattern as *Lmx1b*$^{-/-}$ mice show a duplication of ventral muscle patterning in the limb (Chen et al., 1998).

2.6.4. Condensation and chondrogenesis

Long bones develop by a process of endochondral ossification, during which mesenchymal cells in the limb bud first form condensations, then differentiate into chondrocytes to form the cartilage anlagen (Goldring et al., 2006). Articular cells are formed at the epiphyses, whereas in the shaft, chondrocytes mature, and ultimately the hypertrophic cartilage is replaced by bone.

Wnt signaling is required at specific stages of chondrocyte differentiation, and must be tightly regulated developmentally for normal cartilage formation and skeletogensis. Differentiation of condensed progenitor cells and correct organization of these chondrocytes in skeletal anlagen require canonical Wnt signaling to be relatively inactive (Tamamura et al., 2005). Two independent studies have consistently demonstrated that specific expression of constitutively active *β-catenin* in nascent chondrocytes inhibits further chondrogenic differentiation in mice, judged by lower expression of *Sox9*, the main chondrocyte-specific transcription factor (Akiyama et al., 2004; Guo et al., 2004b). Consequently, the mouse mutants develop severe chondrodysplasia, where all skeletal elements formed by endochondral ossification, including limbs, are extremely small. In a further study using a similar approach to express constitutively active *β-catenin* in early chondrocytes using a type II collagen promoter in mice (Tamamura et al., 2005), severe abnormalities in the organization, structure, and histology of skeletal elements were observed, as well as an inhibition of endochondral ossification. The chondrocytes of these animals fail to mature and assemble into skeletal anlagen with organized growth caps (Tamamura et al., 2005). Conversely, when *β-catenin* is conditionally deleted at early stages of skeletal development, either in mesenchymal condensations in the limb bud before differentiation or in nascent chondrocytes, ectopic chondrocyte differentiation is observed (Day et al., 2005).

Conditional deletion of *β-catenin* in mouse limb mesenchymal condensations or in nascent chondrocytes also causes a delay in chondrocyte maturation (Day et al., 2005).

Tissues surrounding the early mesenchymal condensation express *Wnt14*, and this Wnt protein has been shown to signal through the canonical Wnt pathway *in vivo* (Guo et al., 2004b). When *Wnt14* is overexpressed at high levels in nascent chondrocytes, chrondrocyte differentiation is blocked (Day et al., 2005). However, with lower levels of *Wnt14* expression, chondrocyte differentiation is not inhibited, allowing the effect of canonical Wnt signaling on later developmental stages to be examined. Stimulation of canonical Wnt signaling leads to an acceleration of chondrocyte maturation, indicated by an expanded osteogenic zone and upregulation of *MMP9* and *MMP13*, which are required for ECM remodeling following chondrocyte hypertrophy (Day et al., 2005). Therefore, although at early skeletal development stages, canonical Wnt-signaling levels are required to be low for differentiation of progenitor cells into chondrocytes, later differentiation of mature chondrocytes into hypertrophic cells requires higher levels of canonical Wnt signaling (Day et al., 2005; Tamamura et al., 2005).

In *Wnt5a*$^{-/-}$ mutant mice the skeletal elements formed by endochondral ossification are shortened, with missing distal digits (Yamaguchi et al., 1999a). These mutants show significantly reduced levels of *Sox9*, *Ihh*, which coordinate chondrocyte proliferation and differentiation, and *collagen X*, a maker of chondrocyte hypertrophy (Yang et al., 2003). Through the promotion of β-catenin degradation, *Wnt5a* can inhibit canonical Wnt signaling (Topol et al., 2003). Ectopic canonical Wnt signaling was observed in *Wnt5a*$^{-/-}$ limb buds by using the TOPGAL Wnt-reporter line. Increased protein levels of β-catenin were also found in *Wnt5a*$^{-/-}$ limb buds, suggesting that *Wnt5a* acts in the developing limb to inhibit canonical Wnt signaling (Topol et al., 2003). Mice lacking the Wnt receptor, *Ror2*, which has been shown to bind Wnt5a and transduce noncanonical Wnt signals (Oishi et al., 2003), display a very similar cartilage phenotype to *Wnt5a*$^{-/-}$ mutants and have shortened endochondrally derived bones (DeChiara et al., 2000; Oishi et al., 2003). Human congenital skeletal disorders, including Robinow syndrome and brachydactyly B, have also been associated with mutations in Ror2 (Afzal et al., 2000; Oldridge et al., 2000; Schwabe et al., 2000; van Bokhoven et al., 2000).

2.6.5. Joint formation

In the developing anlagen of long bones, chondrocytes first differentiate across the prospective joint areas, then the noncartilaginous interzone forms 'in a secondary event' (Archer et al., 2003). Cavitation follows, where the developing joint is separated from the cartilage anlagen and finally the synovium and articular cartilage develop. *Wnt4*, *Wnt14*, and *Wnt16* are all expressed in the developing joints of mouse limbs in distinct, but overlapping patterns and using the TOPGAL Wnt-reporter mice, canonical Wnt signaling has been shown to be active in this region (Guo et al., 2004b). Ectopic expression of constitutively active *β-catenin* or *Wnt14* in developing mouse limbs causes ectopic expression of joint markers *Gdf5*, *chordin*, and *Fgf8*, whereas expression of markers of chondrocyte differentiation, including *Noggin* and *Collagen X*, is decreased (Guo et al., 2004b). Conditional deletion of *β-catenin* in early mesenchymal condensations and early differentiating chondrocytes in mouse limb causes joint fusion and in these joints *Gdf5* expression, as well as chondrocyte proliferation is reduced (Guo et al., 2004b). Therefore, it has been proposed that *Wnt14* and canonical Wnt signaling are involved in synovial joint formation in mouse limb development and may function in differentiating chondrocytes to reverse chondrocyte differentiation, which is the first step in the formation of these joints (Guo et al., 2004b).

2.6.6. Bone development

Although flat bones develop through intramembranous ossification, involving direct osteogenic differentiation, most bones, including the long bones of limbs, develop through endochondral ossification, which involves bone formation from a condensed cartilage template (Kronenberg, 2003). Chondrocytes in the center of the model cease proliferation, become hypertrophic, and secrete distinct extracellular matrix. A perichondrium is formed from mesenchymal cells at the border of the condensation and an osteogenic differentiation program is stimulated in perichondrial cells adjacent to the center of the model by Ihh, secreted by hypertrophic chondrocytes. These osteoblasts in the periostium secrete bone matrix and produce a bone collar. Hypertrophic chondrocytes also mineralize the surrounding matrix and release factors to attract blood vessels and chondroclasts, before undergoing apoptosis. Further osteoblasts invade the cartilage model and synthesize bone matrix within it to form the primary spongiosa. Secondary ossification centers form at the ends of the model, and chondrocytes continue to proliferate in growth plates between these two regions. During remodeling, osteoclasts, derived from hematopoietic origins, resorb this woven bone, which is then replaced by lamella bone.

The importance of canonical Wnt signaling in osteogenesis first became apparent when loss-of-function mutations in *Lrp5* were shown to inhibit osteoblast proliferation and differentiation, leading to a reduction in bone density (Gong et al., 2001; Kato et al., 2002). Conversely, an increase in bone mass was associated with gain-of-function mutations in *Lrp5* (Boyden et al., 2002; Little et al., 2002). However, its direct involvement in osteoblast differentiation during embryogenesis has only recently become clear (Hartmann, 2006). Using TOPGAL Wnt-reporter mice, canonical Wnt signaling was found to be active in the perichondrium and osteoblasts at sites of bone formation (Hens et al., 2005), and β-catenin was found to localize in the nuclei of perichondrial osteoblasts (Day et al., 2005; Hill et al., 2005). Ectopic *Wnt14* expression in developing mouse limb has shown to stimulate canonical signaling as *Lef1*, a direct target of this pathway is upregulated in these regions (Day et al., 2005). With low expression levels of *Wnt14* chondrocyte differentiation was not affected, however ossification was enhanced, determined by an upregulation of the osteogenic markers *Runx2*, *Osterix*, and *Osteocalcin*, and increased von Kossa staining (Day et al., 2005). Loss of *β-catenin* in the early limb mesenchyme does not impair early stages of osteogenesis as the early osteoblast markers *alkaline phosphatase* and *Collagen 1α1* are still expressed by periosteal cells (Day et al., 2005; Hill et al., 2005; Hu et al., 2005). However, expression of *Osterix*, which marks osteoblast commitment, and *Osteocalcin*, a mature osteoblast marker, is absent, indicating a block in osteogenesis. Rather than undergoing further osteogenic differentiation, periosteal cells differentiate into chondrocytes, which express the chondrocyte markers *Sox9* and *Collagen 2a1* and secrete a matrix that stains positively with alcian blue (Day et al., 2005; Hill et al., 2005). In summary, recent data suggests that during skeletogenesis canonical Wnt signaling acts cell autonomously in mesenchymal progenitors to repress chondrogenic differentiation and induce osteoblast differentiation.

2.7. Wnt signaling in skin development

Canonical Wnt signaling is essential for normal mammalian skin development (Alonso and Fuchs, 2003), and numerous Wnt and Fz genes are expressed in mouse embryonic skin (Millar et al., 1999; Reddy et al., 2001, 2004). In particular, Wnt signals are critical for the formation of hair follicles (van Genderen et al., 1994; Andl et al., 2002). The canonical Wnt pathway was found to be active in both the epithelium and mesenchyme of developing hair follicles, using the TOPGAL Wnt-reporter transgene (DasGupta and Fuchs, 1999). In postnatal hair follicles Wnt signaling is active in precursor cells of the hair shaft (DasGupta and Fuchs, 1999), matrix cells, and dermal papilla (DP) (Maretto et al., 2003). Expression of a constitutively active form of *β-catenin* in mice under the control of an epidermal promoter, or mis-expression of Lef-1 in epithelium, induces the formation of new hair follicles (Zhou et al., 1995; Gat et al., 1998). Conversely, conditional deletion of *β-catenin* in mouse skin inhibits the formation of placodes that generate hair follicles (Huelsken et al., 2001). *Lef1*$^{-/-}$ mice fail to develop whiskers and have a reduced number of hairs (van Genderen et al., 1994). Ectopic expression of *Dkk1* in basal cells of mouse epidermis completely blocks the development of skin appendages, indicating that Wnt signals precede other inducers of hair follicle formation (Andl et al., 2002). Recent evidence suggests that Wnt10b may be the Wnt ligand that mediates canonical Wnt signaling in skin epithelial cells (Ouji et al., 2006a,b). *Wnt10b* is expressed in developing hair follicles, in particular at early stages, in placodes (Reddy et al., 2001). Addition of Wnt10b to adult mouse-derived primary skin epithelial cells *in vitro* induces differentiation, shown by expression of epithelial cell differentiation markers including *keratin1, keratin2, loricrin, mHa5*, and *mHb5*. These effects were probably mediated through the canonical Wnt pathway, as Wnt10b-stimulated canonical Wnt signaling, demonstrated using a TCF-reporter assay.

Wnt signals in both mesenchymal and epithelial hair follicle progenitors may be regulated by BMP signaling. The BMP inhibitor *noggin* is expressed in the mesenchymal condensation below the developing hair placode and in the developing DP (Botchkarev et al., 1999). Both hair follicle induction and early morphogenesis are retarded in *noggin*-null mice, which also show reduced Lef-1 expression in the hair placode. In contrast, addition of Noggin to embryonic skin organ cultures increases the number of hair follicles, enhances early developmental of hair formation, and increases Lef-1 expression. Simultaneous addition of both Wnt3a and noggin to mouse keratinocytes *in vitro* further activates canonical Wnt signaling (Jamora et al., 2003).

The retinoblastoma family of proteins may also regulate Wnt signaling in the skin (Ruiz et al., 2004). In p107/p130-deficient skin elevated levels of nuclear of *β*-catenin are found in basal keratinocytes (Ruiz et al., 2004). Increased expression of the GSK-3*β*-binding protein Frat and subsequent inhibition of β-catenin phosphorylation was proposed to be the cause of β-catenin accumulation. Wnt signaling may be further regulated in mammalian skin by Wise, which has been found recently to be expressed in rat postnatal DP during the anagen hair–shaft growth phase (O'Shaughnessy et al., 2004). Wise binds LRP6 and its effects on Wnt are context dependent (Itasaki et al., 2003). In 293 cells stably transfected with a Wnt-signaling reporter gene, Wise dramatically inhibits Wnt-signaling activity (Beaudoin et al., 2005). Hairless (Hr) is essential for hair follicle regeneration in both humans and mice, and in $Hr^{-/-}$ mice initial hair growth is normal, but following the regression or catagen phase of the postnatal hair cyclic process, no new hair is produced. It has recently been suggested that Hr is required to reduce expression of Wise and alleviate its inhibition on Wnt signaling at appropriate times in the postnatal hair cycle (Beaudoin et al., 2005).

Several target genes regulated by canonical Wnt signaling in hair follicle development have been identified. A recent study compared the expression profiles of epithelial stem cells isolated from skin of wild-type mice and mutants that expressed a constitutively active form of *β-catenin* (Lowry et al., 2005). Increased expression of proliferation-associated genes, including *cylcin D1*, was found in the epithelial stem cells that expressed stabilized *β-catenin*. Other genes upregulated in response to stabilized β-catenin included *Biglycan*, *Hmgn3*, and *Timp3*. An enhanced proliferation of keratinocytes was observed in response to *Biglycan* and *Timp3*, as well as constitutively active *β-catenin*. These data indicate a role for Wnt signaling in promoting the transition from quiescent epidermal stem cell to proliferating transient amplifying cells, the first stage in differentiation. In embryonic skin Wnt signaling also induces expression of *Eda*, which along with its receptor, *Edar*, plays an important role in cell fate determination in the epidermis and in regulating lineage differentiation in the hair follicle (Headon et al., 2001; Laurikkala et al., 2002; Botchkarev and Fessing, 2005). In addition to its role in converting Lef/Tcf from transcriptional repressors to activators, β-catenin also represses *E-cadherin* transcription through binding to Lef1 in the formation of hair follicle buds (Jamora et al., 2003). Reduced E-cadherin levels and attenuated cell–cell adhesion are likely to be critical for hair follicle morphogenesis (Jamora et al., 2003).

2.8. Wnt signaling in kidney development

In mammals, the urogenital system is derived entirely from intermediate mesoderm (IM) that lies between the somitic and lateral plate mesoderm. Early interaction with paraxial/somitic mesoderm converts adjacent IM to epithelium and results in formation of a pair of nephric ducts, also known as the Wolffian ducts, at the forelimb bud level. Once formed, the Wolffian duct extends caudally and induces medial IM cells to undergo mesenchymal to epithelial conversion to form the mesonephric tubules. By E10.5, the Wolffian duct has reached the cloaca and extends a dorsal diverticulum, the ureteric bud (UB), just rostral to the cloaca. UB continues to extend into the IM at the hindlimb bud level, known as the metanephric mesenchyme (MM). Intricate interactions between UB and MM in turn give rise to the metanephric kidney, the functional postnatal kidney in mammals. On one hand, UB induces MM overlying the UB tip to condense and undergo mesenchymal to epithelial conversion to form epithelial tubules, precursors of the epithelial segments of the nephron. On the other hand, MM supports extensive branching of the UB tip to form the collecting duct network (Perantoni, 2003; Vainio, 2003; Yu et al., 2004).

Several Wnts have been shown to participate in the reciprocal induction between UB and MM during metanephric kidney formation. The first *Wnt* gene identified to be critical for epithelial conversion of MM was *Wnt4*. In *Wnt4*-null mouse embryos, there is little epithelial tubule formation. The defect appears to be specifically due to epithelial conversion, since in *Wnt4* mutants UB is able to invade and branch, and MM condenses and proliferates normally at early stages (Stark et al., 1994; Kispert et al., 1998; Vainio, 2003). Conversely, in an *in vitro* transfilter culture system, Wnt4 expressing cells, when placed adjacent to, but separated from, undifferentiated MM by a porous filter, can induce epithelial conversion. Thus, Wnt4 appears to be both necessary and sufficient for epithelial conversion. Furthermore, Wnt4 expressing cells can also induce epithelial conversion in MM from *Wnt4* mutants, suggesting that Wnt4 may only play a transient role in triggering nephronic specification or that Wnt4 functions through activating a second factor (Kispert et al., 1998; Vainio, 2003). Finally, other Wnts that are known to activate the canonical Wnt pathway, such as Wnt1, can mimic the effect of Wnt4 in the transfilter culture system, suggesting that Wnt4 functions *in vivo* through activating β-catenin-mediated transcription (Kispert et al., 1998). This conclusion is further supported by the fact that treatment with lithium chloride, which stabilizes β-catenin through inhibiting GSK-3β activity, also leads to epithelial conversion (Davies and Garrod, 1995).

However, Wnt4 is unlikely to be the initial signal emanating from the UB that induces epithelial conversion *in vivo* since its expression is limited to the MM upon UB invasion (Stark et al., 1994; Kispert et al., 1998). Thus, Wnt4 functions in an autocrine or paracrine fashion within in the MM. A recent study indicates that another Wnt, Wnt9b, is also required for epithelial conversion and metanephric tubule formation. In contrast to *Wnt4*, *Wnt9b* is expressed throughout the Wolffian duct epithelium, including the UB, but not in the MM, suggesting that it is likely to be part of the initial inductive signal derived from UB. A *Wnt1* transgene expressed exclusively in the Wolffian duct can rescue the kidney defects in *Wnt9b* mutants, strongly suggesting that Wnt9b signals through the canonical Wnt pathway. Furthermore, *Wnt4* expression is never induced in the MM of *Wnt9b*-null embryos, while Wnt4 expressing cells can induce epithelial conversion and tubule formation in *Wnt9b* mutant MM in transfilter culture systems. These studies are consistent with a model in which Wnt9b secreted from the UB activates Wnt4 expression in the MM, either directly or indirectly, to induce epithelial conversion (Carroll et al., 2005).

Conditional removal of *Fgf8* in the mesodermal lineage revealed a complex requirement of Fgf signaling in Wnt-mediated epithelial conversion (Perantoni et al., 2005). In the mutants where *Fgf8* is deleted in most of the mesodermal cells by a *T-Cre* transgene, the MM fails to undergo epithelial conversion, resembling both *Wnt9b* and *Wnt4* mutants. Fgf8 is only expressed in the MM, and its expression is lost in *Wnt9b* mutants. In *Fgf8* mutants, *Wnt4* expression is significantly diminished. Overall, these data support a hypothesis where Wnt9b activates Fgf8, which in turn induces Wnt4 expression. However, in the transfilter culture system, spinal cord tissue, which expresses various Wnts and can induce epithelial tubule formation in *Wnt4* mutant MM, fails to induce tubule formation in *Fgf8* mutant MM. Fgf8-soaked beads, although cannot induce epithelial tubules in explanted MM when present alone, are able to rescue *Fgf8* mutant MM in combination with spinal cord tissue. Thus, these experiments indicate that the role of Fgf8 is not simply to induce *Wnt4* expression, but rather may function in conjunction with Wnt4 during epithelial conversion (Perantoni et al., 2005).

Despite the severe kidney phenotype in all of the mutants described above, the branching of UB appears to be normal initially, suggesting that these signaling ligands, or any other signals secreted upon differentiation of the MM, are not essential for inducing ureteric branching. In contrast, glial cell-derived neurotrophic factor (Gdnf), a ligand of the TGF-β superfamily secreted by the MM, binds to the Ret tyrosine kinase receptor expressed on the ureteric epithelium and plays crucial roles in ureteric branching (Schuchardt et al., 1994; Sanchez et al., 1996; Schuchardt et al., 1996; Airaksinen and Saarma, 2002). Similar to *Ret*, *Wnt11* is also expressed in the tips of UB. Mutation of *Wnt11* results in smaller kidneys in the mouse and detailed analysis revealed clear reduction of early ureteric branching at E12.5. While *Ret* is expressed strongly in the UB of *Wnt11* mutants, a significant reduction of Gdnf in the surrounding MM is observed. Conversely, *Wnt11* expression in the UB is abolished in *Ret* mutants. Finally, *Wnt11* and *Ret* double heterozygotes show reduction of ureteric branching that is not displayed in each single heterozygous mutant (Majumdar et al., 2003). Collectively, these data suggest that *Wnt11* and *Ret/Gdnf* pathways engage in an autoregulatory feedback loop to promote the ureteric branching process. However, since the *Xenopus* and zebrafish homologues of *Wnt11* are only involved in the noncanonical Wnt pathway (Heisenberg et al., 2000; Tada and Smith, 2000) and in the transfilter culture system, Wnt11 fails to induce epithelial conversion as other canonical

Wnts do (Kispert et al., 1998), it is unclear which pathway(s) activated by Wnt11 mediates the branching process.

While the studies described above all indicate essential roles of Wnt signaling in kidney development, analysis of the *ICAT* mutant revealed the intriguing observation that a certain level of suppression of β-catenin transactivation is also important, as a portion of *ICAT*-null mutants lacked kidneys. The cause of this defect is currently unknown (Satoh et al., 2004).

2.9. Wnt signaling in lung development

Similar to kidney development, the formation of the mammalian lung is also accomplished through extensive branching morphogenesis mediated by epithelial–mesenchymal interaction. However, in contrast to kidney epithelium, which is derived from mesoderm through epithelial conversion, the respiratory epithelium is derived from foregut endoderm between the thymus and the stomach. At E9.0–E9.5, the primordial lung buds, formed from outpocketing of the anterior foregut endoderm, invade the splanchnic mesenchyme and undergo initial branching to give rise to the main stem and lobar bronchi around E11. During the subsequent pseudoglandular stages from E11.5 to E16.5, extensive branching and budding of the airways occur to form the proximal airways, which conduct air, and peripheral lung, which mediates gas exchange. Gradual differentiation of the epithelium, in turn, gives rise to various highly specialized cell types such as secretory Clara cells and ciliated epithelium lining the conducting airway and alveolar Types I and II cells on the peripheral airway. During the final stages of lung development, alveolarization of the distal epithelium occurs in parallel with development of the pulmonary vasculature derived from mesoderm, leading to the formation of the functional gas exchange system (Shannon and Hyatt, 2004; Pongracz and Stockley, 2006).

Gene targeting in the mouse again demonstrates that the interplay between the four major signaling pathways, FGF, Hedgehog, TGF-β, and Wnt, is essential for normal lung development. Briefly, highly localized secretion of FGF10 from the surrounding mesenchyme induces lung epithelial tip outgrowth through acting upon FGF receptors, in particular FGFR2IIIb, expressed on epithelial cells (Min et al., 1998; Sekine et al., 1999; De Moerlooze et al., 2000). Concomitantly, FGF signaling increases the expression of BMP4 in the epithelial tip. Distal accumulation of high level of BMP4 ultimately inhibits epithelial cell proliferation and movement, leading to arrest of the advancing airway tip. Coordinated decrease of FGF10 expression at the tip with simultaneous lateral upregulation is believed to initiate budding and thus dichotomous branching of the nascent airway (Weaver et al., 2000). SHH expressed at high level in the distal epithelial tip is proposed to be the signal responsible for suppressing FGF10 expression in the nearby mesenchymal cells (Cardoso, 2001).

Mouse genetics has also revealed crucial roles of both canonical and noncanonical Wnt pathways in mammalian lung formation. Mutation of *Wnt7b*, a canonical Wnt produced in the epithelium, leads to decreased mesenchymal proliferation at E12.5 and severe defects in the smooth muscle component of the major pulmonary vessels at birth (Shu et al., 2002). Using two different Wnt-reporter lines in which LEF/TCF-binding sites are placed in front of *LacZ* gene and a minimal promoter (DasGupta and Fuchs, 1999; Mohamed et al., 2004), the activation of the canonical Wnt pathway is first observed in the trachea as well as the bronchi and bronchioles at E11.5. The canonical Wnt activity can also be visualized in the tracheal mesenchyme at this stage but soon becomes diminished in the mesenchyme. From E12.5 to E15.5, *LacZ* expression is confined to the epithelium, with the highest level in the active branching distal tips. Accompanying the progression of lung development and epithelial differentiation, *LacZ* expression is significantly reduced at E16.5 and completely disappears by E17.5 (Okubo and Hogan, 2004; De Langhe et al., 2005; Dean et al., 2005). Thus, the canonical Wnt pathway is likely to play important roles in lung development.

To determine the function of the canonical Wnt pathway in mammalian lung development, a number of gain- and loss-of-function experiments have been carried out. Specific removal of *β-catenin* in the epithelial cells early in lung morphogenesis results in disruption of the peripheral respiratory structure. However, the proximal conducting airway grows and differentiates normally (Mucenski et al., 2003). Conversely, expression of a stabilized *β-catenin* from E14.5 onwards in both proximal and peripheral epithelial cells leads to ectopic differentiation of alveolar Type II-like cells in the proximal conducting airway (Mucenski et al., 2005). Expression of another activated form of *β*-catenin, a *β*-catenin–Lef1fusion protein, in the epithelium from the beginning of lung development induces a lineage switch to intestinal cells (Okubo and Hogan, 2004). Thus, these experiments all suggest an essential role of the canonical Wnt pathway in cell fate determination or maintenance.

In contrast, *in vitro* manipulation of Wnt-signaling activity in the organ culture system revealed that the canonical Wnt pathway could also regulate branching morphogenesis. Overactivation of Wnt signaling in cultured lung explants, through either Wnt3a or lithium chloride treatment, leads to a decrease in the number of branches formed, while knockdown of *β*-catenin using antisense morpholino increases branch formation (Dean et al., 2005). Intriguingly, in another study blocking Wnt singling using the antagonist Dkk1 inhibits branching morphogenesis (De Langhe et al., 2005). It is unclear why both activation and inhibition of Wnt signaling resulted in similar branching phenotype in these two studies.

The noncanonical Wnt pathway also appears to participate in lung morphogenesis. Mutation of one of the noncanonical *Wnt* genes *Wnt5a* results in truncation of the proximal airway and overexpansion of the peripheral airway. Both epithelium and mesenchyme show increased proliferation and delayed maturation (Li et al., 2002). Conversely, targeted overexpression of *Wnt5a* in the epithelium leads to reduced branching and dilated distal airways, in spite of increased expression of FGF10 in the surrounding mesenchyme (Li et al., 2005). Overexpression of *Wnt5a* in epithelium severely reduces the chemotactic response and shows dilated branching tips when exposed to FGF10 in culture (Li et al., 2005). Although the underlying mechanisms are not known, these results indicate that Wnt5a is critically required for FGF10 induced branching morphogenesis.

2.10. Wnt signaling in gastrointestinal tract development

The formation of the mammalian gastrointestinal tract is initiated by folding of the endoderm at the anterior and posterior ends, creating the anterior intestinal portal (AIP) and caudal intestinal portal (CIP). As the AIP and CIP migrate toward each other, the endoderm fuses ventrally to form a tube. An outer layer of splanchnic mesoderm surrounds the endodermally derived epithelial tube, giving rise to the primitive gut tube. Cross talks between the mesodermal and endodermal lineage subsequently divide the primitive gut tube into three functionally and morphologically distinct segments: foregut, midgut, and hindgut, which in turn form the stomach, intestine, and colon, respectively. Conventional grafting and tissue recombination experiments have clearly demonstrated that early gut endoderm can convert nongut mesoderm to gut-specific smooth muscle cells and have implicated SHH secreted by the endoderm as one of the inductive signals (Apelqvist et al., 1997; Roberts et al., 1998). The mesoderm, in turn, signals back to pattern the endoderm along the A-P axis, leading to the differentiation of many different types of specialized epithelial cells within each segment. While Wnt signaling has long been proposed to play key roles in gastrointestinal tract patterning and a number of studies have been carried out to search for Wnt pathway components during gut development (Wells and Melton, 1999; McBride et al., 2003; Theodosiou and Tabin, 2003; Gregorieff and Clevers, 2005; Gregorieff et al., 2005), the definitive experimental evidence has just started to emerge in the last few years.

Mutations of both *Tcf1* and *Tcf4*, the two genes which encode the essential transcriptional factors in the canonical Wnt pathway and show overlapping expression in the gut, result in severe posterior truncation including the hindlimb, posterior body, and tail in the mouse. At E14.5, while the anterior organs, such as the lung, heart, pancreas, and liver, form and appear normal, the hindgut is missing. In the *Tcf1;Tcf4* double mutants, the paraxial mesoderm is not affected initially, as evidenced by the posterior somite formation and normal *T* expression up to E9.5. In contrast, endoderm markers *Sox17* and *Foxa1* are missing from the posterior region at E8.5. Sectioning of these mutants revealed delayed closure of midgut and a complete lack of hindgut at E9.5. Furthermore, marker analyses suggest that at later stages even the midgut appears to have taken the characteristics of the stomach, with little or no intestine formation (Gregorieff et al., 2004). Thus, canonical Wnt signaling in the endoderm is essential for mid and hindgut patterning.

In contrast, development of the stomach from the foregut requires suppression of the canonical Wnt pathway by secreted Wnt antagonists, reminiscent of the forebrain specification during the A-P patterning of the body axis. A recent study shows that specific expression of a homeobox gene *Barx1* in the stomach mesoderm triggers localized production of sFRP1 and sFRP2 before stomach differentiation. Loss of *Barx1* causes expression of intestine-specific gene *Cdx2* in the gastric epithelium and defective stomach morphogenesis, but intestine development is not affected. These phenotypes are the exact opposite of those in *Tcf1/Tcf4* mutants and suggest that reduced canonical Wnt signaling is critical for stomach-specific epithelial differentiation. In tissue culture ectopic expression of sFRP1, sFRP2, or Dkk1 in stomach mesoderm is sufficient to replace *Barx1* in inducing gastric epithelial differentiation from the cocultured endoderm (Kim et al., 2005).

Promoted by the link between Wnt signaling gain-of-function mutations and human colorectal cancer formation, many studies have been carried out and have revealed interesting stem cell and cancer biology involving the canonical Wnt pathway. During late gestation in the mouse, the highly proliferative, pseudostratified intestinal epithelium has been converted to a single-layered epithelium that forms numerous protrusions, the future villi. Cell proliferation also becomes restricted to the intervillus region. During the first several weeks after birth, the proliferative pockets undergo invagination to form crypts of Lieberkühn (Sancho et al., 2003; Radtke and Clevers, 2005). A small number of stem cells reside close to the bottom of the crypt and give rise to transient progenitors that occupy the rest of the crypt. The transient progenitors undergo limited rounds of proliferation and upon reaching the crypt–villus junction, differentiate into one of the three types of mature epithelium: absorptive epithelial cells, goblet cells, and enteroendocrine cells. The differentiated cells continue an upward migration to replace the villus epithelium. A fourth mature cell type, the Paneth cell, differentiates and remains at the bottom of the crypt.

Canonical Wnt signaling plays multiple roles in intestinal stem cell maintenance and differentiation. In *Tcf4*-null mice, the intervillus proliferative pockets are completely missing (Korinek et al., 1998). A similar phenotype is observed in transgenic mice expressing ectopic *Dkk1* in the intestine epithelial layer (Pinto et al., 2003). Studies in colorectal cancer cell lines and *Dkk1* transgenic mice have further defined an essential role of Wnt signaling in the maintenance of intestinal stem cells through promoting proliferation and blocking differentiation (van de Wetering et al., 2002; Pinto et al., 2003; Sancho et al., 2003). This is achieved, at least in part, through activation of the Tcf-4 target gene *c-Myc* that in turn represses transcription of $p^{21CIP/WAF1}$

(van de Wetering et al., 2002; Pinto et al., 2003). Consequently, shortly after induced inactivation of the Wnt-signaling repressor *APC*, the crypt compartments show dramatic enlargement with increased cell proliferation, impaired cell migration, and elevated *c-Myc* expression. Intriguingly, loss of *APC* in differentiated epithelial cells in the villi does not lead to proliferation or morphological change, suggesting that additional regulatory mechanisms in these cells can override or modify the response induced by increased level of β-catenin (Andreu et al., 2005).

While the above studies indicate that suppression of Wnt signaling is required for the differentiation of epithelial cells fated for the villi, it is important to note that fully differentiated Paneth cells reside in the crypt, where high levels of Wnt ligands are present to sustain stem cell maintenance and proliferation. In fact, a new study has revealed that active Wnt signaling is critical for Paneth cell differentiation. Expression of Fz5 and Tcf4 are both required in the crypt epithelium to activate a number of genes that are essential for the complete maturation and correct localization of Paneth cells (van Es et al., 2005). Conversely, elevated β-catenin levels, due to the loss of *APC*, promote differentiation along the Paneth cell lineage but block terminal differentiation of the absorptive, goblet, and enteroendocrine lineages (Andreu et al., 2005). Overall, these studies have revealed that active Wnt signaling in the crypt can lead to either stem cell maintenance or Paneth cell maturation. How the two drastically different outcomes are achieved in closely related neighboring cells by the same extracellular Wnt signals remains a fascinating biological question to be addressed in the future.

3. The noncanonical Wnt pathway in mammalian morphogenesis

Early *in vitro* experiments have revealed that exposure to certain Wnts, for example Wnt5a, does not induce transformation of certain cultured mammalian cells or stabilization of β-catenin (Wong et al., 1994), which are typical responses to treatment of Wnt1 or Wnt3a. Instead, a number of distinct pathways, including JNK signaling, small GTPase Rho/Rac/Cdc42 signaling, and Ca^{2+} signaling, have been proposed as the downstream targets of these "noncanonical" or Wnt5a class Wnts (Veeman et al., 2003a; Wallingford and Habas, 2005). However, as these targets are also subjected to regulation by numerous other signals, we still do not have a definitive biochemical assay to measure the activation in response to each noncanonical Wnt. Gene targeting in the mouse has revealed a number of phenotypes following mutation of these noncanonical Wnts, such as the axis shortening defect in *Wnt5a* mutants and the kidney defect in *Wnt11* mutants described above (Yamaguchi et al., 1999a; Majumdar et al., 2003; Topol et al., 2003; Yang et al., 2003). However, each phenotype was explained by a different mechanism that is also distinct from the current understanding of the noncanonical Wnt pathway in other model organisms including *Drosophila*, *Xenopus*, and zebrafish (Veeman et al., 2003a; Klein and Mlodzik, 2005).

In flies, the noncanonical Wnt pathway shares two critical components with the canonical Wnt pathway: the Fz receptor and a multifunctional cytoplasmic protein Dishevelled (Dsh in flies, XDsh in frogs, and Dvl in mice) (Klein and Mlodzik, 2005; Wallingford and Habas, 2005). However, all the other members of this pathway are distinct from those of the canonical Wnt pathway and it has been argued that no Wnt ligand is involved (Povelones et al., 2005). The main function of this pathway is to define the PCP, the polarity of a cell within the plane of the epithelium, perpendicular to the apical–basal axis of the cell. Manifestations of this polarity include uniformly oriented trichomes on the wing epithelial cells and bristles on the thorax, abdomen, and leg and precisely coordinated orientation of ommatidial units of the compound eye (Lawrence et al., 2002, 2004; Klein and Mlodzik, 2005). Studies over the last few years have identified a large number of novel cytoplasmic or membrane proteins, including a "core" PCP group consisted of Flamingo, Diego, Prickle, and Stabismus/Van Gogh, in addition to Fz and Dishevelled. Models have been proposed to explain how complex interactions among the proteins, either within the cell or between neighboring cells, set up PCP in individual cells and propagate the polarity information across the entire field of epithelium (Lawrence et al., 2002, 2004; Amonlirdviman et al., 2005; Klein and Mlodzik, 2005).

In *Xenopus* and zebrafish, almost all the homologues of the fly "core" PCP proteins have been found to control convergent extension, a coordinated extension of the A-P axis with concomitant narrowing of the medial–lateral axis (Wallingford et al., 2000; Darken et al., 2002; Goto and Keller, 2002; Keller, 2002; Park and Moon, 2002; Wallingford et al., 2002; Carreira-Barbosa et al., 2003; Kinoshita et al., 2003; Takeuchi et al., 2003b; Veeman et al., 2003b). In contrast to the PCP pathway in fly, both Wnt5a and Wnt11 have been implicated in regulating convergent extension (Heisenberg et al., 2000; Yamanaka et al., 2002). During convergent extension, several PCP proteins have been shown to control the polarity of lamellipodial protrusions that drive polarized cell intercalation (Wallingford et al., 2000; Jessen et al., 2002). In zebrafish, the PCP pathway also appears to regulate convergent extension, at least in part, through determining the orientation of cell division (Gong et al., 2004).

Mouse genetics has provided compelling evidence that a homologous PCP pathway also exists in mammals and regulates a diverse array of morphogenetic processes. In *Loop-tail* (*Lp*) mutant mice, which harbor a loss-of-function point mutation in *Vangl2*, one of the mammalian homologues of the fly PCP gene *Strabismus/Van Gogh* (*Stbm/Vang*), the uniform orientation of stereocilia on the sensory hairs cells is disrupted (Kibar et al., 2001b; Murdoch et al., 2001; Montcouquiol et al., 2003). The uniformly orientated stereocilia on hair cells in the cochlea, reminiscent of the trichome on epithelial cells of the fly wing, has been proposed to be a manifestation of PCP in mammals (Lewis and Davies, 2002). In addition to *Lp*, mutations of several other PCP homologues in the mouse result in misorientation of stereociliary bundles. These include *Crash* and *Spin cycle* (*Crsh* and *Scy*, two distinct mutations of *Celsr1*, a homologue of the PCP pathway member *flamingo*; Curtin et al., 2003) and *Fz3$^{-/-}$;Fz6$^{-/-}$* (Wang et al., 2006b) and *Dvl1$^{-/-}$;Dvl2$^{-/-}$* (Wang et al., 2005) mutants. We have also found that while the stereociliary bundle orientation is normal in either *Dvl2$^{-/-}$* or *Lp/+* mutants, *Dvl2$^{-/-}$;Lp/+* mutants display randomized orientation, suggesting that Dvl and Vangl function together in hair cell PCP establishment (Wang et al., 2005). This argument is also supported by the *in vitro* biochemical assay in which Vangl2 can bind to all three Dvl homologues, Dvl1–3 while the missense point mutation found in *Lp* perturbs Vangl2 interaction with all three Dvls (Torban et al., 2004).

Consistent with the findings in fly where most PCP proteins show an asymmetric plasma membrane distribution, we have found that a transgenic Dvl2–EGFP fusion protein, capable of rescuing the stereocilia orientation defects in *Dvl1$^{-/-}$;Dvl2$^{-/-}$* mutants, displays apically restricted asymmetric plasma membrane localization to the abneural side of the hair cell. The uniform asymmetric localization of Dvl2-EGFP is disrupted in *Lp/Lp* embryos, indicating that like the fly wing epithelium, asymmetric distribution of mammalian PCP protein is dependent on an intact PCP pathway (Wang et al., 2005). More recently, others also reported a similar *Lp*-dependent asymmetric localization of Fz3 and Fz6 in the sensory hair cells, utricles, and cristae (Wang et al., 2006b). Taken together, these data suggest that in mammals, a conserved PCP pathway serves as the underlying mechanism in coordinating stereocilia orientation. Finally, *Fz6$^{-/-}$* mice display misorientation of hairs throughout the body, implying that the mammalian PCP pathway also controls macroscopic hair patterning (Guo et al., 2004a).

In addition to the inner ear polarity defects, *Lp, Crsh, Scy, Fz3$^{-/-}$;Fz6$^{-/-}$*, and *Dvl1$^{-/-}$;Dvl2$^{-/-}$* mutants all result in a unique neural tube closure defect where the entire neural tube from midbrain to tail fails to close, a severe congenital neural tube defect termed craniorachischisis in humans (Kibar et al., 2001a; Murdoch et al., 2001; Hamblet et al., 2002; Curtin et al., 2003; Wang et al., 2006b). On the basis of experiments in *Xenopus* in which overexpression of XDsh and Stbm mutants that blocks convergent extension usually results in similar neural tube closure defects, the cause of craniorachischisis in these mutants has been speculated to be a failure of convergent extension (Darken et al., 2002; Goto and Keller, 2002; Wallingford and Harland, 2002; Copp et al., 2003; Ueno and Greene, 2003). We recently provided supporting evidence for this hypothesis. We have found that during neurulation, the mouse neural plate undergoes concomitant lengthening and narrowing, a morphogenetic process that resembles convergent extension in other vertebrates. Both *Dvl1$^{-/-}$;Dvl2$^{-/-}$* and *Lp/Lp* mutants disrupt the coordinated lengthening and narrowing of the neural plate, as measured by the increase of the length-to-width ratio (LWR). Significantly, the reduction of LWR in $Dvl1^{-/-};Dvl2^{-/-}$ and Lp/Lp mutants is observed from ~4-somite stage, the earliest time point when LWR can be determined and several hours before neural tube closure occurs in control embryos (Wang et al., 2006a). These data suggest that a Dvl/Vangl-mediated homologous PCP pathway is important for a convergent extension-like morphogenetic process during neurulation to reduce the width of the neural plate. A narrowed neural plate, in turn, may facilitate neural tube closure by further shortening the distance between the opposing neural folds, as proposed in *Xenopus* (Wallingford and Harland, 2002). Consistent with this idea, $Dvl2^{-/-};Lp/+$ mutants also show significantly reduced LWR as $Dvl1^{-/-};Dvl2^{-/-}$ and Lp/Lp and fail to close their neural tubes, while $Lp/+$ mutants display moderate reduction of the LWR and delayed neural tube closure (Wang et al., 2006a).

Our studies on *Dvl* have also revealed a high level of conservation between mammalian neural tube closure and the PCP establishment in *Drosophila*. We found that a point mutation identical to the *dsh1* allele in fly, which specifically abolishes the PCP pathway, completely disrupted *Dvl2* function in neural tube closure when introduced in the mouse, suggesting stringent conservation of Dvl function in mammalian convergent extension and PCP (Wang et al., 2006a). Finally, Dvl/Vangl-mediated mammalian PCP pathway may have broad functions in tissue morphogenesis. For example, we and other have also implicated Dvl1/2 and Vangl2 in the polarized, basal to apical elongation of the cochlea (Montcouquiol et al., 2003; Wang et al., 2005).

A role for noncanonical Wnt signaling in mammalian heart development has also been revealed from studies of *Lp* mice. *Lp/Lp* mutants display outflow tract abnormalities including DORV and VSD (Henderson et al., 2001), presumably due to an inability to muscularize of the outflow tract septum caused by defects in myocardialization (Phillips et al., 2005; Henderson et al., 2006). Myocardialization has been argued to require the polarized migration of cells, similar to cell intercalation seen in convergent extension movements. Expression of typically noncanonical Wnts, Wnt5a, and Wnt11, as well as other downstream PCP components, including Rho kinase (ROCK) 1, is also found in the developing outflow tract (Phillips et al., 2005). ROCK1 mediates RhoA signaling, which regulates cytoskeletal organization. In wild-type controls protein levels of ROCK1 are upregulated in myocardializing cells that are extending into the endocardial cushion tissue, but this expression is lost in *Lp/Lp* mutants. The expression of RhoA at the boundary between the cushion cells and the outflow tract myocardium seen in the controls is also reduced in *Lp/Lp* mice. These data suggest that noncanonical Wnt-stimulated ROCK1/RhoA signaling may function in the cytoskeletal rearrangements that are required for the polarized migration of myocardial cells during mammalian heart morphogenesis.

Accumulating evidence has also linked noncanonical Wnt pathway to axonal tract development. Mutation of *Celsr3*, another homologue of *flamingo/starry night* in the mouse, leads to the absence of the internal capsule, anterior commissure, and many of the longitudinal axonal bundles. As a consequence, the cortex is completely disconnected from the subcortical structures (Tissir et al., 2005). While the underlying mechanism is not known, the fact that *Fz3*-null mutants display almost identical abnormalities and that Celsa3 and Fz3 localize to the same neuronal cells during development, strongly implying that the PCP/noncanonical Wnt pathway is involved (Goto and Keller, 2002; Wang et al., 2002).

Mouse genetics has also revealed that the mammalian PCP pathway requires two novel components whose homologues have not been implicated in the fly PCP pathway. The first one, encoded by *mScribble1*, is the mammalian homologue of Scribble that defines the apical–basal polarity in fly (Murdoch et al., 2003). Mice homozygous for *circletail* (*Crc*), a spontaneous mutation in *mScrib1* where a single base insertion leads to a frame shift and premature truncation of the encoded mScrib1 protein, display almost identical neural tube closure and stereocilia polarization defects as *Lp/Lp* mutants. More significantly, the same defects are also observed in mice that are double heterozygous for *Crc* and *Lp*, further arguing that mScrib1 is a component of the mammalian PCP pathway (Murdoch et al., 2003).

The second novel component of the mammalian PCP pathway, identified in a gene-trapping screen, is encoded by the mouse protein tyrosine kinase 7 (*PTK7*) gene, also known as colon carcinoma kinase-4 (*CCK-4*). PTK7 contains an extracellular domain with seven immunoglobulin loops and a C-terminal cytoplasmic domain with homology to protein tyrosine kinase. However, PTK7 does not appear to have any kinase activity and cannot be phosphroylated by itself or other protein tyrosine kinases (Jung et al., 2004; Lu et al., 2004). While the mode of action of this protein is unknown at present, mice homozygous for an insertion–mutation of *PTK7*, which truncates most of the extracellular, transmembrane, and cytoplasmic domains, displays similar neural tube closure and stereocilia orientation defects as *Lp/Lp*. Most *PTK7/+;Lp/+* mutants display spinal bifida, a less severe form of neural tube closure defect. Perturbation of *PTK7* function in *Xenopus* leads to convergent extension defects, suggesting that PTK7 is likely to be a critical component of the PCP pathway in other model organisms as well (Lu et al., 2004). Intriguingly, the fly homologue of *PTK7*, *off-track* (*otk*), has only been implicated as a coreceptor of plexins in semaphoring-mediated axon guidance process (Winberg et al., 2001), although whether *otk* is the functional orthologue of *PTK7* in fly is unclear at present.

While it is clear that a highly conserved PCP pathway in mammals regulates both planar polarization and morphogenesis processes resembling convergent extension, we cannot ignore the significant differences between these two apparently different biological processes. The current models on

PCP establishment in static wing or abdominal cells in *Drosophila* (Lawrence et al., 2002, 2004; Amonlirdviman et al., 2005; Klein and Mlodzik, 2005) do not easily apply to actively intercalating cells during convergent extension in *Xenopus*. In fact, we do not know whether the coordinated lengthening and narrowing of the neural plate in mammals are achieved through medial–lateral intercalation, directly cell migration or oriented cell division. Finally, we note that convergent extension in *Xenopus* and zebrafish does require noncanonical Wnts and also involves Wnt/calcium signaling, a pathway that has not been implicated in PCP determination in *Drosophila* but none-theless requires "core" PCP proteins such as Dsh and Pk (Sheldahl et al., 2003; Veeman et al., 2003a,b). Analyzing how the noncanonical Wnts and the Wnt/calcium pathways are integrated with the PCP pathway in the mouse may hold the key for a better understanding of how the core PCP proteins have been adapted to regulate both convergent extension and planar polarization.

4. Summary

In this chapter, we illustrated some of the most recent progresses in our understanding of the canonical and noncanonical Wnt-signaling pathways in mammalian development. The studies we outlined above have demonstrated clearly that, as in other model systems, the mammalian canonical Wnt pathway is an integral component of a large signaling network that involves many other pathways, including the FGF, TGF-β, hedgehog, and Notch-signaling path-ways. The intricate interactions among these signaling components ultimately determine the output of Wnt signaling, for example, either as morphogens to specify different cell lineages along the A-P axis or as mitogens to promote cell proliferation of a defined population. The noncanonical Wnt pathway, in contrast, does not appear to affect either cell fate or proliferation but rather impacts morphogenesis through remodeling cell arrangements along different axes. Nevertheless, emerging evidence in *Xenopus* indicates that proper rear-rangement of mesodermal cells during convergent extension also depends on A-P patterning specified by Activin, a member of the TGF-β family (Ninomiya et al., 2004). Understanding how the individual pathways are coordinately regulated and orchestrated to achieve the desired pattern and morphology will be one of the greatest challenges in developmental biology in the future.

References

Adamska, M., Billi, A.C., Cheek, S., Meisler, M.H. 2005. Genetic interaction between Wnt7a and Lrp6 during patterning of dorsal and posterior structures of the mouse limb. Dev. Dyn. 233(2), 368–372.

Afzal, A.R., Rajab, A., Fenske, C.D., Oldridge, M., Elanko, N., Ternes-Pereira, E., Tuysuz, B., Murday, V.A., Patton, M.A., Wilkie, A.O., Jeffery, S. 2000. Recessive Robinow syndrome, allelic to dominant brachydactyly type B, is caused by mutation of ROR2. Nat. Genet. 25(4), 419–422.

Agarwal, P., Wylie, J.N., Galceran, J., Arkhitko, O., Li, C., Deng, C., Grosschedl, R., Bruneau, B.G. 2003. Tbx5 is essential for forelimb bud initiation following patterning of the limb field in the mouse embryo. Development 130(3), 623–633.

Ai, Z., Fischer, A., Spray, D.C., Brown, A.M., Fishman, G.I. 2000. Wnt-1 regulation of connexin43 in cardiac myocytes. J. Clin. Invest. 105(2), 161–171.

Airaksinen, M.S., Saarma, M. 2002. The GDNF family: Signalling, biological functions and therapeutic value. Nat. Rev. Neurosci. 3(5), 383–394.

Akiyama, H., Lyons, J.P., Mori-Akiyama, Y., Yang, X., Zhang, R., Zhang, Z., Deng, J.M., Taketo, M.M., Nakamura, T., Behringer, R.R., McCrea, P.D., de Crombrugghe, B. 2004. Interactions between Sox9 and beta-catenin control chondrocyte differentiation. Genes Dev. 18(9), 1072–1087.

Alonso, L., Fuchs, E. 2003. Stem cells in the skin: Waste not, Wnt not. Genes Dev. 17(10), 1189–1200.

Amonlirdviman, K., Khare, N.A., Tree, D.R., Chen, W.S., Axelrod, J.D., Tomlin, C.J. 2005. Mathematical modeling of planar cell polarity to understand domineering nonautonomy. Science 307(5708), 423–426.

Andl, T., Reddy, S.T., Gaddapara, T., Millar, S.E. 2002. WNT signals are required for the initiation of hair follicle development. Dev. Cell 2(5), 643–653.

Andreu, P., Colnot, S., Godard, C., Gad, S., Chafey, P., Niwa-Kawakita, M., Laurent-Puig, P., Kahn, A., Robine, S., Perret, C., Romagnolo, B. 2005. Crypt-restricted proliferation and commitment to the Paneth cell lineage following Apc loss in the mouse intestine. Development 132(6), 1443–1451.

Apelqvist, A., Ahlgren, U., Edlund, H. 1997. Sonic hedgehog directs specialised mesoderm differentiation in the intestine and pancreas. Curr. Biol. 7(10), 801–804.

Archer, C.W., Dowthwaite, G.P., Francis-West, P. 2003. Development of synovial joints. Birth Defects Res. C Embryo Today 69(2), 144–155.

Augustine, K., Liu, E.T., Sadler, T.W. 1993. Antisense attenuation of Wnt-1 and Wnt-3a expression in whole embryo culture reveals roles for these genes in craniofacial, spinal cord, and cardiac morphogenesis. Dev. Genet. 14(6), 500–520.

Austin, T.W., Solar, G.P., Ziegler, F.C., Liem, L., Matthews, W. 1997. A role for the Wnt gene family in hematopoiesis: Expansion of multilineage progenitor cells. Blood 89(10), 3624–3635.

Bachiller, D., Klingensmith, J., Kemp, C., Belo, J.A., Anderson, R.M., May, S.R., McMahon, J.A., McMahon, A.P., Harland, R.M., Rossant, J., De Robertis, E.M. 2000. The organizer factors Chordin and Noggin are required for mouse forebrain development. Nature 403(6770), 658–661.

Backman, M., Machon, O., Mygland, L., van den Bout, C.J., Zhong, W., Taketo, M.M., Krauss, S. 2005. Effects of canonical Wnt signaling on dorso-ventral specification of the mouse telencephalon. Dev. Biol. 279(1), 155–168.

Baron, M.H. 2003. Embryonic origins of mammalian hematopoiesis. Exp. Hematol. 31(12), 1160–1169.

Barrow, J.R., Thomas, K.R., Boussadia-Zahui, O., Moore, R., Kemler, R., Capecchi, M.R., McMahon, A.P. 2003. Ectodermal Wnt3/beta-catenin signaling is required for the establishment and maintenance of the apical ectodermal ridge. Genes Dev. 17(3), 394–409.

Beaudoin, G.M., 3rd, Sisk, J.M., Coulombe, P.A., Thompson, C.C. 2005. Hairless triggers reactivation of hair growth by promoting Wnt signaling. Proc. Natl. Acad. Sci. USA 102(41), 14653–14658.

Borello, U., Coletta, M., Tajbakhsh, S., Leyns, L., De Robertis, E.M., Buckingham, M., Cossu, G. 1999. Transplacental delivery of the Wnt antagonist Frzb1 inhibits development of caudal paraxial mesoderm and skeletal myogenesis in mouse embryos. Development 126(19), 4247–4255.

Botchkarev, V.A., Fessing, M.Y. 2005. Edar signaling in the control of hair follicle development. J. Invest. Dermatol. Symp. Proc. 10(3), 247–251.

Botchkarev, V.A., Botchkareva, N.V., Roth, W., Nakamura, M., Chen, L.H., Herzog, W., Lindner, G., McMahon, J.A., Peters, C., Lauster, R., McMahon, A.P., Paus, R. 1999. Noggin is a mesenchymally derived stimulator of hair-follicle induction. Nat. Cell Biol. 1(3), 158–164.

Boyden, L.M., Mao, J., Belsky, J., Mitzner, L., Farhi, A., Mitnick, M.A., Wu, D., Insogna, K., Lifton, R.P. 2002. High bone density due to a mutation in LDL-receptor-related protein 5. N. Engl. J. Med. 346(20), 1513–1521.

Buendia, M.A. 2000. Genetics of hepatocellular carcinoma. Semin. Cancer Biol. 10(3), 185–200.

Cadoret, A., Ovejero, C., Saadi-Kheddouci, S., Souil, E., Fabre, M., Romagnolo, B., Kahn, A., Perret, C. 2001. Hepatomegaly in transgenic mice expressing an oncogenic form of beta-catenin. Cancer Res. 61(8), 3245–3249.

Cadoret, A., Ovejero, C., Terris, B., Souil, E., Levy, L., Lamers, W.H., Kitajewski, J., Kahn, A., Perret, C. 2002. New targets of beta-catenin signaling in the liver are involved in the glutamine metabolism. Oncogene 21(54), 8293–8301.

Capdevila, J., Izpisua Belmonte, J.C. 2001. Patterning mechanisms controlling vertebrate limb development. Annu. Rev. Cell Dev. Biol. 17, 87–132.

Cardoso, W.V. 2001. Molecular regulation of lung development. Annu. Rev. Physiol. 63, 471–494.

Carreira-Barbosa, F., Concha, M.L., Takeuchi, M., Ueno, N., Wilson, S.W., Tada, M. 2003. Prickle 1 regulates cell movements during gastrulation and neuronal migration in zebrafish. Development 130(17), 4037–4046.

Carroll, T.J., Park, J.S., Hayashi, S., Majumdar, A., McMahon, A.P. 2005. Wnt9b plays a central role in the regulation of mesenchymal to epithelial transitions underlying organo-genesis of the mammalian urogenital system. Dev. Cell 9(2), 283–292.

Chen, H., Lun, Y., Ovchinnikov, D., Kokubo, H., Oberg, K.C., Pepicelli, C.V., Gan, L., Lee, B., Johnson, R.L. 1998. Limb and kidney defects in Lmx1b mutant mice suggest an involvement of LMX1B in human nail patella syndrome. Nat. Genet. 19(1), 51–55.

Chiang, C., Litingtung, Y., Lee, E., Young, K.E., Corden, J.L., Westphal, H., Beachy, P.A. 1996. Cyclopia and defective axial patterning in mice lacking Sonic hedgehog gene function. Nature 383(6599), 407–413.

Church, V.L., Francis-West, P. 2002. Wnt signalling during limb development. Int. J. Dev. Biol. 46(7), 927–936.

Colnot, S., Decaens, T., Niwa-Kawakita, M., Godard, C., Hamard, G., Kahn, A., Giovannini, M., Perret, C. 2004. Liver-targeted disruption of Apc in mice activates beta-catenin signaling and leads to hepatocellular carcinomas. Proc. Natl. Acad. Sci. USA 101(49), 17216–17221.

Copp, A.J., Greene, N.D., Murdoch, J.N. 2003. The genetic basis of mammalian neurulation. Nat. Rev. Genet. 4(10), 784–793.

Cossu, G., Borello, U. 1999. Wnt signaling and the activation of myogenesis in mammals. EMBO J. 18(24), 6867–6872.

Curtin, J.A., Quint, E., Tsipouri, V., Arkell, R.M., Cattanach, B., Copp, A.J., Henderson, D.J., Spurr, N., Stanier, P., Fisher, E.M., Nolan, P.M., Steel, K.P., et al. 2003. Mutation of Celsr1 disrupts planar polarity of inner ear hair cells and causes severe neural tube defects in the mouse. Curr. Biol. 13(13), 1129–1133.

Cygan, J.A., Johnson, R.L., McMahon, A.P. 1997. Novel regulatory interactions revealed by studies of murine limb pattern in Wnt-7a and En-1 mutants. Development 124(24), 5021–5032.

Darken, R.S., Scola, A.M., Rakeman, A.S., Das, G., Mlodzik, M., Wilson, P.A. 2002. The planar polarity gene strabismus regulates convergent extension movements in *Xenopus*. EMBO J. 21(5), 976–985.

DasGupta, R., Fuchs, E. 1999. Multiple roles for activated LEF/TCF transcription complexes during hair follicle development and differentiation. Development 126(20), 4557–4568.

Davies, J.A., Garrod, D.R. 1995. Induction of early stages of kidney tubule differentiation by lithium ions. Dev. Biol. 167(1), 50–60.

Day, T.F., Guo, X., Garrett-Beal, L., Yang, Y. 2005. Wnt/beta-catenin signaling in mesenchymal progenitors controls osteoblast and chondrocyte differentiation during vertebrate skeletogenesis. Dev. Cell 8(5), 739–750.

De Langhe, S.P., Sala, F.G., Del Moral, P.M., Fairbanks, T.J., Yamada, K.M., Warburton, D., Burns, R.C., Bellusci, S. 2005. Dickkopf-1 (DKK1) reveals that fibronectin is a major target of Wnt signaling in branching morphogenesis of the mouse embryonic lung. Dev. Biol. 277(2), 316–331.

De Moerlooze, L., Spencer-Dene, B., Revest, J., Hajihosseini, M., Rosewell, I., Dickson, C. 2000. An important role for the IIIb isoform of fibroblast growth factor receptor 2 (FGFR2) in mesenchymal-epithelial signalling during mouse organogenesis. Development 127(3), 483–492.

Dean, C.H., Miller, L.A., Smith, A.N., Dufort, D., Lang, R.A., Niswander, L.A. 2005. Canonical Wnt signaling negatively regulates branching morphogenesis of the lung and lacrimal gland. Dev. Biol. 286(1), 270–286.

DeChiara, T.M., Kimble, R.B., Poueymirou, W.T., Rojas, J., Masiakowski, P., Valenzuela, D.M., Yancopoulos, G.D. 2000. Ror2, encoding a receptor-like tyrosine kinase, is required for cartilage and growth plate development. Nat. Genet. 24(3), 271–274.

del Barco Barrantes, I., Davidson, G., Grone, H.J., Westphal, H., Niehrs, C. 2003. Dkk1 and noggin cooperate in mammalian head induction. Genes Dev. 17(18), 2239–2244.

Ding, J., Yang, L., Yan, Y.T., Chen, A., Desai, N., Wynshaw-Boris, A., Shen, M.M. 1998. Cripto is required for correct orientation of the anterior-posterior axis in the mouse embryo. Nature 395(6703), 702–707.

Dreyer, S.D., Zhou, G., Baldini, A., Winterpacht, A., Zabel, B., Cole, W., Johnson, R.L., Lee, B. 1998. Mutations in LMX1B cause abnormal skeletal patterning and renal dysplasia in nail patella syndrome. Nat. Genet. 19(1), 47–50.

Duprez, D. 2002. Signals regulating muscle formation in the limb during embryonic development. Int. J. Dev. Biol. 46(7), 915–925.

Francis-West, P.H., Antoni, L., Anakwe, K. 2003. Regulation of myogenic differentiation in the developing limb bud. J. Anat. 202(1), 69–81.

Galceran, J., Farinas, I., Depew, M.J., Clevers, H., Grosschedl, R. 1999. Wnt3a-/-like phenotype and limb deficiency in Lef1(-/-)Tcf1(-/-) mice. Genes Dev. 13(6), 709–717.

Galceran, J., Miyashita-Lin, E.M., Devaney, E., Rubenstein, J.L., Grosschedl, R. 2000. Hippocampus development and generation of dentate gyrus granule cells is regulated by LEF1. Development 127(3), 469–482.

Gat, U., DasGupta, R., Degenstein, L., Fuchs, E. 1998. *De Novo* hair follicle morphogenesis and hair tumors in mice expressing a truncated beta-catenin in skin. Cell 95(5), 605–614.

Gitler, A.D., Lu, M.M., Jiang, Y.Q., Epstein, J.A., Gruber, P.J. 2003. Molecular markers of cardiac endocardial cushion development. Dev. Dyn. 228(4), 643–650.

Goldring, M.B., Tsuchimochi, K., Ijiri, K. 2006. The control of chondrogenesis. J. Cell. Biochem. 97(1), 33–44.

Gong, Y., Slee, R.B., Fukai, N., Rawadi, G., Roman-Roman, S., Reginato, A.M., Wang, H., Cundy, T., Glorieux, F.H., Lev, D., Zacharin, M., Oexle, K., et al. 2001. LDL receptor-related protein 5 (LRP5) affects bone accrual and eye development. Cell 107(4), 513–523.

Gong, Y., Mo, C., Fraser, S.E. 2004. Planar cell polarity signalling controls cell division orientation during zebrafish gastrulation. Nature 430(7000), 689–693.

Goto, T., Keller, R. 2002. The planar cell polarity gene strabismus regulates convergence and extension and neural fold closure in Xenopus. Dev. Biol. 247(1), 165–181.

Gregorieff, A., Clevers, H. 2005. Wnt signaling in the intestinal epithelium: From endoderm to cancer. Genes Dev. 19(8), 877–890.

Gregorieff, A., Grosschedl, R., Clevers, H. 2004. Hindgut defects and transformation of the gastro-intestinal tract in Tcf4(−/−)/Tcf1(−/−) embryos. EMBO J. 23(8), 1825–1833.

Gregorieff, A., Pinto, D., Begthel, H., Destree, O., Kielman, M., Clevers, H. 2005. Expression pattern of Wnt signaling components in the adult intestine. Gastroenterology 129(2), 626–638.

Guha, U., Gomes, W.A., Kobayashi, T., Pestell, R.G., Kessler, J.A. 2002. *In vivo* evidence that BMP signaling is necessary for apoptosis in the mouse limb. Dev. Biol. 249(1), 108–120.

Gunhaga, L., Marklund, M., Sjodal, M., Hsieh, J.C., Jessell, T.M., Edlund, T. 2003. Specification of dorsal telencephalic character by sequential Wnt and FGF signaling, Nat. Neurosci. 6(7), 701–707.

Guo, N., Hawkins, C., Nathans, J. 2004a. Frizzled 6 controls hair patterning in mice. Proc. Natl. Acad. Sci. USA 101(25), 9277–9281.

Guo, X., Day, T.F., Jiang, X., Garrett-Beal, L., Topol, L., Yang, Y. 2004b. Wnt/beta-catenin signaling is sufficient and necessary for synovial joint formation. Genes Dev. 18(19), 2404–2417.

Hamblet, N.S., Lijam, N., Ruiz-Lozano, P., Wang, J., Yang, Y., Luo, Z., Mei, L., Chien, K.R., Sussman, D.J., Wynshaw-Boris, A. 2002. Dishevelled 2 is essential for cardiac outflow tract development, somite segmentation and neural tube closure. Development 129(24), 5827–5838.

Haramis, A.G., Brown, J.M., Zeller, R. 1995. The limb deformity mutation disrupts the SHH/ FGF-4 feedback loop and regulation of 5′ HoxD genes during limb pattern formation. Development 121(12), 4237–4245.

Hartmann, C. 2006. A Wnt canon orchestrating osteoblastogenesis. Trends Cell Biol. 16(3), 151–158.

Hasegawa, S., Sato, T., Akazawa, H., Okada, H., Maeno, A., Ito, M., Sugitani, Y., Shibata, H., Miyazaki Ji, J., Katsuki, M., Yamauchi, Y., Yamamura Ki, K., et al. 2002. Apoptosis in neural crest cells by functional loss of APC tumor suppressor gene. Proc. Natl. Acad. Sci. USA 99(1), 297–302.

Headon, D.J., Emmal, S.A., Ferguson, B.M., Tucker, A.S., Justice, M.J., Sharpe, P.T., Zonana, J., Overbeek, P.A. 2001. Gene defect in ectodermal dysplasia implicates a death domain adapter in development. Nature 414(6866), 913–916.

Heisenberg, C.P., Tada, M., Rauch, G.J., Saude, L., Concha, M.L., Geisler, R., Stemple, D.L., Smith, J.C., Wilson, S.W. 2000. Silberblick/Wnt11 mediates convergent extension movements during zebrafish gastrulation. Nature 405(6782), 76–81.

Henderson, D.J., Conway, S.J., Greene, N.D., Gerrelli, D., Murdoch, J.N., Anderson, R.H., Copp, A.J. 2001. Cardiovascular defects associated with abnormalities in midline development in the Loop-tail mouse mutant. Circ. Res. 89(1), 6–12.

Henderson, D.J., Phillips, H.M., Chaudhry, B. 2006. Vang-like 2 and noncanonical Wnt signaling in outflow tract development. Trends Cardiovasc. Med. 16(2), 38–45.

Hens, J.R., Wilson, K.M., Dann, P., Chen, X., Horowitz, M.C., Wysolmerski, J.J. 2005. TOPGAL mice show that the canonical Wnt signaling pathway is active during bone development and growth and is activated by mechanical loading *in vitro*. J. Bone Miner. Res. 20(7), 1103–1113.

Hill, T.P., Spater, D., Taketo, M.M., Birchmeier, W., Hartmann, C. 2005. Canonical Wnt/beta-catenin signaling prevents osteoblasts from differentiating into chondrocytes. Dev. Cell 8(5), 727–738.

Hill, T.P., Taketo, M.M., Birchmeier, W., Hartmann, C. 2006. Multiple roles of mesenchymal {beta}-catenin during murine limb patterning. Development.

Hirabayashi, Y., Gotoh, Y. 2005. Stage-dependent fate determination of neural precursor cells in mouse forebrain. Neurosci. Res. 51(4), 331–336.

Hirabayashi, Y., Itoh, Y., Tabata, H., Nakajima, K., Akiyama, T., Masuyama, N., Gotoh, Y. 2004. The Wnt/beta-catenin pathway directs neuronal differentiation of cortical neural precursor cells. Development 131(12), 2791–2801.

Hsieh, J.C., Lee, L., Zhang, L., Wefer, S., Brown, K., DeRossi, C., Wines, M.E., Rosenquist, T., Holdener, B.C. 2003. Mesd encodes an LRP5/6 chaperone essential for specification of mouse embryonic polarity. Cell 112(3), 355–367.

Hu, H., Hilton, M.J., Tu, X., Yu, K., Ornitz, D.M., Long, F. 2005. Sequential roles of Hedgehog and Wnt signaling in osteoblast development. Development 132(1), 49–60.

Huelsken, J., Birchmeier, W. 2001. New aspects of Wnt signaling pathways in higher vertebrates. Curr. Opin. Genet. Dev. 11(5), 547–553.

Huelsken, J., Vogel, R., Brinkmann, V., Erdmann, B., Birchmeier, C., Birchmeier, W. 2000. Requirement for beta-catenin in anterior-posterior axis formation in mice. J. Cell Biol. 148(3), 567–578.

Huelsken, J., Vogel, R., Erdmann, B., Cotsarelis, G., Birchmeier, W. 2001. beta-Catenin controls hair follicle morphogenesis and stem cell differentiation in the skin. Cell 105(4), 533–545.

Hurlstone, A.F., Haramis, A.P., Wienholds, E., Begthel, H., Korving, J., Van Eeden, F., Cuppen, E., Zivkovic, D., Plasterk, R.H., Clevers, H. 2003. The Wnt/beta-catenin pathway regulates cardiac valve formation. Nature 425(6958), 633–637.

Hussain, S.Z., Sneddon, T., Tan, X., Micsenyi, A., Michalopoulos, G.K., Monga, S.P. 2004. Wnt impacts growth and differentiation in ex vivo liver development. Exp. Cell Res. 292(1), 157–169.

Ikeya, M., Takada, S. 1998. Wnt signaling from the dorsal neural tube is required for the formation of the medial dermomyotome. Development 125(24), 4969–4976.

Ikeya, M., Lee, S.M., Johnson, J.E., McMahon, A.P., Takada, S. 1997. Wnt signalling required for expansion of neural crest and CNS progenitors. Nature 389(6654), 966–970.

Ishikawa, T., Tamai, Y., Zorn, A.M., Yoshida, H., Seldin, M.F., Nishikawa, S., Taketo, M.M. 2001. Mouse Wnt receptor gene Fzd5 is essential for yolk sac and placental angiogenesis. Development 128(1), 25–33.

Ishikawa, T.O., Tamai, Y., Li, Q., Oshima, M., Taketo, M.M. 2003. Requirement for tumor suppressor Apc in the morphogenesis of anterior and ventral mouse embryo. Dev. Biol. 253(2), 230–246.

Itasaki, N., Jones, C.M., Mercurio, S., Rowe, A., Domingos, P.M., Smith, J.C., Krumlauf, R. 2003. Wise, a context-dependent activator and inhibitor of Wnt signalling. Development 130(18), 4295–4305.

Jamora, C., DasGupta, R., Kocieniewski, P., Fuchs, E. 2003. Links between signal transduction, transcription and adhesion in epithelial bud development. Nature 422(6929), 317–322.

Jessen, J.R., Topczewski, J., Bingham, S., Sepich, D.S., Marlow, F., Chandrasekhar, A., Solnica-Krezel, L. 2002. Zebrafish trilobite identifies new roles for Strabismus in gastrulation and neuronal movements. Nat. Cell Biol. 4(8), 610–615.

Jung, J.W., Shin, W.S., Song, J., Lee, S.T. 2004. Cloning and characterization of the full-length mouse Ptk7 cDNA encoding a defective receptor protein tyrosine kinase. Gene 328, 75–84.

Kato, M., Patel, M.S., Levasseur, R., Lobov, I., Chang, B.H., Glass, D.A., 2nd, Hartmann, C., Li, L., Hwang, T.H., Brayton, C.F., Lang, R.A., Karsenty, G., et al. 2002. Cbfa1-independent decrease in osteoblast proliferation, osteopenia, and persistent embryonic eye vascularization in mice deficient in Lrp5, a Wnt coreceptor. J. Cell Biol. 157(2), 303–314.

Kawakami, Y., Capdevila, J., Buscher, D., Itoh, T., Rodriguez Esteban, C., Izpisua Belmonte, J.C. 2001. WNT signals control FGF-dependent limb initiation and AER induction in the chick embryo. Cell 104(6), 891–900.

Keller, R. 2002. Shaping the vertebrate body plan by polarized embryonic cell movements. Science 298(5600), 1950–1954.

Kelly, O.G., Pinson, K.I., Skarnes, W.C. 2004. The Wnt co-receptors Lrp5 and Lrp6 are essential for gastrulation in mice. Development 131(12), 2803–2815.

Kemp, C., Willems, E., Abdo, S., Lambiv, L., Leyns, L. 2005. Expression of all Wnt genes and their secreted antagonists during mouse blastocyst and postimplantation development. Dev. Dyn. 233(3), 1064–1075.

Kengaku, M., Capdevila, J., Rodriguez-Esteban, C., De La Pena, J., Johnson, R.L., Belmonte, J.C., Tabin, C.J. 1998. Distinct WNT pathways regulating AER formation and dorsoventral polarity in the chick limb bud. Science 280(5367), 1274–1277.

Khokha, M.K., Hsu, D., Brunet, L.J., Dionne, M.S., Harland, R.M. 2003. Gremlin is the BMP antagonist required for maintenance of Shh and Fgf signals during limb patterning. Nat. Genet. 34(3), 303–307.

Kibar, Z., Vogan, K.J., Groulx, N., Justice, M.J., Underhill, D.A., Gros, P. 2001a. Ltap, a mammalian homolog of Drosophila Strabismus/Van Gogh, is altered in the mouse neural tube mutant Loop-tail. Nat. Genet. 28(3), 251–255.

Kibar, Z., Vogan, K.J., Groulx, N., Justice, M.J., Underhill, D.A., Gros, P. 2001b. Ltap, a mammalian homolog of Drosophila Strabismus/Van Gogh, is altered in the mouse neural tube mutant Loop-tail. Nat. Genet. 28(3), 251–255.

Kim, B.M., Buchner, G., Miletich, I., Sharpe, P.T., Shivdasani, R.A. 2005. The stomach mesenchymal transcription factor Barx1 specifies gastric epithelial identity through inhibition of transient Wnt signaling. Dev. Cell 8(4), 611–622.

Kinoshita, N., Iioka, H., Miyakoshi, A., Ueno, N. 2003. PKC delta is essential for Dishevelled function in a noncanonical Wnt pathway that regulates Xenopus convergent extension movements. Genes Dev. 17(13), 1663–1676.

Kispert, A., Vainio, S., McMahon, A.P. 1998. Wnt-4 is a mesenchymal signal for epithelial transformation of metanephric mesenchyme in the developing kidney. Development 125(21), 4225–4234.

Klein, T.J., Mlodzik, M. 2005. Planar cell polarization: An emerging model points in the right direction. Annu. Rev. Cell Dev. Biol. 21, 155–176.

Korinek, V., Barker, N., Moerer, P., van Donselaar, E., Huls, G., Peters, P.J., Clevers, H. 1998. Depletion of epithelial stem-cell compartments in the small intestine of mice lacking Tcf-4. Nat. Genet. 19(4), 379–383.

Kronenberg, H.M. 2003. Developmental regulation of the growth plate. Nature 423(6937), 332–336.

Kuhlman, J., Niswander, L. 1997. Limb deformity proteins: Role in mesodermal induction of the apical ectodermal ridge. Development 124(1), 133–139.

Lagutin, O.V., Zhu, C.C., Kobayashi, D., Topczewski, J., Shimamura, K., Puelles, L., Russell, H.R., McKinnon, P.J., Solnica-Krezel, L., Oliver, G. 2003. Six3 repression of Wnt signaling in the anterior neuroectoderm is essential for vertebrate forebrain development. Genes Dev. 17(3), 368–379.

Laurikkala, J., Pispa, J., Jung, H.S., Nieminen, P., Mikkola, M., Wang, X., Saarialho-Kere, U., Galceran, J., Grosschedl, R., Thesleff, I. 2002. Regulation of hair follicle development by the TNF signal ectodysplasin and its receptor Edar. Development 129(10), 2541–2553.

Lawrence, P.A., Casal, J., Struhl, G. 2002. Towards a model of the organisation of planar polarity and pattern in the Drosophila abdomen. Development 129(11), 2749–2760.

Lawrence, P.A., Casal, J., Struhl, G. 2004. Cell interactions and planar polarity in the abdominal epidermis of Drosophila. Development 131(19), 4651–4664.

Lee, S.M., Tole, S., Grove, E., McMahon, A.P. 2000. A local Wnt-3a signal is required for development of the mammalian hippocampus. Development 127(3), 457–467.

Lewandoski, M., Sun, X., Martin, G.R. 2000. Fgf8 signalling from the AER is essential for normal limb development. Nat. Genet. 26(4), 460–463.

Lewis, J., Davies, A. 2002. Planar cell polarity in the inner ear: How do hair cells acquire their oriented structure? J. Neurobiol. 53(2), 190–201.

Li, C., Xiao, J., Hormi, K., Borok, Z., Minoo, P. 2002. Wnt5a participates in distal lung morphogenesis. Dev. Biol. 248(1), 68–81.

Li, C., Hu, L., Xiao, J., Chen, H., Li, J.T., Bellusci, S., Delanghe, S., Minoo, P. 2005. Wnt5a regulates Shh and Fgf10 signaling during lung development. Dev. Biol. 287(1), 86–97.

Lickert, H., Kutsch, S., Kanzler, B., Tamai, Y., Taketo, M.M., Kemler, R. 2002. Formation of multiple hearts in mice following deletion of beta-catenin in the embryonic endoderm. Dev. Cell 3(2), 171–181.

Little, R.D., Carulli, J.P., Del Mastro, R.G., Dupuis, J., Osborne, M., Folz, C., Manning, S.P., Swain, P.M., Zhao, S.C., Eustace, B., Lappe, M.M., Spitzer, L., et al. 2002. A mutation in the LDL receptor-related protein 5 gene results in the autosomal dominant high-bone-mass trait. Am. J. Hum. Genet. 70(1), 11–19.

Liu, P., Wakamiya, M., Shea, M.J., Albrecht, U., Behringer, R.R., Bradley, A. 1999. Requirement for Wnt3 in vertebrate axis formation. Nat. Genet. 22(4), 361–365.

Loomis, C.A., Harris, E., Michaud, J., Wurst, W., Hanks, M., Joyner, A.L. 1996. The mouse Engrailed-1 gene and ventral limb patterning. Nature 382(6589), 360–363.

Lowry, W.E., Blanpain, C., Nowak, J.A., Guasch, G., Lewis, L., Fuchs, E. 2005. Defining the impact of beta-catenin/Tcf transactivation on epithelial stem cells. Genes Dev. 19(13), 1596–1611.

Lu, X., Borchers, A.G., Jolicoeur, C., Rayburn, H., Baker, J.C., Tessier-Lavigne, M. 2004. PTK7/CCK-4 is a novel regulator of planar cell polarity in vertebrates. Nature 430(6995), 93–98.

Maclean, K., Dunwoodie, S.L. 2004. Breaking symmetry: A clinical overview of left-right patterning. Clin. Genet. 65(6), 441–457.

Majumdar, A., Vainio, S., Kispert, A., McMahon, J., McMahon, A.P. 2003. Wnt11 and Ret/Gdnf pathways cooperate in regulating ureteric branching during metanephric kidney development. Development 130(14), 3175–3185.

Maretto, S., Cordenonsi, M., Dupont, S., Braghetta, P., Broccoli, V., Hassan, A.B., Volpin, D., Bressan, G.M., Piccolo, S. 2003. Mapping Wnt/beta-catenin signaling during mouse development and in colorectal tumors. Proc. Natl. Acad. Sci. USA 100(6), 3299–3304.

Marvin, M.J., Di Rocco, G., Gardiner, A., Bush, S.M., Lassar, A.B. 2001. Inhibition of Wnt activity induces heart formation from posterior mesoderm. Genes Dev. 15(3), 316–327.

McBride, H.J., Fatke, B., Fraser, S.E. 2003. Wnt signaling components in the chicken intestinal tract. Dev. Biol. 256(1), 18–33.

Micsenyi, A., Tan, X., Sneddon, T., Luo, J.H., Michalopoulos, G.K., Monga, S.P. 2004. Beta-catenin is temporally regulated during normal liver development. Gastroenterology 126(4), 1134–1146.

Millar, S.E., Willert, K., Salinas, P.C., Roelink, H., Nusse, R., Sussman, D.J., Barsh, G.S. 1999. WNT signaling in the control of hair growth and structure. Dev. Biol. 207(1), 133–149.

Min, H., Danilenko, D.M., Scully, S.A., Bolon, B., Ring, B.D., Tarpley, J.E., DeRose, M., Simonet, W.S. 1998. Fgf-10 is required for both limb and lung development and exhibits striking functional similarity to Drosophila branchless. Genes Dev. 12(20), 3156–3161.

Mohamed, O.A., Clarke, H.J., Dufort, D. 2004. Beta-catenin signaling marks the prospective site of primitive streak formation in the mouse embryo. Dev. Dyn. 231(2), 416–424.

Mohamed, O.A., Jonnaert, M., Labelle-Dumais, C., Kuroda, K., Clarke, H.J., Dufort, D. 2005. Uterine Wnt/beta-catenin signaling is required for implantation. Proc. Natl. Acad. Sci. USA 102(24), 8579–8584.

Monga, S.P., Monga, H.K., Tan, X., Mule, K., Pediaditakis, P., Michalopoulos, G.K. 2003. Beta-catenin antisense studies in embryonic liver cultures: Role in proliferation, apoptosis, and lineage specification. Gastroenterology 124(1), 202–216.

Montcouquiol, M., Rachel, R.A., Lanford, P.J., Copeland, N.G., Jenkins, N.A., Kelley, M.W. 2003. Identification of Vangl2 and Scrb1 as planar polarity genes in mammals. Nature 423(6936), 173–177.

Moon, A.M., Capecchi, M.R. 2000. Fgf8 is required for outgrowth and patterning of the limbs. Nat. Genet. 26(4), 455–459.

Morkel, M., Huelsken, J., Wakamiya, M., Ding, J., van de Wetering, M., Clevers, H., Taketo, M.M., Behringer, R.R., Shen, M.M., Birchmeier, W. 2003. Beta-catenin regulates Cripto- and Wnt3-dependent gene expression programs in mouse axis and mesoderm formation. Development 130(25), 6283–6294.

Mucenski, M.L., Wert, S.E., Nation, J.M., Loudy, D.E., Huelsken, J., Birchmeier, W., Morrisey, E.E., Whitsett, J.A. 2003. Beta-Catenin is required for specification of proximal/distal cell fate during lung morphogenesis. J. Biol. Chem. 278(41), 40231–40238.

Mucenski, M.L., Nation, J.M., Thitoff, A.R., Besnard, V., Xu, Y., Wert, S.E., Harada, N., Taketo, M.M., Stahlman, M.T., Whitsett, J.A. 2005. Beta-catenin regulates differentiation of respiratory epithelial cells *in vivo*. Am. J. Physiol. Lung Cell. Mol. Physiol. 289(6), L971–L999.

Mukhopadhyay, M., Shtrom, S., Rodriguez-Esteban, C., Chen, L., Tsukui, T., Gomer, L., Dorward, D.W., Glinka, A., Grinberg, A., Huang, S.P., Niehrs, C., Belmonte, J.C. 2001. Dickkopf1 is required for embryonic head induction and limb morphogenesis in the mouse. Dev. Cell 1(3), 423–434.

Mulroy, T., McMahon, J.A., Burakoff, S.J., McMahon, A.P., Sen, J. 2002. Wnt-1 and Wnt-4 regulate thymic cellularity. Eur. J. Immunol. 32(4), 967–971.

Murdoch, J.N., Doudney, K., Paternotte, C., Copp, A.J., Stanier, P. 2001. Severe neural tube defects in the loop-tail mouse result from mutation of Lpp1, a novel gene involved in floor plate specification. Hum. Mol. Genet. 10(22), 2593–2601.

Murdoch, J.N., Henderson, D.J., Doudney, K., Gaston-Massuet, C., Phillips, H.M., Paternotte, C., Arkell, R., Stanier, P., Copp, A.J. 2003. Disruption of scribble (Scrb1) causes severe neural tube defects in the circletail mouse. Hum. Mol. Genet. 12(2), 87–98.

Muroyama, Y., Fujihara, M., Ikeya, M., Kondoh, H., Takada, S. 2002. Wnt signaling plays an essential role in neuronal specification of the dorsal spinal cord. Genes Dev. 16(5), 548–553.

Naiche, L.A., Papaioannou, V.E. 2003. Loss of Tbx4 blocks hindlimb development and affects vascularization and fusion of the allantois. Development 130(12), 2681–2693.

Nakamura, T., Sano, M., Songyang, Z., Schneider, M.D. 2003. A Wnt- and beta-catenin-dependent pathway for mammalian cardiac myogenesis. Proc. Natl. Acad. Sci. USA 100(10), 5834–5839.

Nakaya, M.A., Biris, K., Tsukiyama, T., Jaime, S., Rawls, J.A., Yamaguchi, T.P. 2005. Wnt3a links left-right determination with segmentation and anteroposterior axis elongation. Development 132(24), 5425–5436.

Nelson, W.J., Nusse, R. 2004. Convergence of Wnt, beta-catenin, and cadherin pathways. Science 303(5663), 1483–1487.

Ninomiya, H., Elinson, R.P., Winklbauer, R. 2004. Antero-posterior tissue polarity links mesoderm convergent extension to axial patterning. Nature 430(6997), 364–367.

Niswander, L., Martin, G.R. 1993. FGF-4 and BMP-2 have opposite effects on limb growth. Nature 361(6407), 68–71.

Nordstrom, U., Jessell, T.M., Edlund, T. 2002. Progressive induction of caudal neural character by graded Wnt signaling. Nat. Neurosci. 5(6), 525–532.

O'Shaughnessy, R.F., Yeo, W., Gautier, J., Jahoda, C.A., Christiano, A.M. 2004. The WNT signalling modulator, Wise, is expressed in an interaction-dependent manner during hair-follicle cycling. J. Invest. Dermatol. 123(4), 613–621.

Oishi, I., Suzuki, H., Onishi, N., Takada, R., Kani, S., Ohkawara, B., Koshida, I., Suzuki, K., Yamada, G., Schwabe, G.C., Mundlos, S., Shibuya, H., et al. 2003. The receptor tyrosine kinase Ror2 is involved in non-canonical Wnt5a/JNK signalling pathway. Genes Cells 8(7), 645–654.

Okamura, R.M., Sigvardsson, M., Galceran, J., Verbeek, S., Clevers, H., Grosschedl, R. 1998. Redundant regulation of T cell differentiation and TCRalpha gene expression by the transcription factors LEF-1 and TCF-1. Immunity 8(1), 11–20.

Okubo, T., Hogan, B.L. 2004. Hyperactive Wnt signaling changes the developmental potential of embryonic lung endoderm. J. Biol. 3(3), 11.

Oldridge, M., Fortuna, A.M., Maringa, M., Propping, P., Mansour, S., Pollitt, C., DeChiara, T.M., Kimble, R.B., Valenzuela, D.M., Yancopoulos, G.D., Wilkie, A.O. 2000. Dominant mutations in ROR2, encoding an orphan receptor tyrosine kinase, cause brachydactyly type B. Nat. Genet. 24(3), 275–278.

Ouji, Y., Yoshikawa, M., Shiroi, A., Ishizaka, S. 2006a. Wnt-10b promotes differentiation of skin epithelial cells *in vitro*. Biochem. Biophys. Res. Commun. 342(1), 28–35.

Ouji, Y., Yoshikawa, M., Shiroi, A., Ishizaka, S. 2006b. Wnt-10b secreted from lymphocytes promotes differentiation of skin epithelial cells. Biochem. Biophys. Res. Commun. 342(4), 1063–1069.

Park, M., Moon, R.T. 2002. The planar cell-polarity gene stbm regulates cell behaviour and cell fate in vertebrate embryos. Nat. Cell Biol. 4(1), 20–25.

Parr, B.A., McMahon, A.P. 1995. Dorsalizing signal Wnt-7a required for normal polarity of D-V and A-P axes of mouse limb. Nature 374(6520), 350–353.

Parr, B.A., Shea, M.J., Vassileva, G., McMahon, A.P. 1993. Mouse Wnt genes exhibit discrete domains of expression in the early embryonic CNS and limb buds. Development 119(1), 247–261.

Perantoni, A.O. 2003. Renal development: Perspectives on a Wnt-dependent process, Semin. Cell Dev. Biol. 14(4), 201–208.

Petropoulos, H., Skerjanc, I.S. 2002. Beta-catenin is essential and sufficient for skeletal myogenesis in P19 cells. J. Biol. Chem. 277(18), 15393–15399.

Perantoni, A.O., Timofeeva, O., Naillat, F., Richman, C., Pajni-Underwood, S., Wilson, C., Vainio, S., Dove, L.F., Lewandoski, M. 2005. Inactivation of FGF8 in early mesoderm reveals an essential role in kidney development. Development 132(17), 3859–3871.

Phillips, H.M., Murdoch, J.N., Chaudhry, B., Copp, A.J., Henderson, D.J. 2005. Vangl2 acts via RhoA signaling to regulate polarized cell movements during development of the proximal outflow tract. Circ. Res. 96(3), 292–299.

Pinson, K.I., Brennan, J., Monkley, S., Avery, B.J., Skarnes, W.C. 2000. An LDL-receptor-related protein mediates Wnt signalling in mice. Nature 407(6803), 535–538.

Pinto, D., Gregorieff, A., Begthel, H., Clevers, H. 2003. Canonical Wnt signals are essential for homeostasis of the intestinal epithelium. Genes Dev. 17(14), 1709–1713.

Pongracz, J.E., Stockley, R.A. 2006. Wnt signalling in lung development and diseases. Respir. Res. 7, 15.

Popperl, H., Schmidt, C., Wilson, V., Hume, C.R., Dodd, J., Krumlauf, R., Beddington, R.S. 1997. Misexpression of Cwnt8C in the mouse induces an ectopic embryonic axis and causes a truncation of the anterior neuroectoderm. Development 124(15), 2997–3005.

Povelones, M., Howes, R., Fish, M., Nusse, R. 2005. Genetic evidence that *Drosophila* frizzled controls planar cell polarity and Armadillo signaling by a common mechanism. Genetics 171(4), 1643–1654.

Radtke, F., Clevers, H. 2005. Self-renewal and cancer of the gut: Two sides of a coin. Science 307(5717), 1904–1909.

Ranheim, E.A., Kwan, H.C., Reya, T., Wang, Y.K., Weissman, I.L., Francke, U. 2005. Frizzled 9 knock-out mice have abnormal B-cell development. Blood 105(6), 2487–2494.

Reddy, S., Andl, T., Bagasra, A., Lu, M.M., Epstein, D.J., Morrisey, E.E., Millar, S.E. 2001. Characterization of Wnt gene expression in developing and postnatal hair follicles and identification of Wnt5a as a target of Sonic hedgehog in hair follicle morphogenesis. Mech. Dev. 107(1–2), 69–82.

Reddy, S.T., Andl, T., Lu, M.M., Morrisey, E.E., Millar, S.E. 2004. Expression of Frizzled genes in developing and postnatal hair follicles. J. Invest. Dermatol. 123(2), 275–282.

Reya, T., Duncan, A.W., Ailles, L., Domen, J., Scherer, D.C., Willert, K., Hintz, L., Nusse, R., Weissman, I.L. 2003. A role for Wnt signalling in self-renewal of haematopoietic stem cells. Nature 423(6938), 409–414.

Robb, L., Tam, P.P. 2004. Gastrula organiser and embryonic patterning in the mouse. Semin. Cell Dev. Biol. 15(5), 543–554.

Roberts, D.J., Smith, D.M., Goff, D.J., Tabin, C.J. 1998. Epithelial-mesenchymal signaling during the regionalization of the chick gut. Development 125(15), 2791–2801.

Ruiz, S., Segrelles, C., Santos, M., Lara, M.F., Paramio, J.M. 2004. Functional link between retinoblastoma family of proteins and the Wnt signaling pathway in mouse epidermis. Dev. Dyn. 230(3), 410–418.

Sanchez, M.P., Silos-Santiago, I., Frisen, J., He, B., Lira, S.A., Barbacid, M. 1996. Renal agenesis and the absence of enteric neurons in mice lacking GDNF. Nature 382(6586), 70–73.

Sancho, E., Batlle, E., Clevers, H. 2003. Live and let die in the intestinal epithelium. Curr. Opin. Cell Biol. 15(6), 763–770.

Satoh, K., Kasai, M., Ishidao, T., Tago, K., Ohwada, S., Hasegawa, Y., Senda, T., Takada, S., Nada, S., Nakamura, T., Akiyama, T. 2004. Anteriorization of neural fate by inhibitor of beta-catenin and T cell factor (ICAT), a negative regulator of Wnt signaling. Proc. Natl. Acad. Sci. USA 101(21), 8017–8021.

Schilham, M.W., Wilson, A., Moerer, P., Benaissa-Trouw, B.J., Cumano, A., Clevers, H.C. 1998. Critical involvement of Tcf-1 in expansion of thymocytes. J. Immunol. 161(8), 3984–3991.

Schneider, V.A., Mercola, M. 2001. Wnt antagonism initiates cardiogenesis in *Xenopus laevis*. Genes. Dev. 15(3), 304–315.

Schuchardt, A., D'Agati, V., Larsson-Blomberg, L., Costantini, F., Pachnis, V. 1994. Defects in the kidney and enteric nervous system of mice lacking the tyrosine kinase receptor Ret. Nature 367(6461), 380–383.

Schuchardt, A., D'Agati, V., Pachnis, V., Costantini, F. 1996. Renal agenesis and hypody-splasia in ret-k- mutant mice result from defects in ureteric bud development. Development 122(6), 1919–1929.

Schwabe, G.C., Tinschert, S., Buschow, C., Meinecke, P., Wolff, G., Gillessen-Kaesbach, G., Oldridge, M., Wilkie, A.O., Komec, R., Mundlos, S. 2000. Distinct mutations in the receptor tyrosine kinase gene ROR2 cause brachydactyly type B. Am. J. Hum. Genet. 67(4), 822–831.

Sekine, K., Ohuchi, H., Fujiwara, M., Yamasaki, M., Yoshizawa, T., Sato, T., Yagishita, N., Matsui, D., Koga, Y., Itoh, N., Kato, S. 1999. Fgf10 is essential for limb and lung formation. Nat. Genet. 21(1), 138–141.

Shannon, J.M., Hyatt, B.A. 2004. Epithelial-mesenchymal interactions in the developing lung. Annu. Rev. Physiol. 66, 625–645.

Sheldahl, L.C., Slusarski, D.C., Pandur, P., Miller, J.R., Kuhl, M., Moon, R.T. 2003. Dishevelled activates Ca^{2+} flux, PKC, and CamKII in vertebrate embryos. J. Cell Biol. 161(4), 769–777.

Shu, W., Jiang, Y.Q., Lu, M.M., Morrisey, E.E. 2002. Wnt7b regulates mesenchymal proliferation and vascular development in the lung. Development 129(20), 4831–4842.

Soshnikova, N., Zechner, D., Huelsken, J., Mishina, Y., Behringer, R.R., Taketo, M.M., Crenshaw, E.B., 3rd, Birchmeier, W. 2003. Genetic interaction between Wnt/beta-catenin and BMP receptor signaling during formation of the AER and the dorsal-ventral axis in the limb. Genes Dev. 17(16), 1963–1968.

Staal, F.J., Clevers, H.C. 2003. Wnt signaling in the thymus. Curr. Opin. Immunol. 15(2), 204–208.

Staal, F.J., Clevers, H.C. 2005. WNT signalling and haematopoiesis: A WNT-WNT situation. Nat. Rev. Immunol. 5(1), 21–30.

Staal, F.J., Meeldijk, J., Moerer, P., Jay, P., van de Weerdt, B.C., Vainio, S., Nolan, G.P., Clevers, H. 2001. Wnt signaling is required for thymocyte development and activates Tcf-1 mediated transcription. Eur. J. Immunol. 31(1), 285–293.

Stark, K., Vainio, S., Vassileva, G., McMahon, A.P. 1994. Epithelial transformation of metanephric mesenchyme in the developing kidney regulated by Wnt-4. Nature 372(6507), 679–683.

Tada, M., Smith, J.C. 2000. Xwnt11 is a target of Xenopus Brachyury: Regulation of gastrulation movements via Dishevelled, but not through the canonical Wnt pathway. Development 127(10), 2227–2238.

Tajbakhsh, S., Borello, U., Vivarelli, E., Kelly, R., Papkoff, J., Duprez, D., Buckingham, M., Cossu, G. 1998. Differential activation of Myf5 and MyoD by different Wnts in explants of mouse paraxial mesoderm and the later activation of myogenesis in the absence of Myf5. Development 125(21), 4155–4162.

Takeuchi, J.K., Koshiba-Takeuchi, K., Suzuki, T., Kamimura, M., Ogura, K., Ogura, T. 2003a. Tbx5 and Tbx4 trigger limb initiation through activation of the Wnt/Fgf signaling cascade. Development 130(12), 2729–2739.

Takeuchi, M., Nakabayashi, J., Sakaguchi, T., Yamamoto, T.S., Takahashi, H., Takeda, H., Ueno, N. 2003b. The prickle-related gene in vertebrates is essential for gastrulation cell movements. Curr. Biol. 13(8), 674–679.

Tamamura, Y., Otani, T., Kanatani, N., Koyama, E., Kitagaki, J., Komori, T., Yamada, Y., Costantini, F., Wakisaka, S., Pacifici, M., Iwamoto, M., Enomoto-Iwamoto, M. 2005. Developmental regulation of Wnt/beta-catenin signals is required for growth plate assembly, cartilage integrity, and endochondral ossification. J. Biol. Chem. 280(19), 19185–19195.

Tan, X., Apte, U., Micsenyi, A., Kotsagrelos, E., Luo, J.H., Ranganathan, S., Monga, D.K., Bell, A., Michalopoulos, G.K., Monga, S.P. 2005. Epidermal growth factor receptor: A novel target of the Wnt/beta-catenin pathway in liver. Gastroenterology 129(1), 285–302.

Theodosiou, N.A., Tabin, C.J. 2003. Wnt signaling during development of the gastrointestinal tract. Dev. Biol. 259(2), 258–271.

Tissir, F., Bar, I., Jossin, Y., De Backer, O., Goffinet, A.M. 2005. Protocadherin Celsr3 is crucial in axonal tract development. Nat. Neurosci. 8(4), 451–457.

Topol, L., Jiang, X., Choi, H., Garrett-Beal, L., Carolan, P.J., Yang, Y. 2003. Wnt-5a inhibits the canonical Wnt pathway by promoting GSK-3-independent beta-catenin degradation. J. Cell Biol. 162(5), 899–908.

Torban, E., Wang, H.J., Groulx, N., Gros, P. 2004. Independent mutations in mouse vangl2 that cause neural tube defects in looptail mice impair interaction with members of the dishevelled family. J. Biol. Chem. 279(50), 52703–52713.

Ueno, N., Greene, N.D. 2003. Planar cell polarity genes and neural tube closure. Birth Defects Res. C Embryo Today 69(4), 318–324.

Vainio, S.J. 2003. Nephrogenesis regulated by Wnt signaling. J. Nephrol. 16(2), 279–285.

van Bokhoven, H., Celli, J., Kayserili, H., van Beusekom, E., Balci, S., Brussel, W., Skovby, F., Kerr, B., Percin, E.F., Akarsu, N., Brunner, H.G. 2000. Mutation of the gene encoding the ROR2 tyrosine kinase causes autosomal recessive Robinow syndrome. Nat. Genet. 25(4), 423–426.

van de Wetering, M., Sancho, E., Verweij, C., de Lau, W., Oving, I., Hurlstone, A., van der Horn, K., Batlle, E., Coudreuse, D., Haramis, A.P., Tjon-Pon-Fong, M., Moerer, P., et al. 2002. The beta-catenin/TCF-4 complex imposes a crypt progenitor phenotype on colorectal cancer cells. Cell 111(2), 241–250.

Van Den Berg, D.J., Sharma, A.K., Bruno, E., Hoffman, R. 1998. Role of members of the Wnt gene family in human hematopoiesis. Blood 92(9), 3189–3202.

van der Heyden, M.A., Rook, M.B., Hermans, M.M., Rijksen, G., Boonstra, J., Defize, L.H., Destree, O.H. 1998. Identification of connexin43 as a functional target for Wnt signalling. J. Cell Sci. 111(Pt. 12), 1741–1749.

van Es, J.H., Jay, P., Gregorieff, A., van Gijn, M.E., Jonkheer, S., Hatzis, P., Thiele, A., van den Born, M., Begthel, H., Brabletz, T., Taketo, M.M., Clevers, H. 2005. Wnt signalling induces maturation of Paneth cells in intestinal crypts. Nat. Cell Biol. 7(4), 381–386.

van Genderen, C., Okamura, R.M., Farinas, I., Quo, R.G., Parslow, T.G., Bruhn, L., Grosschedl, R. 1994. Development of several organs that require inductive epithelial-mesenchymal interactions is impaired in LEF-1-deficient mice. Genes Dev. 8(22), 2691–2703.

Veeman, M.T., Axelrod, J.D., Moon, R.T. 2003a. A second canon. Functions and mechanisms of beta-catenin-independent Wnt signaling. Dev. Cell 5(3), 367–377.

Veeman, M.T., Slusarski, D.C., Kaykas, A., Louie, S.H., Moon, R.T. 2003b. Zebrafish prickle, a modulator of noncanonical Wnt/Fz signaling, regulates gastrulation movements. Curr. Biol. 13(8), 680–685.

Wallingford, J.B., Habas, R. 2005. The developmental biology of Dishevelled: An enigmatic protein governing cell fate and cell polarity. Development 132(20), 4421–4436.

Wallingford, J.B., Harland, R.M. 2002. Neural tube closure requires Dishevelled-dependent convergent extension of the midline. Development 129(24), 5815–5825.

Wallingford, J.B., Rowning, B.A., Vogeli, K.M., Rothbacher, U., Fraser, S.E., Harland, R.M. 2000. Dishevelled controls cell polarity during Xenopus gastrulation. Nature 405(6782), 81–85.

Wallingford, J.B., Fraser, S.E., Harland, R.M. 2002. Convergent extension: The molecular control of polarized cell movement during embryonic development. Dev. Cell 2(6), 695–706.

Wang, J., Mark, S., Zhang, X., Qian, D., Yoo, S.J., Radde-Gallwitz, K., Zhang, Y., Lin, X., Collazo, A., Wynshaw-Boris, A., Chen, P. 2005. Regulation of polarized extension and planar cell polarity in the cochlea by the vertebrate PCP pathway. Nat. Genet. 37(9), 980–985.

Wang, J., Hamblet, N.S., Mark, S., Dickinson, M.E., Segil, N., Fraser, S., Chen, P., Wallingford, J.B., Wynshaw-Boris, A. 2006a. Dishevelled genes mediate a conserved mammalian PCP pathway to regulate convergent extension during neurulation. Development 133(9), 1767–1778.

Wang, Y., Guo, N., Nathans, J. 2006b. The role of Frizzled3 and Frizzled6 in neural tube closure and in the planar polarity of inner-ear sensory hair cells. J. Neurosci. 26(8), 2147–2156.

Wang, Y., Thekdi, N., Smallwood, P.M., Macke, J.P., Nathans, J. 2002. Frizzled-3 is required for the development of major fiber tracts in the rostral CNS. J. Neurosci. 22(19), 8563–8573.

Weaver, M., Dunn, N.R., Hogan, B.L. 2000. Bmp4 and Fgf10 play opposing roles during lung bud morphogenesis. Development 127(12), 2695–2704.

Wells, J.M., Melton, D.A. 1999. Vertebrate endoderm development. Annu Rev Cell Dev. Biol. 15, 393–410.

Willert, K., Brown, J.D., Danenberg, E., Duncan, A.W., Weissman, I.L., Reya, T., Yates, J.R., 3rd, Nusse, R. 2003. Wnt proteins are lipid-modified and can act as stem cell growth factors. Nature 423(6938), 448–452.

Winberg, M.L., Tamagnone, L., Bai, J., Comoglio, P.M., Montell, D., Goodman, C.S. 2001. The transmembrane protein Off-track associates with Plexins and functions downstream of Semaphorin signaling during axon guidance. Neuron 32(1), 53–62.

Wong, G.T., Gavin, B.J., McMahon, A.P. 1994. Differential transformation of mammary epithelial cells by Wnt genes. Mol. Cell. Biol. 14(9), 6278–6286.

Xu, X., Weinstein, M., Li, C., Naski, M., Cohen, R.I., Ornitz, D.M., Leder, P., Deng, C. 1998. Fibroblast growth factor receptor 2 (FGFR2)-mediated reciprocal regulation loop between FGF8 and FGF10 is essential for limb induction. Development 125(4), 753–765.

Yamaguchi, TP. 2001. Heads or tails: Wnts and anterior-posterior patterning. Curr. Biol. 11(17), R713–R724.

Yamaguchi, T.P., Bradley, A., McMahon, A.P., Jones, S. 1999a. A Wnt5a pathway underlies outgrowth of multiple structures in the vertebrate embryo. Development 126(6), 1211–1223.

Yamaguchi, T.P., Takada, S., Yoshikawa, Y., Wu, N., McMahon, A.P. 1999b. T (Brachyury) is a direct target of Wnt3a during paraxial mesoderm specification. Genes Dev. 13(24), 3185–3190.

Yamanaka, H., Moriguchi, T., Masuyama, N., Kusakabe, M., Hanafusa, H., Takada, R., Takada, S., Nishida, E. 2002. JNK functions in the non-canonical Wnt pathway to regulate convergent extension movements in vertebrates. EMBO Rep. 3(1), 69–75.

Yang, Y. 2003. Wnts and wing: Wnt signaling in vertebrate limb development and musculo-skeletal morphogenesis. Birth Defects Res. C Embryo Today 69(4), 305–317.

Yang, Y., Topol, L., Lee, H., Wu, J. 2003. Wnt5a and Wnt5b exhibit distinct activities in coordinating chondrocyte proliferation and differentiation. Development 130(5), 1003–1015.

Yoshikawa, Y., Fujimori, T., McMahon, A.P., Takada, S. 1997. Evidence that absence of Wnt-3a signaling promotes neuralization instead of paraxial mesoderm development in the mouse. Dev. Biol. 183(2), 234–242.

Yu, J., McMahon, A.P., Valerius, M.T. 2004. Recent genetic studies of mouse kidney development. Curr. Opin. Genet. Dev. 14(5), 550–557.

Zechner, D., Fujita, Y., Hulsken, J., Muller, T., Walther, I., Taketo, M.M., Crenshaw, E.B., 3rd, Birchmeier, W., Birchmeier, C. 2003. beta-Catenin signals regulate cell growth and the balance between progenitor cell expansion and differentiation in the nervous system. Dev. Biol. 258(2), 406–418.

Zeller, R., Jackson-Grusby, L., Leder, P. 1989. The limb deformity gene is required for apical ectodermal ridge differentiation and anteroposterior limb pattern formation. Genes Dev. 3(10), 1481–1492.

Zhou, P., Byrne, C., Jacobs, J., Fuchs, E. 1995. Lymphoid enhancer factor 1 directs hair follicle patterning and epithelial cell fate. Genes Dev. 9(6), 700–713.

Zuniga, A., Haramis, A.P., McMahon, A.P., Zeller, R. 1999. Signal relay by BMP antagonism controls the SHH/FGF4 feedback loop in vertebrate limb buds. Nature 401(6753), 598–602.

Cell migration under control of Wnt-signaling in the vertebrate embryo

Almut Köhler, Alexandra Schambony and Doris Wedlich

Universitaet Karlsruhe (TH), Zoologisches Institut II,
D-76131 Karlsruhe, Germany

Contents

Advances in Developmental Biology
Volume 17 ISSN 1574-3349
DOI: 10.1016/S1574-3349(06)17005-2

During embryonic development, cell fate specification and morphogenetic movements are tightly controlled by a network of signaling cascades. Among these, Wnt-signaling historically has predominantly been associated with cell fate decisions, however, during the last years evidence is accumulating that Wnt-signaling pathways also play an essential role in cell migration as part of morphogenetic processes. Wnt signals control and coordinate cell polarity and proper formation of lamellipodia and filopodia in a large number of morphogenetic movements, including gastrulation, neural crest migration, and eye field formation. More recently, Wnt ligands and Wnt antagonists have been reported to function as cell guidance molecules, for example, in the migration of the anterior visceral endoderm of mouse embryos and in axon pathfinding during establishment of retinotectal projection. In this chapter, we aim to provide an overview of Wnt-regulated cell migration processes, the different downstream signaling pathways and the newly identified receptors and coreceptors involved in Wnt-stimulated signal transduction. Deciphering the growing complexity of the Wnt-signaling network and its multiple roles will be the challenge of the next future.

1. Introduction

Cell migration is a crucial and recurring process in embryonic morphogenesis, which is tightly coupled with cell differentiation. Signaling cascades important for axis formation, patterning of germ layers, and organ differentiation also control cell motility. Therefore, a clear distinction between differentiation and migration is sometimes difficult to make. Before the planar cell polarity (PCP) pathway was identified the main function of Wnt-signaling was seen in regulating cell fate decisions, and less attention was given to its role in cell migration. Meanwhile, a growing number of reports describe noncanonical and canonical Wnt-signaling effects on cell migration in early development and also in organogenesis of the vertebrate embryo. This chapter provides an overview of cell migration processes in which Wnt-signaling plays an important role. It also aims to help in sorting the different Wnt-signaling pathways and branches and to allocate them to specific cell behaviors. Since cell migration in embryogenesis and organogenesis shows many facets, which to some extent are organism specific and have to be introduced, this chapter concentrates on vertebrate development.

The first section compares different cell migration types. Cellular aspects of cell migration have been extensively studied in single cell organisms like *Dictyostelium* or cell culture cells like fibroblasts, keratinocytes, or osteoblasts. The complexity of the embryo and the very fast changes of its shape make additional regulatory mechanisms necessary, especially when embryonic cells migrate as cell sheets or cell clusters.

The following sections are arranged in chronology to the developmental program starting with migration processes in the early embryo and ending up with morphogenetic movements in heart development. Wnt-dependent cell movements in early embryogenesis have been reported for anterior visceral endoderm (AVE) formation in mice, gastrulation, neural tube closure, and neural crest migration. Gastrulation is a central part of this section because the different Wnt-signaling pathways and their branches have been identified and intensively studied in this morphogenetic process. They can be assigned to specific cell behaviors emphasizing similarities but also differences in gastrulation between different species. Therefore, a detailed description of the Wnt-signaling cascades in regard to cell migration is found in this section. More indirect evidence for a role of canonical Wnt-signaling has been reported in neuronal migration of the cortex and the hindbrain. Eye development is the topic of the following section. A sequence of morphogenetic movements characterizes eye development. Important contributions of Wnt-signaling in eye field separation, retinal progenitor movement, and axon guidance in retinotectal projection have recently been described. An antagonistic influence of canonical and noncanonical Wnt-signaling seems to be crucial for proper eye development. The last section reviews Wnt-dependent morphogenetic movements, which form the multichambered heart. Noncanonical Wnt/PCP signaling is crucial for the midline convergence of myocardial precursor cells. The latter leads to the fusion of the two lateral heart field primordia. A different branch of the noncanonical Wnt-signaling cascade seems to be important for heart tube formation. Myocardialization, a more recently identified migration process, describes the de-epithelialization and invasion of myocardiocytes into the mesenchymal cushion. This process might also be controlled by noncanonical Wnt-signaling; however, the pathway remains to be clarified. There are some hints that the invasion of the cardiac neural crest, which forms the aorticopulmonary septum, also depends on Wnt-signaling. But it is unclear whether canonical or noncanonical Wnt-signaling or both are required and if so, at which time point.

2. Modes of cell migration in the vertebrate embryo

Cell migration comprises a complex program of cellular behavior including orientation, cell polarization, cytoskeletal rearrangement, and changes in adhesion and shape. As long as a cell migrates, these cellular processes underlie a continuous regulation and feedback control to direct cell migration. The basic principles of cell motility are best studied in single cells like fibroblasts or *Dictyostelium*. Accordingly, single cell migration is classified into amoeba-like or fibroblast-like movements depending on the more fluidic or fibrillar cellular surrounding, for example, leukocytes in the blood versus macrophages in connective tissue (Fig. 1A, B). These cells establish a front–rear polarity. In *Dictyostelium*, this cell polarity is mediated by the recruitment of PI3K to the front and PTEN to the rear end initiated by binding of cAMP to the receptor (reviewed in Janetopoulos et al., 2005). In fibroblasts the focal adhesion kinase seems to be crucial in forming the leading edge (Tilghman et al., 2005).

In the embryo, single cell migration is observed in a minority of cases. Instead, clusters of cells or cell sheets are on move (Fig. 1C and E). They use other cells as substrates for migration and orientation. This requires that migratory cells communicate with their neighbors and that individual cell movements within a group or tissue layer have to be coordinated. When the functions of Wnt-signaling in different cell migration processes during embryonic development are compared, it seems that the noncanonical pathway is required for cell communication, modulation of cell–cell adhesion,

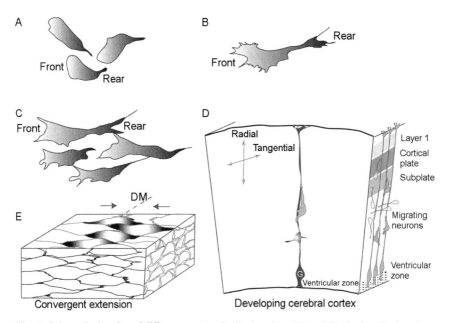

Fig. 1. Schematic drawing of different modes of cell migration. (A) and (B) single cell migration, (A) amoebic-like migration of leukocytes or *Dictyostelium*, (B) migrating fibroblast, (C) cells migrating in clusters, (D) migrating neurons in the cerebral cortex guided by glia cells (G), and (E) convergent extension movement, arrows indicate the direction of the migration toward the dorsal midline (DM).

cell polarity formation, and coordination of motile cells. Recent reports discuss a novel function of Wnt-family members as repelling guidance cues in migration of the future AVE and in axon outgrowth (Kimura-Yoshida et al., 2005; Ouchi et al., 2005; Schmitt et al., 2006).

Convergent extension movement in *Xenopus* gastrulation represents an extreme mass cell movement as a cell sheet is rearranged. Apart from acquiring bipolarity cells move in the direction of the dorsal midline in a coor-dinated fashion (Fig. 1E; Keller et al., 2000; Keller, 2002). Noncanonical Wnt-signaling controls the establishment of polarity (Wallingford et al., 2000), and together with XPAPC it coordinates the cell movements (Unterseher et al., 2004).

Neural crest cells are of fibroblast-like shape with the typical front–rear polarity (Fig. 1B). Filopodia and lamellipodia formation at the front depends on noncanonical Wnt-signaling (De Calisto et al., 2005). Cranial neural crest cells express classical type II cadherins (Borchers et al., 2001; Takahashi and Osumi, 2005; Taneyhill and Bronner-Fraser, 2005; for review see Pla et al., 2001; Ruan et al., 2003) and migrate in loose cell clusters. *In vivo* time-lapse studies reveal that even when they migrate as single cells they repeatedly contact directly their neural crest companions (Kashef and Köhler, unpublished data).

Neurons seem to migrate as individuals. They choose glia cells as guidance cues in the locomotion mode or still unknown factors in the translocation mode during radial cortex migration (Nadarajah et al., 2001, Fig. 1D). Although migrating as individuals, neurons form cell junctions with glia cells. These adhesion complexes consist of integrins and junctional glia proteins (Anton et al., 1996, 1999; Cameron et al., 1997). Thus, single cell migration in the developing cerebral cortex requires cell communication and cell–cell adhesion, which has more in common with the migratory behavior of cell clusters rather than with fibroblasts.

3. Cell migration in the early embryo

3.1. Migration of the AVE in mouse development

The earliest morphogenetic movement observed in mouse development following implantation of the blastocyst is the active migration of a specified region in the visceral endoderm by which the future AVE is formed. This process takes place prior to gastrulation. After implantation the blastocyst elongates into a cylindrical structure, the egg cylinder, which consists of the extraembryonic ectoderm, the epiblast, and the surrounding visceral endoderm (Fig. 2). The embryo only develops from the epiblast but its forebrain development depends on signals emanating from the AVE. Apart from this role in anterior patterning of the embryo, the visceral endoderm has been

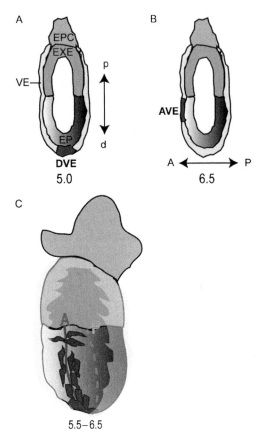

Fig. 2. Proximodistal to anteroposterior axis conversion and formation of the anterior visceral endoderm (AVE) in the mouse embryo. (A) Premigratory stage 5.0, the antecedents of the AVE are located distally forming the DVE. (B) Postmigratory stage 6.5, AVE is found in anterior position at the junction between extraembryonic ectoderm (EXE) and epiblast (EP). (C) Schematic drawing of the migrating cells, anterior view facing the reader. Color code: light gray: ectoplacental cone (EPC), yellow: visceral endoderm (VE), blue: extraembryonic ectoderm (EXE), gray: epiblast (EP), dark blue: DVE/AVE, p: proximal, d: distal, A: anterior, P: posterior. (See Color Insert.)

reported to function in nutrient uptake and delivery, retinol transfer, and controlling the formation of the proamniotic cavity (reviewed in Bielinska et al., 1999).

The AVE region derives from the distal tip of the egg cylinder (stage 5.0 dpc) and is therefore called distal visceral endoderm (DVE) before migration (Fig. 2A). The DVE is distinguishable from the residual visceral endoderm by the expression of specific genes, for example, *Hex, Hesx1, lim1,* and *cer1* (for review see Martinez-Barbera and Beddington, 2001). The expression of these genes is maintained during migration and finally localized in the AVE (Fig. 2A, B). Around stage 5.5 dpc the DVE cells start to migrate to the future anterior pole of the embryo. They stop migration when they reach the junction of the epiblast and the extraembryonic ectoderm and form the AVE. At the end of this migration process the proximodistal axis has been converted into the anteroposterior axis, which becomes obvious in the asymmetry in marker gene expression (Thomas et al., 1998; Bielinska et al., 1999). The requirement of the AVE for the forebrain differentiation was first demonstrated by extirpation experiments. Removal of the AVE resulted in loss of anterior head structures (Thomas and Beddington, 1996). Gene knockout studies performed later demonstrate that the AVE might function in repressing posterior genes in the anterior epiblast (Kimura et al., 2000; Perea-Gomez et al., 2001a; see also review by Perea-Gomez et al., 2001b). With the acquirement to express GFP in the visceral endoderm by injecting *in vitro*-synthesized transcripts or GFP constructs driven by the *cerl*- or *Hex*-promoter, the movement of these cells could be traced *in vivo* when embryos were taken in culture (Weber et al., 1999; Srinivas et al., 2004; Kimura-Yoshida et al., 2005). Detailed time-lapse studies by Srinivas et al. (2004) revealed that visceral endoderm cells from the distal tip of the egg cylinder migrate within 4–5 hours over a distance of 100 μm to the anterior– proximal position. The cells migrate unilaterally and project filopodia in the direction of their motion. They seem to move as a monolayered cell sheet in direct contact with the epiblast. When they reach the border of epiblast and extraembryonic ectoderm they spread laterally along the tissue border before they stop migration (Fig. 2C). The molecular mechanisms regulating this migratory behavior are still unknown. Canonical Wnt-signaling seems to play an important role because in the *β-catenin* knockout mouse the DVE-specific genes *cerl* and *lim1* are expressed, but the DVE cells fail to move to the anterior position (Huelsken et al., 2000). A similar phenotype is observed in *Otx2* and *cripto* knockout mice. AVE/DVE marker expression is maintained in the distal tip of the egg cylinder at stage 6.5 dpc indicating a defect in cell migration (Ding et al., 1998; Kimura et al., 2000). The latter is confirmed by lineage tracing of DVE cells in *Otx2* knockout mice. In these mutants the DVE cells fail to migrate anteriorly (Perea-Gomez et al., 2001a). In *otx2/cripto* double knockout mice expression of *cerl, lim1*, and *Wnt3* is completely suppressed giving evidence that both genes act in parallel pathways and are required for induction and migration of the DVE, and thus, in anteroposterior axis formation (Kimura et al., 2001).

Cripto seems to be a direct target gene of *β*-catenin, which explains the similarity in the phenotype of the *cripto* and the *β-catenin* knockout mice. *Cripto* is identified as strongest downregulated gene in the *β*-catenin$^{-/-}$ embryo at stage 6.0 dpc. Furthermore, Morkel et al. (2003) identify a *β*-catenin/ TCF-responsive element in the *Cripto* enhancer, which is found to be active in a *β*-catenin/TCF-dependent reporter gene assay. These data indicate that canonical Wnt-signaling is required for anterior migration of the DVE.

But the appropriate ligand of the β-catenin/cripto pathway remains elusive. Wnt3 expression is detected in the proximal epiblast in the prestreak stage (6.25 dpc) and concentrated higher at the posterior side. However, it does not influence AVE formation, since the Wnt3$^{-/-}$ mouse shows normal AVE marker expression and localization while the primitive streak, the mesoderm, and the node are not formed (Liu et al., 1999).

Recently, also a link between Otx2 function and canonical Wnt-signaling is reported by Kimura-Yoshida et al. (2005) emphasizing an opposite relation: attenuation of canonical Wnt-signaling by dkk1 is required for guiding the DVE cells to the anterior position. Dkk1 binds the Wnt coreceptor LRP5/6 thereby blocking Wnt/β-catenin signaling (Glinka et al., 1998; Li et al., 2002). Otx2$^{-/-}$ embryos lack *dkk1* expression in the egg cylinder stage. Kimura-Yoshida et al. (2005) demonstrated that dkk1 is a downstream target of Otx2 by inserting the *dkk1* gene into the *Otx2* locus, which results in a partial rescue of the anteroposterior axis defect observed in *Otx2* knockout embryos. To mimic the function of dkk1 as repressor of Wnt/β-catenin signaling Kimura-Yoshida et al. (2005) deleted one copy of the *β-catenin* gene in Otx2$^{-/-}$ animals. Lowering the β-catenin concentration in the *Otx2* knockout mouse also restores correct localization of AVE marker genes. Conversely, ectopic expression of murine Wnt8A destroys proximodistal to anteroposterior axis conversion. AVE marker genes are found mislocalized in the distal position of the egg cylinder stage 6.5 as seen in Otx2$^{-/-}$ embryos.

In stage 5.5 dpc wild-type embryos *dkk1* transcripts are localized in the visceral endoderm proximal to the adjacent DVE. In the following stages the belt-like *dkk1* expression shifts in anterior direction, remaining directly in front of the migrating DVE cells. The ligand Wnt3, instead, forms an opposite expression gradient with highest concentration in the posterior proximal epiblast, the future primitive streak region (Liu et al., 1999). The expression patterns led Kimura-Yoshida et al. (2005) to hypothesize that dkk1 might act as an attractant and Wnt3 as a repellent guidance cue for the migrating DVE cells. The direction of the migrating cells changes when beads soaked with dkk1 or Wnt3A are applied to embryos opposite to the endogenous expression area. From these results Kimura-Yoshida et al. (2005) conclude that Otx2 directs DVE migration to the anterior position by local repression of canonical Wnt-signaling through dkk1 expression. How do their data correspond with the observations by Huelsken et al. (2000) and Morkel et al. (2003) that β-catenin and its target gene *cripto* are required for proper localization of the AVE? It seems most likely, that the morphogenetic movement of AVE formation is under tight spatial and temporal control of canonical Wnt-signaling. Wnt/β-catenin signaling might be required to initiate cell motility in DVE cells, for example by altering cell–cell adhesion, organization of the cytoskeleton, initiating filo-podia formation or by polarizing the DVE cells. When cells have acquired motility, they escape from the Wnt influence following a dkk1 gradient, which also may prevent their posteriorization. Some evidence for such a tight spatial regulation of canonical Wnt-signaling is provided by the nuclear localization of β-catenin. Kimura-Yoshida et al. (2005) state that nuclear β-catenin is found in the posterior visceral endoderm but only to a lower degree in the AVE. Additional studies are required that further dissect the migration process into different cellular behaviors and correlate them with Wnt and other growth factor-signaling cascades to understand the formation of the AVE and by this, axis conversion. With the improved embryo cultivation techniques and the GFP-labeled DVE cells perfect tools have been developed to address this question in the next future.

3.2. Gastrulation

During gastrulation the three germ layers are formed and positioned relative to each other and the embryos body axes. Correct positioning of the germ layers requires massive tissue rearrangements, which are dominated by the morphogenetic movements of the mesoderm. These cell movements are characteristic for different species and range from individual cell migration in the chicken and mouse to the mass cell movement of an intact tissue in *Xenopus*. However, a prerequisite in all vertebrates is the specification of dorsal mesoderm capable to undergo morphogenetic movements. An early event in mesoderm induction and dorsalization is the formation of an organizing center, which requires combinatorial signaling of TGF-β and Wnt/β-catenin pathways. The organizer induces the expression of genes that specify dorsal mesoderm and neuroectoderm, and of genes required for gastrulation movements (reviewed in Boettger et al., 2001; Bouwmeester, 2001; Robb and Tam, 2004; Schier and Talbot, 2005). However, in this context the canonical Wnt/β-catenin pathway has rather an inductive function and is not directly related to cell movements, therefore organizer formation is not within the scope of this article.

Gastrulation movements depend on the activity of the noncanonical Wnt/PCP pathway. In both, canonical and noncanonical Wnt pathways (Fig. 3), a secreted Wnt ligand binds to its receptor of the Frizzled family. In the canonical pathway a coreceptor of the LRP5/6 family (Mao et al., 2001; He et al., 2004) is required to transmit the signal to Disheveled and achieve the inhibition of the APC/axin/GSK-3β complex that targets β-catenin for proteasomal degradation. As a result of active canonical Wnt-signaling, β-catenin accumulates in the cytoplasm, translocates into the nucleus and there acts as transactivator of Lef/TCF transcription factors, and stimulates target gene expression (Wodarz and Nusse, 1998). Noncanonical Wnt-signaling resembles the planar polarity pathway of *Drosophila* (Fanto and McNeill, 2004) and is therefore termed Wnt/PCP signaling. In the Wnt/PCP pathway, Frizzled together with strabismus (stbm) (Darken et al., 2002;

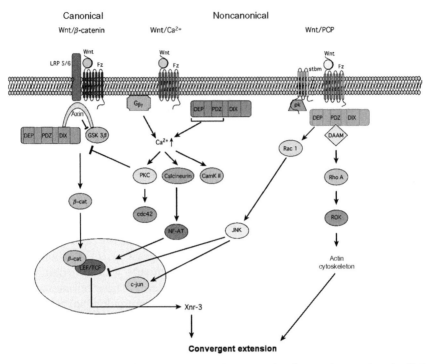

Fig. 3. Overview of canonical and noncanonical Wnt-signaling pathways. For detailed descriptions refer to the main text (Section 3.2). (See Color Insert.)

Park and Moon, 2002) and prickle (pk) (Takeuchi et al., 2003; Veeman et al., 2003) recruits Disheveled to the plasma membrane (Axelrod et al., 1998) and stimulates the Disheveled-mediated activation of the small GTPases Rho A and Rac 1 and their downstream effectors ROK and JNK (Marlow et al., 2002; Yamanaka et al., 2002; Habas et al., 2003; Kim and Han, 2005). The differential subcellular localization of Disheveled is a criterion to distinguish these two branches of Wnt pathways. Disheveled is localized in the cytoplasm of cells with an active canonical Wnt pathway and associates with the cell membrane in response to Wnt/PCP signaling. However, a study demonstrated that although membrane localization of Disheveled is a prerequisite for activation of Wnt/PCP signaling, Disheveled mutants targeted to the cell membrane are still capable to activate the canonical Wnt/β-catenin pathway (Park et al., 2005). Disheveled contains three main protein–protein interaction domains: an N-terminal Disheveled/Axin (DIX) domain, a central PSD-95/DLG/ZO1 (PDZ) domain, and a Disheveled/ EGL-10/Pleckstrin (DEP) domain. Effectors of the canonical and the non-canonical Wnt pathway interact with distinct domains of Disheveled. The interaction of Axin with the DIX domain of Disheveled is essential for the canonical pathway, but canonical signaling also requires the PDZ domain

(Axelrod et al., 1998; Boutros et al., 1998; Moriguchi et al., 1999). Wnt/PCP signaling on the other hand does not involve the DIX domain but depends on the PDZ and the DEP domain (Axelrod et al., 1998; Boutros et al., 1998; Moriguchi et al., 1999; Heisenberg et al., 2000; Tada and Smith, 2000; Wallingford et al., 2000). Especially the DEP domain is essential for membrane recruitment and activation of Rac 1 in the Wnt/PCP pathway (Habas et al., 2003; Park et al., 2005). The second noncanonical Wnt pathway, the Wnt/Ca^{2+} pathway also acts through the PDZ and DEP domains and does not require the DIX domain (Sheldahl et al., 2003); however the specific protein–protein interactions within this pathway have not been identified yet. Besides Disheveled, Wnt/Ca^{2+} signaling involves activation of a trimeric G-protein, which triggers Ca^{2+} release and activation of the Ca^{2+}-responsive protein kinases PKC and CamK II (Sheldahl et al., 1999; Kuehl et al., 2000a,b). PKC triggers activation of the small GTPase cdc42 (Choi and Han, 2002), and inhibits the canonical Wnt-pathway upstream of β-catenin, while CamK II modulates canonical signaling at the level of the Lef/TCF transcription factor complex (Kuehl et al., 2001). In a parallel response to Wnt/Ca^{2+} signaling, NF-AT becomes activated via calcineurin (Saneyoshi et al., 2002).

Wnt signals are modified by the availability of Frizzled receptors and coreceptors (LRP) (He et al., 2004; Hsieh, 2004) and by secreted Wnt antagonists including FrzB, sizzled, Dickkopf (Dkk), Cerberus (Niehrs et al., 2001; Kawano and Kypta, 2003), and Crescent (Shibata et al., 2005). The first three block canonical Wnt-signaling, while Crescent modulates noncanonical Wnt activity in a dual mode: it stimulates Wnt/PCP signaling in the mesoderm and blocks the same pathway in the neuroectoderm (Shibata et al., 2005). In addition, Wnt ligands associate with extracellular matrix proteins (Bradley and Brown, 1990). Heparan sulfate proteoglycans (HSPGs) (Itoh and Sokol, 1994; Lin and Perrimon, 2002; Liu et al., 2003) and HSPG-modifying enzymes (Baeg et al., 2001; Dhoot et al., 2001; Giraldez et al., 2002; Ai et al., 2003; De Cat et al., 2003) have been shown to modulate Wnt-signaling activity. Zebrafish knypek (Topczewski et al., 2001) and its *Xenopus* homologue glypican-4 (Ohkawara et al., 2003) selectively stimulate noncanonical Wnt/PCP signaling and are required for correct gastrulation movements, while *Xenopus* glypican-3 modulates both canonical and noncanonical Wnt-signaling (De Cat et al., 2003). Depletion of functional glypicans results in the same phenotypes as inhibition of the Wnt/PCP pathway. Together with the observation that these glypicans bind Wnt ligands with diverging specificity (De Cat et al., 2003; Ohkawara et al., 2003), it is assumed that glypicans act to modulate Wnt-signaling activity.

Gastrulation movements are best studied in zebrafish and frogs where the embryos are easily accessible for *in vivo* observation. Although in both species noncanonical Wnt-signaling is essential for proper gastrulation movements, there are specific differences in the types of cell movements and their regulation by Wnt-signaling.

3.2.1. Gastrulation movements in zebrafish

In zebrafish embryos gastrulation starts about 6 hour after fertilization (hpf) with the involution of cells at the margin of the blastoderm that form the hypoblast. The thickening at the involution zone is called the germ ring. Although involution seems to be a movement of individual cells, it is also considered as the coordinated movement of a sheet of cells. The hypoblast cells originally migrate toward the animal pole of the embryo, while the epiblast cells continue to move toward the vegetal pole. Thus, at early gastrulation hypoblast and epiblast cells move in opposite directions. The hypoblast cells at that stage migrate as individual cells on the yolk cell and the inner side of the epiblast.

At the dorsal side of the embryo the germ ring is thicker, and the cells are more compact than at the lateral and ventral side. This thickening, the embryonic shield, represents the organizer region of zebrafish embryos and is homologous to Spemann's organizer in *Xenopus* and Hensen's node in chicken. At midgastrula the migration direction of the cells of both, the epiblast and the hypoblast changes, and the cells start to move dorsally until finally hypoblast and epiblast cells move in parallel toward the dorsal midline. Convergence of the dorsal hypoblast cells seems to be driven by mediolateral intercalation (Glickman et al., 2003).

A number of genetic mutants revealed that noncanonical Wnt/PCP signaling controls convergence and extension movements during zebrafish gastrulation. The first hints were obtained from two mutant fish lines with nonfunctional genes for *strabismus* (*stbm*)/*trilobite* (*tri*) (Sepich et al., 2000) and *Wnt11/silberblick* (*slb*) (Heisenberg et al., 2000). In both mutants, cell movements toward the dorsal midline were disturbed without changing mesodermal patterning. The *silberblick* phenotype was not identical to phenotypes obtained after inhibition of canonical Wnt-signaling and could be rescued by a Disheveled mutant that lacks the DIX domain required for activation of the canonical Wnt pathway (Heisenberg et al., 2000). Stbm/tri has been shown to interact with Disheveled via its PDZ domain, to recruit Disheveled to the cell membrane, and to activate JNK (Park and Moon, 2002). In zebrafish the pathway is not only activated by Wnt11/slb; the Wnt5a/pipetail (ppt) mutant shows similar phenotypes. Furthermore, both Wnt ligands are at least partially redundant in the anterior mesoderm (Kilian et al., 2003). The resemblance between the noncanonical pathway downstream of Wnt11 and the PCP pathway of *Drosophila* was further supported by the identification of additional homologues of *Drosophila* PCP genes that caused similar phenotypes and act downstream of Wnt11 in zebrafish namely prickle, Frizzled, JNK (Carreira-Barbosa et al., 2003; Veeman et al., 2003), and Rho-activated kinase ROK2 (Marlow et al., 2002).

Mutants or suppression of these genes lead to migration defects during gastrulation. In *slb* mutants shape and migration of the epiblast cells are not affected, but hypoblast cells migrate slower and in random directions at early gastrula (Ulrich et al., 2003). In contrast mutation of *Wnt5a/ppt* affects both ectodermal and mesodermal cells in a non cell-autonomous way. The cells in these embryos fail to elongate, do not align, and consequently migrate slower and in multiple directions (Kilian et al., 2003). The mechanism of Wnt11/slb and Wnt5a/ppt, however, seems to be similar because in both *slb* and *ppt* mutants the formation of protrusions in the migrating cells was affected (Kilian et al., 2003; Ulrich et al., 2003). The same inhibition of cell elongation, randomized migration directions, and decreased velocity has been observed in embryos injected with pk antisense morpholino oligonucleotide (Veeman et al., 2003), in *stbm/tri* mutants (Jessen et al., 2002) and in embryos injected with a dominant-negative ROK2 construct (Marlow et al., 2002). Interestingly, in all experiments a non cell-autonomous function of the PCP genes was observed, but stbm/tri and ROK2 showed an additional cell autonomous effect on cell elongation (Jessen et al., 2002; Marlow et al., 2002).

PCP signaling is further modulated by a growing number of proteins. Diversin is an ankyrin-repeat protein that shows similarity to *Drosophila* Diego and is capable to act as a switch between canonical and noncanonical Wnt-signaling. It stimulates the Wnt/PCP pathway and JNK activity via a still unknown mechanism and recruits casein kinase CKIε to the axin/ conductin complex and thereby triggers degradation of β-catenin (Schwarz-Romond et al., 2002). Other proteins have been reported that selectively modulate the noncanonical pathway. The seven-pass transmembrane cadherin flamingo (Fmi) is part of the *Drosophila* PCP-signaling pathway. The zebrafish homologues zFmi1a and zFmi1b show functional interaction with Wnt11 and stbm in the control of anterior migration of hypoblast cells in early gastrula and in dorsal convergence of the lateral mesoderm in a dose-dependent manner (Formstone and Mason, 2005). An activator of protein phosphatase 2A, widerborst (wdb), is also part of the PCP pathway in *Drosophila* and zebrafish. Suppression of wdb in zebrafish embryos leads to an inhibition of convergence and extension movements. Studies in *Drosophila* revealed that wdb shows a polar subcellular localization independent of Frizzled, but influences the localization of dsh and Fmi (Hannus et al., 2002). The Fyn/Yes protein kinases activate Rho A and thereby influence gastrulation movements and interact with Wnt/PCP signaling; however they are not part of this pathway (Jopling and den Hertog, 2005).

A prerequisite for moving cells is the modulation of cell–cell adhesion. In the context of gastrulation, cells have to loosen their contacts with neighboring cells to move, but still tissue stability has to be preserved. In zebrafish embryos adhesion of blastomeres is mediated by E-cadherin, which is expressed maternally and in all cells during the first cleavages. In gastrulating embryos E-cadherin is found in the anterior mesoderm and the developing epithelia (Babb et al., 2001). Mutations in the *E-cadherin* gene as found in the mutants *half-baked* (*hab*), *weg, lawine* (*law*), and *avalanche* (*ava*) display slower epiboly and in homozygous mutants a retraction of the blastoderm followed by dissociation of the cells (Kane et al., 1996; McFarland et al., 2005). The defective epiboly is caused by disturbed radial intercalation behavior and the cells' failure to stably integrate into the epiblast layer (Kane et al., 2005). In addition to these defects in epiboly, heterozygous embryos showed widened dorsal axes, which indicated an inhibition of convergent extension movements. The inhibition of gastrulation movements was stronger in the epiblast than in the hypoblast and consistently ectodermal structures were more affected than mesodermal tissues (McFarland et al., 2005). Morpholino knock-down studies of E-cadherin induce comparable phenotypes. It has been observed that epiboly is inhibited, while gastrulation is initiated, but convergent extension movements are blocked. The embryos displayed wider notochord and somites, the latter indicating that also mesoderm-derived structures are affected (Babb and Marrs, 2004). In a second study it was observed that at shield stage, E-cadherin is upregulated in hypoblast cells shortly after their internalization. Injection of E-cadherin MO did not inhibit internalization, but elongation, intercalation, and migration of the hypoblast cells. Simultaneously, epiboly of the epiblast cells was inhibited as has been reported for the *E-cadherin* mutants and in the earlier study using E-cadherin MO (Babb and Marrs, 2004; Kane et al., 2005; McFarland et al., 2005).

Interestingly, the strongest defects caused by E-cadherin MO injection have been observed in the development of the prechordal plate and its derivatives. In embryos injected with E-cadherin MO the prechordal plate forms, but later most cells become apoptotic, which results in severe defects in the forebrain and, although weaker, in the mid- and hindbrain (Babb and Marrs, 2004). Prechordal plate progenitor migration is controlled by Wnt11/slb (Heisenberg et al., 2000; Ulrich et al., 2003). In a study (Ulrich et al., 2005) it has been shown that *slb* mutants show a reduced cohesion of hypoblast cells similar to the situation in E-cadherin MO injected embryos. The authors further show that Wnt11/slb affects the subcellular localization of E-cadherin and that this activity involves Rab 5c. Knock-down of Rab 5c phenocopies the *slb* mutant and injection of *slb* embryos with a constitutively active Rab 5c mutant rescues this phenotype. The control of cell adhesion via Rab 5c-mediated endocytosis of E-cadherin (Ulrich et al., 2005) represents a new mechanism of Wnt11/slb function in the control of gastrulation movements.

3.2.2. Gastrulation movements in Xenopus

The first visible indicator for gastrulation in *Xenopus* is the formation of the blastopore at the dorsal vegetal side of the embryo. Apical constriction of the bottle cells results in the formation of a groove that is accompanied by a concentration of pigment and thus in the formation of a darker arc, the dorsal blastoporous lip. The blastoporous lip demarcates the area through which the prospective mesoderm involutes. During gastrulation the blastoporous lip extends ventrally until it forms a circle and involution of the ventral mesoderm commences. In parallel, the vegetal endoderm moves to the dorsal animal region of the embryo and causes a vegetal translocation of the involuting marginal zone. This movement called vegetal rotation initiates involution of the mesoderm, and it places the dorsal mesendoderm adjacent to the blastocoel roof (BCR). The mesendoderm cells make contact with the extracellular matrix that lines the inner surface of the BCR and start to migrate directionally toward the animal pole.

The involuted mesoderm that forms the mesodermal structures of the trunk, notochord, and somitic mesoderm, undergoes massive rearrangements. Initially, the involuting marginal zone thins by radial intercalation processes. By midgastrulation the originally multipolar cells acquire a bipolar shape with their long axes aligned along the mediolateral axis of the embryo. The subsequent mediolateral intercalation movements narrow the tissue along the mediolateral axis and lead to elongation along the anteroposterior body axis, a process referred to as convergent extension movements. For a detailed description of cell movements during *Xenopus* gastrulation (see Keller and Winklbauer, 1992; Keller et al., 2000, 2003; Keller, 2002).

As in zebrafish, noncanonical Wnt-signaling controls convergent extension movements in *Xenopus*. Xwnt11 (Tada and Smith, 2000) activates the β-catenin independent Wnt/PCP pathway that involves Disheveled (Tada and Smith, 2000), stbm (Park and Moon, 2002), prickle (Takeuchi et al., 2003), the small GTPases Rho A and Rac 1 (Habas et al., 2003), and the kinases ROK and JNK (Kim and Han, 2005). The role of the small GTPases Rho A and Rac 1 has been studied in greater detail in *Xenopus*. While Rac 1 activation is mediated by the DEP domain of Disheveled, Rho A becomes activated via Daam 1 bound to the PDZ domain of Disheveled (Habas et al., 2003). Rho A and Rac 1 play distinct roles in the regulation of cellular behavior during convergent extension movements. Rac 1 stimulates filopodia formation at the elongated sides of intercalating cells, while acquiring of the typical bipolar morphology essentially depends on Rho A. In addition, both small GTPases are involved in the formation of lamellipodia at the short sides of the cells. It has been proposed that Rho A and Rac 1 are activated independently downstream of Disheveled and Daam 1 (Tahinci and Symes, 2003). Similarly, Rho A and Rac 1 modulate cell polarity, directed migration, and protrusive activity in mesendoderm cells migrating on the inner BCR surface (Ren et al., 2006), although Wnt/PCP signaling so far has not been reported to influence migration behavior in these cells.

In contrast to zebrafish, an additional role of the canonical Wnt/β-catenin pathway in the regulation of convergent extension movements has been demonstrated in *Xenopus*. Canonical Wnt-signaling controls gene expression in Spemann's organizer including the direct canonical target genes *siamois, twin,* and *Xnr-3*. The first hint that Xnr-3 is required for convergent extension movements of the dorsal mesoderm has been obtained in overexpression studies of Xwnt5a. Xwnt5a activates the noncanonical Wnt/Ca^{2+} pathway (Kuehl et al., 2000b), and its downstream effectors PKC and CamK II inhibit the canonical pathway on the level of Disheveled and the transcription factor complex, respectively (Kuehl et al., 2001). The inhibition of canonical signaling by Xwnt5a is further enhanced by the observation that Wnt/Ca^{2+} signaling activates NF-AT via calcineurin, which stimulates the expression of target genes that antagonize canonical Wnt-signaling downstream of Disheveled and upstream of β-catenin (Saneyoshi et al., 2002). Xwnt5a overexpression inhibits CE movements and downregulates the canonical target gene *Xnr-3*. The inhibition of gastrulation movements can be rescued by coexpression of Xnr-3 (Kuehl et al., 2001). An Xnr-3 knockdown study confirmed the role of Xnr-3 in gastrulation. Xnr-3 is indeed necessary for CE movements and acts in cooperation with FGFR1 to promote morphogenetic movements (Yokota et al., 2003).

The influence of Xwnt5a on CE movements is not only due to the downregulation of Xnr-3 but also to ectopic activation of the small GTPase cdc 42 downstream of Xwnt5a and PKC (Choi and Han, 2002). It is discussed, whether Wnt/Ca^{2+} and Wnt/PCP signaling can be considered as distinct signaling pathways or whether they represent alternative branches of the same pathway. There is accumulating evidence that in *Xenopus* Xwnt5a and Xwnt11 can both activate effectors of the Wnt/Ca^{2+} and the Wnt/PCP pathway. During gastrulation, the activity of PKCδ is required for the subcellular translocation of Disheveled to the cell membrane and activation of JNK signaling (Kinoshita et al., 2003), which both are required for the cellular rearrangements during convergent extension. Furthermore, Xwnt5a has been shown to activate MKK 7/JNK signaling, which is essential for CE movements in *Xenopus* (Yamanaka et al., 2002), while Xwnt11 via Frizzled 7 and a trimeric G-protein stimulates cdc 42 activation (Penzo-Mendez et al., 2003).

The requirement of both, canonical and noncanonical signaling for convergent extension movements in *Xenopus* and the mutual interactions between these pathways raises the question, how the signaling activities are fine-tuned to allow coordinated cell movements. Molecules that can modulate both pathways are interesting candidates for future investigations. Like the related protein Diversin in zebrafish, *Xenopus* Inversin acts as a molecular switch between canonical and noncanonical Wnt-signaling pathways (Simons et al., 2005); however its role in the control of gastrulation movements has not been investigated.

The *Xenopus* Fmi homologue XFmi is expressed in the involuting cells during gastrulation, and its overexpression impairs gastrulation movements. Mouse Fmi inhibits canonical Wnt-signaling and blocks posteriorization of the neuroectoderm in *Xenopus*, but an interaction of XFmi with the Wnt/PCP pathway so far has not been reported (Morgan et al., 2003). Another member of the cadherin superfamily, paraxial protocadherin (PAPC) is also expressed in the involuting mesoderm during gastrulation and overexpression of a dominant–negative mutant impairs gastrulation movements in *Xenopus* (Kim et al., 1998) and zebrafish (Yamamoto et al., 1998). Later studies showed that PAPC acts upstream of Rho A in *Xenopus* and mouse (Hukriede et al., 2003). Closer characterization of PAPC in *Xenopus* revealed a role in tissue separation (Medina et al., 2004) and coordination of gastrulation movements (Unterseher et al., 2004) that both depend on the signaling activity of XPAPC. Furthermore, the XPAPC knock-down phenotypes are also observed in Frizzled 7 loss-of-function, which indicates a functional interaction between XPAPC and noncanonical Wnt-signaling (Medina et al., 2004; Unterseher et al., 2004). The extracellular domains of XPAPC and Frizzled 7 have been shown to interact (Medina et al., 2004), thus XPAPC could be part of the noncanonical Wnt pathway. However, unlike PCP-signaling downstream of Disheveled (Habas et al., 2003), XPAPC activates Rho A and simultaneously inhibits Rac 1, and activates JNK through Rho A (Medina et al., 2004; Unterseher et al., 2004). As XPAPC is asymmetrically distributed in the bipolar mesodermal cells that undergo mediolateral intercalation movements (Unterseher et al., 2004), XPAPC could act as a local modulator or coreceptor of the Wnt/PCP pathway (Fig. 4).

Like in zebrafish, classical cadherins also play a role in *Xenopus* gastrulation. The predominant cadherin that mediates adhesion between blastomeres in *Xenopus* is not E-cadherin but XB-cadherin, also known as C-cadherin. Modulation of C/XB-cadherin-mediated cell–cell adhesion leads to inhibition of gastrulation movements (Kuehl et al., 1996; Zhong et al., 1999). The cytoplasmic domain of classical cadherins binds to members of the catenin family including *β*-catenin and p120 catenin. Unlike *β*-catenin, p120 catenin does not activate the canonical Wnt pathway. However, overexpression of p120 catenin disturbs gastrulation movements in *Xenopus* (Geis et al., 1998; Paulson et al., 1999), probably by decreasing C-cadherin mediated cell–cell adhesion (Paulson et al., 1999). Knock-down of p120 catenin and the related ARVCF, another member of the p120ctn subfamily of armadillo repeat proteins, results in decreased protein levels of C-cadherin and a modulation of Rho A and Rac 1 activity. ARVCF and p120 catenin bind to Rho A, and act as Rho GDIs that bind inactive GDP-bound Rho A. In parallel both proteins bind to Vav2, a Guanine-nucleotide exchange factor of Rho-family GTPases, and stimulate Rac 1 activity (Fang et al., 2004). Similar to XPAPC, p120 catenins modulate Rho A and Rac 1 activity, although antagonistically, and thereby influence gastrulation movements. Besides its function in small GTPase signaling p120 catenin acts as cofactor for the transcriptional repressor XKaiso (Kim et al., 2002). Both overexpression and downregulation of XKaiso inhibit gastrulation movements by transcriptional regulation of effectors of the Wnt/PCP signaling pathway. XKaiso directly represses Xwnt11 expression and indirectly modulates the expression of Xstbm. The repression of Xwnt11 expression by XKaiso is reverted by p120 catenin. Binding of p120 catenin to XKaiso dissociates the repressor from the *Xwnt11* promotor and

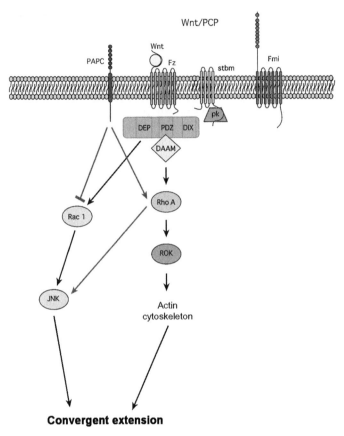

Convergent extension

Fig. 4. Interactions between XPAPC- and Wnt/PCP-pathways. For detailed description refer to the main text (Section 3.2.2). (See Color Insert.)

allows transcription (Kim et al., 2004). Thus, p120 catenin can modulate PCP signaling on the level of the small GTPases Rho A and Rac 1 and directly by regulation of *Xwnt11* gene expression in cooperation with XKaiso. Additional transcriptional downregulation of Xwnt11 has been observed by insulin-like growth factor signaling via the upregulation of Otx2, which indicates antagonistic roles of IGF and noncanonical Wnt-signaling (Carron et al., 2005).

3.3. Convergent extension of the neuroectoderm

The posterior neuroectoderm, hindbrain and spinal chord, converges and extends similar to the mesoderm. In zebrafish, epiblast cells change their direction by midgastrula until they move parallel with the hypoblast cells toward the dorsal midline (Glickman et al., 2003). In *Xenopus*, the neuroectoderm in contrast to the dorsal mesoderm forms medially biased protrusions and thus shows a monopolar cell shape oriented toward the dorsal midline. Mediolateral intercalation behavior and movement toward the dorsal midline of these neuroectoderm cells essentially depends on the presence of the notochord/notoplate that demarcates the dorsal midline (Elul and Keller, 2000; Keller, 2002).

Although convergence and extension movements of the neuroectoderm are studied less intensively as mesodermal morphogenetic movements, there is accumulating evidence that noncanonical Wnt-signaling controls cell polarity and migration in both tissues. The zebrafish Wnt5a/ppt mutant described above affects cell migration in hypoblast and epiblast (Kilian et al., 2003). In *Xenopus*, genes expressed in the neuroectoderm including Xror2, PTK7, and Xsyndecan 4 (Xsyn4) have been shown to modulate Wnt/PCP signaling (Hikasa et al., 2002; Lu et al., 2004b; Munoz et al., 2006). Overexpression of Xror2 inhibits convergent extension movements in both the mesoderm and the neuroectoderm and acts synergistically with Xwnt11 and Xfz7. Coinjection of dominant-negative cdc42 with Xror2 rescued elongation of animal cap explants, which further supports a role of Xror2 in noncanonical Wnt-signaling (Hikasa et al., 2002). The recently identified Xsyn4 is a cell surface transmembrane heparan sulfate proteoglycan, which is expressed in the dorsal mesoderm and the anterior neuroectoderm. It has been shown that both, gain and loss of function of Xsyn4 blocked convergent-extension movements in mesodermal and neural tissue. Xsyn4 interacts with Xfz7 and is sufficient to recruit Disheveled to the cell membrane, thus activates Wnt/PCP signaling. Interestingly, Xsyn4 activity seems to be regulated by fibronectin (Elul and Keller, 2000; Keller, 2002). These findings indicate that Xsyn4 could be the link between the well-studied Wnt/PCP pathway and the accumulating evidence that fibronectin and integrins are likewise required for proper convergent extension movements (Marsden and DeSimone, 2001; Davidson et al., 2002; Marsden and DeSimone, 2003) and that fibronectin and Wnt/PCP signaling mutually interact (Goto et al., 2005).

Although these data indicate that CE in the neuroectoderm is controlled by noncanonical Wnt pathways, one should take into account that CE of the neuroectoderm depends on the correct formation of and CE in the notochord (Elul and Keller, 2000). Xror2 and Xsyndecan 4 are expressed in the notochord and the neuroectoderm (Hikasa et al., 2002; Munoz et al., 2006) and thus most likely affect morphogenetic movements in both tissues. Evidence for an additional independent role of noncanonical Wnt-signaling in CE of the neuroectoderm comes from studies of genes that are either expressed only in neural tissue or differentially affect mesodermal and neural tissue. In contrast to Xror2 and Xsyn4, PTK7 is not expressed in mesodermal tissue, and its expression in the neuroectoderm starts at late gastrula stages (Lu et al., 2004b), while CE of the mesoderm has already begun at mid-gastrula. Knock-down studies using antisense morpholino oligonucleotides in *Xenopus* resulted in an inhibition of convergent extension movements in the neuroectoderm and delayed or impaired neural tube closure, indicating that PTK7 acts in the control of CE in the neuroectoderm. A relationship to Wnt/PCP signaling has been found in PTK7 knockout mice. These mice show inhibited neural tube closure and disorganized polarity of hair cells in the cochlea similar to phenotypes of PCP-defective mouse strains including stbm/Vangl2. PTK7 genetically interacts with stbm/Vangl2 in the mouse. While single heterozygote animals show only weak phenotypes, severe defects in neural tube closure are observed in the double heterozygous line (Lu et al., 2004b). Differential modulation of Wnt/PCP signaling in the mesoderm and neuroectoderm has been reported for the secreted protein Crescent. Crescent acts synergistically with Xwnt11 in the mesoderm in positive regulation of Wnt/PCP signaling, but antagonizes Xwnt11 in the neuroectoderm. Moreover, Crescent knock-down by antisense morpholino oligonucleotides resulted in defective convergent extension of the neuroectoderm and inhibition of head formation (Shibata et al., 2005).

3.4. Neural tube closure

Neural tube closure defects (NTD) are common birth defects in humans with an incidence of 1 out of every 1000 births. The process of neural fold formation, initiation of hinge points, and intermingling of the two folds in the dorsal midline needs a close concert of interacting molecules. To form the neural folds and the hinge points the cells of the neural plate undergo morphological changes. They show an apical constriction and a basal fixation to the surrounding tissue thus resulting in a wedge-shaped form. By fixation to the notochord ventrally and to the surface ectoderm laterally the neural plate folds in to form the neural tube. For the final closure the cells from the neural folds of both sides have to adhere to each other and to merge. The process of neural tube closure does not occur along the whole neural plate simultaneously but starts either at the area of the future midbrain and moves cranially and caudally (birds) or begins at different points along the anteroposterior body axis (mammals).

Publications demonstrate the influence of noncanonical Wnt-signaling especially on this last process of neural tube closure. The zebrafish mutant *trilobite* which harbors a knock-out of *Vangl2* (a s*trabismus* homologue) leads to a loss of polarity in the neural keel cells (Ciruna et al., 2006). During cell divisions the cells usually round up and loose polarity before they separate in two cells. Both cells then retain polarity before they reintegrate in the neuroepithelium. In the mutant the daughter cells do not recover the polarity. Additionally, the daughter cell that usually intercalates across the midline to the contralateral side of its birth is no longer able to cross the midline but accumulates at the ipsilateral side. The PCP pathway is directly able to influence the distribution of effector molecules in the cells since an EGFP-tagged prickle can be localized at the anterior membrane of the neural keel cells. In the mutant this asymmetric distribution cannot be observed (Ciruna et al., 2006). Interestingly, blocking of cell divisions is able to reduce the phenotype thus indicating that PCP signaling is downstream of cell division events. These effects seem to be the result of a close balance of PCP activation since activation of Vangl2 also leads to NTD (Shariatmadari et al., 2005). Overexpression of Wnt7a in mice, which can function in the canonical as well as the PCP pathway results in an upregulation of Vangl2. The mice show an impaired closure of the neural tube in contrast to the zebrafish mutants (Ciruna et al., 2006). In contrast to the *Vangl2* mutant, the neural plate is not elevated and the formation of hinge points is disturbed in severe phenotypes. In mildly affected animals the neural tube cannot stabilize the hinge points and collapses outside or inside. The authors can demonstrate that the PCP pathway is affected since β-catenin is not trans-located to the nucleus, which would be expected in case of canonical Wnt-signaling activation. Instead, a mislocation of N-cadherin leads to loss of adherens junctions. The Wnt7 overexpression increases apoptosis but does not influence proliferation, which is consistent with the idea of independent proliferation prior to PCP pathway activation. Thus, daughter cells of a cell division are not able to reestablish their cell polarity. Another regulatory level might be between Wnt and Disheveled. A morpholino approach to knock-down the expression of the protein phosphatase 2A (PP2A:B56ε) leads also to an impaired neural tube closure in *Xenopus* (Yang et al., 2003). The authors demonstrate that this molecule is placed between Wnt and Disheveled in the canonical pathway. The embryos treated with morpholino lose dorsal head structures and show a decrease in cytosolic β-catenin. The NTD might also be a result of a disturbed PCP pathway since in the *drosophila* mutant *widerborst* PP2A:B56 was found to be involved in noncanonical signaling (Hannus et al., 2002).

3.5. Influence of Wnt molecules on the migration behavior of the neural crest

In contrast to induction or differentiation of the neural crest where a role of Wnt molecules is well defined, the migration of these cells out of the neural plate border toward their target regions is mostly influenced by FGF, Notch, or BMP pathways (LaBonne and Bronner-Fraser, 1999; Raible and Ragland, 2005; Steventon et al., 2005). Nevertheless, Wnt molecules are also involved in the process of delamination and migration. It is yet unclear whether intrinsic or extrinsic cues define the final migration pathways of the cells. There seems to be evidence that mostly extrinsic cues can guide the cranial neural crest cells in the different branchial arches (Kulesa et al., 2005). The cells delaminate from the neural tube during or after the neural tube closure. For this process they have to undergo epithelial-to-mesenchymal transition and lose their adherens junctions with neighbor cells. The activation of RhoB leads to a reorganization of the cytoskeleton thus introducing a polarity to the migrating cell with formation of lamellipodia (Hall, 1998).

First ideas to connect the canonical Wnt pathway to the migratory behavior of neural crest cells occurred by expression studies of cadherins, which are thought to interact with the Wnt/β-catenin pathway via competition for β-catenin. Overexpression of extracellularly truncated Xcadherin-11, a neural crest-specific type II cadherin, leads to premature emigration and a reduced marker expression of twist that can be rescued by β-catenin (Borchers et al., 2001). Nevertheless, the migration in *Xenopus* seems not to be regulated directly by the canonical pathway since overexpression of both, wild type and a cytoplasmatically truncated Xcadherin-11 without the β-catenin-binding site inhibit migration. Similarly, the β-*catenin* knockout mouse does not show any defects in migration behavior of neural crest cells, but later differentiation into dorsal root ganglia is affected (Hari et al., 2002). Still, a competition for β-catenin between signaling and adhesion is discussed but with a temporal aspect. With onset of migration, β-catenin is recruited to the adhesion sites and diminishes from the nuclear localization (de Melker et al., 2004). A nuclear localization due to Wnt1 overexpression inhibits migration, indicating a competition between the two functions of β-catenin in signaling and adhesion. Nevertheless, the *Wnt1* knockout mouse does not show any defect in migration of neural crest cells but an influence on gap junction communication (Xu et al., 2001a,b). This indirect stimulation may occur via induction of neural crest-specific genes including *FoxD3*. This gene is upregulated by Wnt8 or β-catenin and inhibits neural crest formation by an autoinhibitory loop. This leads to downregulation of neural crest formation as well as migration (Pohl and Knoechel, 2001). Other neural crest-specific genes like *slug* that contains a β-catenin as well as a Lef-binding site in its promoter have been shown to promote the delamination of the cells from the neural tube via RhoB (Nieto et al., 1994; Liu and Jessell, 1998; Carl et al., 1999; LaBonne and Bronner-Fraser, 2000). The direct influence of Wnt on RhoB expression was described in real time PCR studies (Taneyhill and Bronner-Fraser, 2005), but the authors observe a positive as well as a negative interaction depending on culture conditions. Explants from chicken neural crest directly cultured in a Wnt3A-containing medium show a decrease while those precultured for 6 hour in serum-free media before adding the Wnt upregulate RhoB. The influence might be triggered by even upstream signaling molecules including BMPs. An inhibition of the BMP pathway results in a reduced expression of Wnt1, causing a downregulation of RhoB that results in a reduced migration (Burstyn-Cohen et al., 2004; reviewed in Kalcheim and Burstyn-Cohen, 2005). Since β-catenin is able to rescue this phenotype there is strong evidence for a direct cooperation between BMP and canonical

Wnt-signaling. Further evidence is the promotion of G1 to S phase transition by *β*-catenin and Lef. Cell division is a prerequisite for delamination from the neural tube (Burstyn-Cohen and Kalcheim, 2002; Burstyn-Cohen et al., 2004). Some discrepancy between real time PCR data (Taneyhill and Bronner-Fraser, 2005) and immunohistologic data (Burstyn-Cohen et al., 2004) in the chick embryo might be explained by technical or temporal differences in the experiments. It will be interesting to follow up the identification of other modulators of Wnt-signaling. One new candidate might be Frodo, a mediator of Wnt, which can bind to the PDZ domain of Disheveled as well as to TCF. It is involved in several morphogenetic processes in animal development and is expressed in the neural crest in mouse and *Xenopus* (Hunter et al., 2006).

The involvement of the PDZ domain can also point into the direction of noncanonical signaling, which seems more likely to be involved in the direct migratory behavior of neural crest cells. Using the different Disheveled mutants, de Calisto et al. (2005) distinguish between the induction of neural crest by the canonical and the migration of the neural crest by the PCP pathway (De Calisto et al., 2005). Wnt11 and its receptor Fz7 are shown to influence the formation of filopodia and lamellipodia in neural crest, a prerequisite for migratory behavior. Overexpression of dominant-negative Wnt11 leads to a reduction in cell movement that can be rescued by the Disheveled mutant DshΔDIX, an activator of the noncanonical Wnt pathway. Until today it is not possible to distinguish precisely between the Wnt/PCP and the Wnt/Ca^{2+} pathway with these mutants, so this question remains open. But some evidences lead to the idea that the Wnt/PCP pathway might be more important.

4. Neuronal migration

4.1. Cerebellar cortex formation by migrating neurons

The development of the brain strongly depends on the correct migration of neurons, which are born in the germinal cell layer lining the ventricular system. They migrate over long distances before reaching their final destination. In the primate brain migration routes up to 7 mm in length have been reported (Rakic, 2003). In the developing brain the distances grow over time with the formation of new cell layers. In the beginning only a preplate is formed by postmitotic neurons, which with time is segregated into a superficial layer, the marginal zone, and the subplate by the invasion of the later born neurons. All three layers build the so-called cortical plate. Further waves of migrating neurons bypass the earlier generated neurons in the cortical plate forming additional cell layers by inside to out sequence (reviewed in Hatten, 2002; Honda et al., 2003; Kriegstein and Noctor, 2004). Different migration routes are described for different neuron populations depending on their place of birth. Cortical pyramidal neurons deriving from the dorsal telencephalon migrate radially to their laminar position. The interneurons, which are generated in the ventral telencephalon, migrate tangentially in the cortex and later take the radial path to enter the cortical plate. Neurons migrate more rapidly on the tangential (ca 50 μm/hour) than on the radial routes (ca 10 μm/hour) (Kriegstein and Noctor, 2004). The kind of the substrate on which the cells migrate is still a matter of discussion. Radial glia fibers may guide the interneurons on their radial path, but the substrate used during the tangential migration is obscure. Two types of migration are distinguished when neurons migrate radially in the cortical plate: somal translocation and locomotion (Nadarajah et al., 2001). In case of somal translocation the cells form long pia-directed processes of 60–95 μm in length that get attached to the pial surface of the marginal zone. The cell moves by shortening the leading processes, which remain fixed to the pia surface, thereby drawing the cell body behind. Locomotion describes the glia-guided movement. The leading processes of the cells are shorter but more or less constant in their length. The migration behavior is discontinuous. Cells switch between periods of rapid and slow forward movements. Hatten (2002) states that the mode of neurons guided by glia fibers strongly differs from that of axial growth cones because the extension and retraction of the leading processes are dominated by the reorganization of the microtubule system and less influenced by the actin cytoskeleton.

Although there are only few reports about a putative role of Wnt-signaling in the migration of neurons they should be mentioned since a growing number of neuronal migration disorders have been described. Their molecular nature is not well understood yet (Gleeson and Walsh, 2000; Ross and Walsh, 2001).

Machon et al. (2003) inactivated β-catenin in the cerebral cortex and hippocampus by crossing the mouse *D6-Cre* line to the *β-catenin* flox/flox line (Machon et al., 2003). This conditional *β-catenin* knockout resulted in a severe damage of the brain showing reduction in size and disorganized morphology of hippocampus and cortex. Since cell–cell adhesion but also Wnt-dependent gene transcription might be affected in these mice the malformations can be explained by defects in proliferation, differentiation, migration, and loss of tissue integrity. However, the phenotype resembles that of *Wnt3a* and *Lef-1* knockout mice (Galceran et al., 2000; Lee and Frasch, 2000) so that there is strong evidence, that inhibition of Wnt-signaling strongly contributes to the *D6-Cre/β-catenin*$^{flox/flox}$ phenotype. In their detailed immunohistologic analysis of this mutant Machon et al. (2003) detect ectopically formed ventricular tissue that consists of undifferentiated and differentiated neurons, which they explain with impaired migration of maturing neurons. They also observe that early born cortical neurons exhibit a defect in radial migration because they do not migrate from the germinal ventricular zone into the preplate. Migration of late-born cortical neurons might also be affected in these mice because the number of these neurons is reduced and they cluster abnormally in the superficial layers. Since loss of β-catenin inhibits cadherin-mediated cell–cell adhesion but also Wnt-signaling, this study, at least, does not clearly answer the role of Wnt-signaling for cortical cell migration. Further support to a role of Wnt-signaling in neuronal migration is given by the analysis of a Wnt1 enhancer/lacZ transgenic mice (Nichols and Bruce, 2006). Although functional data are missing in this study, the β-gal staining in migrating cells of the murine hindbrain points to a requirement of Wnt-signaling in neuronal cell movement which remains to be clarified by additional investigations.

4.2. Migration of oligodendrocytes in the spinal chord

Oligodendrocytes produce the myelin-related proteins, which are required for myelinization of axons to improve their conductivity. During development oligodendrocyte progenitor cells proliferate in the floor plate and the ventral region of the neural tube and from there they migrate throughout the developing spinal chord. With the identification of the oligodendrocyte-specific transcription factor Oligo-1–3 these cells can be traced on their migratory routes (Kakinuma et al., 2004a). There is some evidence that Wnt-signaling seems to promote the motility of the oligodendrocyte progenitors. Kakinuma et al. (2004a) identify an extracellular sulfatase (RsulfIFP1), which is expressed in the floor plate and required for migration of the immature oligodendrocytes. In cell culture studies they can show that expression of this sulfatase leads to tyrosine phosphorylation of β-catenin and strong activation of TOPFLASH, the artificial Wnt-responsive promoter reporter construct. A further link to canonical Wnt-signaling is observed when premature oligodendrocytes are cultured on tenascin C (Kakinuma et al., 2004b). Their motility decreases on this substrate, which is accompanied by the translocation of β-catenin from the nuclei to the cytoplasm. Accordingly, expression of the Wnt-signaling antagonist Dapper in oligodendrocytes inhibits their migration (Kakinuma et al., 2004b). A closer look on the role of Wnt-signaling to understand its effect on oligodendrocyte migration will be promising.

5. Cell migration in eye development

Eye development is a process that involves many morphogenetic steps from eye field formation and separation to invagination of the eye vesicle, lens induction, and retina lamination. And finally the newly developed axons of the retinal ganglion cell (RGC) layer have to be connected to the visual center of the brain. The eye field is formed as part of the anterior neural plate (ANP) during the gastrulation process (Fig. 5). It is characterized by the expression of a subset of transcription factors, which specify the eye field against telencephalon rostrally and the diencephalon caudally (Fig. 5A).

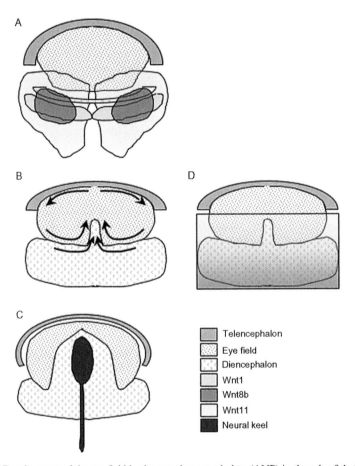

Fig. 5. Development of the eye field in the anterior neural plate (ANP) in the zebrafish. During the gastrulation the ANP forms at the rostral part of the embryo. (A) Behind the future telencephalon the eye field has a lozenge-shape. The future diencephalon is caudal to the eye field. The expression pattern of Wnt1 does not overlap with the eye field as does Wnt8b. But the latter is quite close to the caudal edge of the eye field. Wnt11 overlaps with the caudal part of the future eye field. (B) The diencephalon cells move toward the midline during gastrulation and push in the caudal edge of the eye field. Also, cells in the eye field migrate to the middle, but the pushing forces from the diencephalon let the anterior cells move laterally and the eye field starts to separate. (C) The separation of the eye field moves on as the neural keel moves more and more rostrally. (D) An antagonistic gradient of canonical and noncanonical Wnt signaling triggers the formation of the eye field in comparison to telencephalon and diencephalon. (See Color Insert.)

During gastrulation and neurulation the whole neural plate undergoes convergent extension movements thus condensing the cells in the dorsal midline. By that the diencephalon forms a protrusion and causes an indentation in the caudal border of the eye field region. The cells of the eye region are pressed against the telencephalon and escape laterally. That movement separates the single eye field in two lateral eye fields. The movement of the eye field cells themselves toward the midline strengthens this process (Fig. 5B). The condensation of cells in the midline causes the development of the neural keel, which further promotes the separation of the eye fields (Fig. 5C; for reviews see Chuang and Raymond, 2002; Wilson and Houart, 2004).

5.1. Eye field formation

Wnt-signaling plays not only a role in eye field formation during gastrulation but also in neurulation separating the eye progenitors from the residual brain compartments.

Fate-map analysis of the D1.1.1 blastomere derivatives, which contribute about 50% to the retina in *Xenopus*, reveal that Disheveled is required for efficient targeting of these cells to the retina (Lee et al., 2006). The migration of these progenitor cells into the eye field is also controlled by ephrinB1. Ephrin B1 functionally interacts with the Wnt/PCP pathway by binding to the DEP domain of Disheveled. This causes the recruitment of Disheveled to the cell membrane and results in activation of PKCδ, RhoA, and JNK thus reflecting the activation of PCP signaling. Morpholinos against components of the ephrinB1/PCP pathway result in a condensation of D1.1.1 derivatives at the dorsal midline and prevent the cells from moving into the eye field. Wnt11 is able to rescue the ephrinB1 morpholino phenotype, but dominant-negative Wnt11 expression does not mimic Ephrin B1 depletion. It seems that there are two molecules using the same pathway, but it remains elusive how the different responses between Wnt11 and ephrinB1 are mediated. Since Wnt/PCP signaling is the main pathway that controls CE movements during gastrulation (see page 168) one might suspect that eye field positioning also depends on CE movements of the mesoderm and the overlying neural tissue.

Later in neural plate development, Wnts also trigger the movements of cells inside the ANP as has been shown by time-lapse movies in zebrafish (Cavodeassi et al., 2005). Detailed analysis shows that cells expressing Wnt11 show a stronger adhesion and tend to form clusters when transplanted into a host region lacking Wnt11. Cells overexpressing Wnt11 show a stronger directionality indicating a requirement of Wnt11 for the regulation of morphogenetic movements. Interestingly, distinct expression patterns are detected for different Wnt molecules in the anterior neural plate. While Wnt11 expression overlaps with the caudal area of the eye field, the anterior boundaries of Wnt1 and Wnt8b expression areas lie behind the eye field.

Wnt8b expression ends quite close to the eye region while Wnt1 is found more caudally (Fig. 5A). An overexpression of Wnt11 results in an enlarged eye field. In contrast, the Wnt/β-catenin pathway (Wnt8b) promotes the expansion of the diencephalon on expense of the eye field. The authors identify Fz5 as possible receptor for Wnt11 and Fz8a for Wnt8b. Given that Wnt11 is able to suppress canonical signaling, it can be assumed that an opposing gradient of canonical versus noncanonical pathways shapes the eye field.

After the formation of the single median eye field the separation in the two lateral eye fields is the next critical event. It is well described that shh secreted from the prechordal plate is the factor separating the eye field. The noncanonical Wnt antagonist Crescent can induce cyclopia by inhibiting the formation of the dorsal midline (Pera and De Robertis, 2000). By this, no movement of the axial mesoderm is possible. This effect is specific for Crescent, as other Wnt antagonists do not show the same phenotype. Crescent overexpression inhibits the formation of the prechordal plate. Two different mechanisms may explain this phenotype. (1) Absence of the prechordal plate and thereby loss of the shh source. (2) The inhibition of CE movements and neural tube closure (Shibata et al., 2005), which might prevent the formation of the diencephalic protrusion that separates the eye field (Fig. 5B).

5.2. Lens development

During lens development then also some cells show migratory behavior. The cells below the equator region move along the posterior capsule toward the sutures. An enrichment of Fz molecules at the more basal ends of these cells may indicate an influence of Wnt-signaling (Chen et al., 2004). This induction is triggered by FGF signaling that promotes the formation of filopodia and leads to an upregulation of Fz.

5.3. Retinotectal projection

Recently, a novel function of Wnt ligands has been discovered in axon guidance. Like in eye field formation, an interaction between the canonical and the noncanonical pathway seems to regulate the correct pathfinding of RGC axons to the tectum or the superior colliculus. On the one hand, the Wnt/β-catenin pathway inhibits the axonal outgrowth of RGC axons and changes the distribution of RGC cell bodies from the outer nuclear layer more to the inner nuclear layer (Ouchi et al., 2005). Additionally, Wnt can also influence the dorsal to ventral gradient of RGC axon growth. In cell culture studies, it was found that high levels of Wnt3 block the axon growth completely, but low concentrations promote RGC axon growth dorsally and inhibit it ventrally (Schmitt et al., 2006). The low-concentration answer is mediated via Fz, as it is already known for Wnt-signaling. The high-concentration reaction is transformed via Ryk, which is a coreceptor of canonical Wnt-signaling (Lu et al., 2004a). The same gradient exists not only in the eye, but also in the tectum, where Wnt3 is expressed in a medial to lateral decreasing gradient in chicken and mouse (Schmitt et al., 2006). Ryk and EphB form a dorsoventral gradient in the eye, while Wnt3 and Ephrin B1 form a mediolateral gradient in the tectum (McLaughlin et al., 2003; Schmitt et al., 2006). Wnt3 and Ryk seem to act as repellents, while EphB and Ephrin B1 are attractants in axon guidance. Thus, the gradient of Wnt/Ryk signaling counterbalances the EphB/Ephrin B1-mediated attraction.

Antagonists of noncanonical Wnt pathways have been reported to function as repellents in axon guidance. The secreted Fz related protein 1 (SFRP1) interacts with Fz2 and probably activates noncanonical signaling via G(alpha) protein. SFRP1 repels axon outgrowth on a laminin surface in chicken and *Xenopus* (Rodriguez et al., 2005), but promotes it on a fibronectin surface. Thus, the cellular response can also be modulated by the extracellular matrix.

6. Heart development

The development of the multichambered vertebrate heart is one of the most fascinating processes in organogenesis. A sequence of morphogenetic movements consisting of migration and fusion of cardiogenic tissue, heart tube formation and looping, septa growth, and myocardialization leads to the formation of atrial and ventricular chambers and the outflow tract (for review see Mohun et al., 2000, 2003; Stainier, 2001; Harvey, 2002; Zaffran and Frasch, 2002). In addition, different cell populations contribute to heart formation like cardiogenic (primary heart field) and pharyngeal (secondary heart field) mesoderm as well as neural crest (reviewed in Schoenwolf and Garcia-Martinez, 1995; Harvey, 2002; Mohun et al., 2003; Stoller and Epstein, 2005). Therefore, it is not surprising that congenital heart disease is the most frequent abnormality at birth (Edmonds and James, 1993).

During heart formation the different Wnt-signaling pathways play important roles in determining the cell fate and in regulating the cell movements. Importantly, inhibition of canonical Wnt-signaling (Marvin et al., 2001; Schneider and Mercola, 2001) and activation of noncanonical Wnt-signaling are essential to induce beating myocard tissue in ventral tissue explants from *Xenopus* gastrulae or murine embryonic carcinoma stem cell line P19 (Yamaguchi et al., 1999; Pandur et al., 2002). The mesenchymal heart tissue, on the other hand, requires canonical Wnt-signaling because a balanced expression of Wnt9a and the antagonist Frzb has been reported to control proliferation in the atrioventricular endocardial cushion (Person et al., 2005). A role of Wnt-signaling in morphogenetic movements has been addressed to different migratory phases: (1) fusion of the two heart fields at the ventral midline, (2) heart tube formation, (3) the formation of the outflow tract by myocardialization of the endocardial cushion, and (4) the invasion of neural crest cells into heart tissue.

6.1. Midline convergence of myocardial precursor cells

Defects in the fusion of myocardial precursors at the ventral midline result in partially bilaterally duplicated heart structures, which are termed cardia bifida. This phenotype indicates that cells fail to migrate but are still able to differentiate properly. Fusion of the initially bilateral induced heart anlagen by migration of cells toward the midline is termed midline convergence and observed for many unpaired organs. Matsui et al. (2005) demonstrate that noncanonical Wnt-signaling is required for the midline convergence of the bilateral heart primordia. In zebrafish, cardia bifida is observed upon dominant-negative expression of the Disheveled mutant lacking the DEP domain, which is required in Wnt/PCP signaling (DvlΔDEP, also termed dshΔDEP in other organisms) (Fig. 3). Importantly, the heart defect becomes obvious at very low doses of DvlΔDEP expression that do not affect gastrulation. This heart phenotype is rescued by constitutive active RhoA, but not by constitutive active Rac1 or cdc42. Conversely, expression of dominant-negative RhoA alone also results in cardia bifida, which is not observed when dominant-negative forms of Rac1 or cdc42 are injected into the zebrafish embryo. Since no influence on fusion is observed for CamKII, JNK, and PKC, Matsui et al. (2005) postulate that a noncanonical Wnt-signaling pathway acting via Disheveled and RhoA is regulating migration of myocardial precursor cells toward the ventral midline. In search for a Wnt ligand they identified Wnt11, Wnt4a, and Wnt11-related (Wnt11-R), which partially overlap in their expression domains in the correct time window. Importantly, loss-of-function analyses using antisense morpholino oligonucleotides reveal that all three Wnt molecules are required in this process. The cellular behavior during midline convergence is not well characterized. The migration behavior of the myocard precursor cells is recorded by in vivo time-lapse imaging and migration defects in DvlΔDEP embryos are obvious (Matsui et al., 2005), but it remains to be clarified, which cellular process is regulated by the Wnt/Dlv/RhoA pathway, for example, stress fiber formation, cell adhesion or cell polarity.

6.2. Heart tube formation

Wnt11-R has also been identified in *Xenopus* and assigned to the noncanonical group of Wnt ligands because it does not induce a secondary axis (Garriock et al., 2005). Its expression in the heart anlage starts just before heart tube formation. Loss-of-function studies by antisense morpholino oligonucleotide injections led to partially bilaterally duplicated hearts.

In contrast to the previously described phenotype in zebrafish (Matsui et al., 2005), cardiac primordia migrate to the ventral midline in the Wnt11-R depleted *Xenopus* embryos but fails to fuse and to form a proper heart tube. Since a broadening of the intercellular spaces is seen in the myocardium wall of the morphants, the authors assume that Wnt11-R controls cell adhesion although they could not detect a change in N-cadherin expression (Garriock et al., 2005). The Wnt11-R knock-down phenotype is mimicked by the JNK inhibitor (SP600125) but not by the CamKII inhibitor KN-93 indicating to a noncanonical Wnt/JNK pathway.

It seems amazing that the heart tube formation phenotype is not observed in the zebrafish *silberblick* (*Wnt11*) mutant or in the zebrafish Wnt11-R knock-down embryo (Matsui et al., 2005), and migration of the precardiac cells to the ventral midline is not affected in Wnt11-R depleted *Xenopus* embryos. Furthermore, in zebrafish an influence of JNK in midline convergence is excluded. In *Xenopus*, however, JNK activation is required for heart tube formation. Modifications in the migratory behavior of myocardial precursor cells between the organisms or different contributions of other noncanonincal Wnt-signaling pathways or branches in frog and fish might explain the variations in the morphant phenotypes.

6.3. Formation of the outflow tract

The naturally occurring mouse mutant *Loop-tail* (*Lp*) develops severe heart abnormalities, which are visible in double outlet right ventricle with parallel atrial trunks and ventricular septal defects (Phillips et al., 2005; Henderson et al., 2006). Genetic characterization of the mutant revealed a defect in the *strabismus/Van Gogh* gene (also termed *Vangl2, Ltap, Lpp1*), which is a component of the PCP pathway (Fig. 3). In *Loop-tail* mice the *Vangl2* gene mutation results in a serine to asparagine substitution (Kibar et al., 2001; Murdoch et al., 2001). Vangl2 is expressed in the outflow myocardium and accumulates in those cadiomyocytes that invade the proximal outlet septum (Phillips et al., 2005). Myocardialization of the initially mesenchymal outflow tract cushion is achieved by migration and in growth of cardiomyocytes into the cushion mesenchyme (for review, see van den Hoff et al., 2004). During this process cardiomyocytes form lamellipodia and filopodia extending into the mesenchymal cushion tissue. Phillips et al. (2005) demonstrate that cardiomyocytes do not form such cellular protrusions in Loop-tail mice. As a consequence a large part of the proximal outlet septum remains mesenchymal and appears broader on expense of the myocard. When they analyzed the expression domains of Wnt-signaling components they found Wnt11, Dvl2, ROCK1 and 2 coexpressed with Vangl2 in the myocardium, while Wnt5a transcripts were detected in the endocardial cushions of the outflow tact. The expression patterns of these genes were not altered in *Lp* mice. RhoA, however, which is normally found in the mesenchymal cushion cells adjacent to the cardiomyocytes is lost in the *Lp* mutants indicating a requirement of the PCP/RhoA pathway for myocardialization. JNK seems not to be involved, as it is not found expressed in the outflow tract (Phillips et al., 2005). Loss of RhoA protein in the endocardial cushion cells is difficult to explain in context of Wnt/PCP signaling because alterations in RhoA activity are expected to occur in *Lp*-myocardial cells, which are unable to form cellular protrusions. Further experiments are required to clarify the specific Wnt-signaling transduction pathway. Apart from that, cardiomyocyte motility has to be characterized in more detail. The mentioned "de-epithelialization" of the invading myocardial cells addresses the question whether Wnt-signaling, which has been reported for cardiomyocytes in culture (Toyofuku et al., 2000), modifies cadherin function.

6.4. Migration of neural crest cells into heart tissue

More than 20 years ago Kirby et al. (1983) demonstrated by fate mapping and transplantations of neural crest cells that a subpopulation of these cells contributes to normal aorticopulmonary septation. Since these first transplantation experiments different transgenic mice have been created to identify cardiac neural crest (Waldo et al., 1999; Jiang et al., 2000; Li et al., 2000). In one of the transgenic lines a 5.5-kb fragment of the *Wnt1*-promoter is successfully used to direct *Cre* expression in neural crest cells, which after crossing with the *R26R* allele allows to fate-map cardiac neural crest cells by staining heart tissue for β-galactosidase (Jiang et al., 2000). This indicates a function of Wnt1 for neural crest development including the cardiac subpopulation. The fate-map studies reveal that two streams of neural crest cells invade the outflow tract which later fuse to form the septum separating the mature aorta and the pulmonary arteries (reviewed in Stoller and Epstein, 2005). Canonical Wnt-signaling is crucial for neural crest induction (for review see Raible and Ragland, 2005), while noncanonical Wnt-signaling is important for their delamination and migration (De Calisto et al., 2005). There is some evidence that Wnt-signaling may also play a role for invasion of the neural crest cells into the developing heart. *Disheveled* (*Dvl2*) knockout mice show pleiotropic defects including malformation of the outflow tract (Hamblet et al., 2002). The phenotype is more severe and affects more tissues than the *Vangl2* mutation in the *Loop-tail* mouse (Phillips et al., 2005). Furthermore, in *Dvl1* knockout mice neural crest cells are strongly reduced in the heart tissue (Hamblet et al., 2002). Since Disheveled is a component of all Wnt-signaling pathways, further studies are necessary to answer whether block of NC induction or migration and delamination causes the outflow tract phenotype in *Dvl1* knockout mice.

7. Conclusions

Our understanding of the specific mechanisms of Wnt-signaling in cell migration has evolved rapidly over the last few years. Important contributions are still coming from the analyses of gastrulation processes in *Xenopus* and zebrafish. The advantages of these systems rely on the basic knowledge of different cellular behaviors, the specific migration assays developed in *Xenopus* and the numerous zebrafish mutants. Although the involuting mesoderm in both embryos undergoes convergent extension movements, which are controlled by noncanonical Wnt-signaling, significant differences are observed in some details, which reflect dissimilarities in tissue morphology and cell-migration modes. For example, in *Xenopus* Wnt11 and Wnt5a activate different branches of noncanonical pathways while the corresponding zebrafish mutants point to redundancy in function. Furthermore, in *Xenopus* canonical Wnt-signaling and the canonical target gene Xnr-3 are crucial for convergent extension movement. However, Xnr-3 seems to be unique to *Xenopus* as so far no Xnr-3 homologue has been identified in any other vertebrate. These species-specific characteristics might reflect morphological differences of the migrating cell sheets; the *Xenopus* involuting mesoderm appears more compact than the hypoblast in zebrafish. Nevertheless, both organisms have in common that noncanonical Wnt-signaling in combination with cadherins and protocadherins (XPAPC, Fmi) is required to establish cell polarity, to control protrusive activity, and to regulate cell–cell communication and adhesion of migrating cells. These might be general functions of noncanonical Wnt-signaling in cell movements because similar cellular behaviors are controlled by this pathway during neural crest cell migration, eye field formation or myocardialization of the heart cushion. It is assumed that with upcoming knowledge about the specific behavior of migrating cells in organogenesis more parallels in molecular regulation of cell motility to those observed in gastrulation will be found. A similar general role for canonical Wnt-signaling in cell migration is less obvious. Most likely, this pathway is required to guide cells via a Wnt gradient. It is amazing that this mechanism is used to repel *AVE* cells from settling at posterior position and also to repel axons in the eye retina from ventral positions and in the tectum from caudal positions in retinotectal projection. We also learned from studying Wnt-signaling in cell migration that coreceptors seem to specify the response to different Wnt ligands as it has been shown for LRP5/6 in the Wnt/β-catenin pathway. It will be extremely important to investigate the role of ephrinB, ephB, XPAPC, and flamingo as well as Ryk in this context. In general, the now accessible genetic and molecular tools and the steadily improving digital imaging techniques will bring a great breakthrough in the understanding of cell movements. A fundamental knowledge, especially about Wnt-signaling in cell migration, will have important bearing on various therapy and repair strategies including wound healing, tumor invasion, immune response, tissue regeneration, and stem cell therapy.

Acknowledgments

This work was supported by grants from the German Research Foundation (DFG SCHA 965/2 and DFG We 1208/9) and a scientific award from the state government Baden-Wuerttemberg.

References

Ai, X., Do, A.T., Lozynska, O., Kusche-Gullberg, M., Lindahl, U., Emerson, C.P., Jr. 2003. QSulf1 remodels the 6-O sulfation states of cell surface heparan sulfate proteoglycans to promote Wnt signaling. J. Cell Biol. 162, 341–351.

Anton, E.S., Cameron, R.S., Rakic, P. 1996. Role of neuron-glial junctional domain proteins in the maintenance and termination of neuronal migration across the embryonic cerebral wall. J. Neurosci. 16, 2283–2293.

Anton, E.S., Kreidberg, J.A., Rakic, P. 1999. Distinct functions of alpha3 and alpha(v) integrin receptors in neuronal migration and laminar organization of the cerebral cortex. Neuron 22, 277–289.

Axelrod, J.D., Miller, J.R., Shulman, J.M., Moon, R.T., Perrimon, N. 1998. Differential recruitment of Dishevelled provides signaling specificity in the planar cell polarity and Wingless signaling pathways. Genes Dev. 12, 2610–2622.

Babb, S.G., Marrs, J.A. 2004. E-cadherin regulates cell movements and tissue formation in early zebrafish embryos. Dev. Dyn. 230, 263–277.

Babb, S.G., Barnett, J., Doedens, A.L., Cobb, N., Liu, Q., Sorkin, B.C., Yelick, P.C., Raymond, P.A., Marrs, J.A. 2001. Zebrafish E-cadherin: Expression during early embryogenesis and regulation during brain development. Dev. Dyn. 221, 231–237.

Baeg, G.H., Lin, X., Khare, N., Baumgartner, S., Perrimon, N. 2001. Heparan sulfate proteoglycans are critical for the organization of the extracellular distribution of Wingless. Development 128, 87–94.

Bielinska, M., Narita, N., Wilson, D.B. 1999. Distinct roles for visceral endoderm during embryonic mouse development. Int. J. Dev. Biol. 43, 183–205.

Boettger, T., Knoetgen, H., Wittler, L., Kessel, M. 2001. The avian organizer. Int. J. Dev. Biol. 45, 281–287.

Borchers, A., David, R., Wedlich, D. 2001. Xenopus cadherin-11 restrains cranial neural crest migration and influences neural crest specification. Development 128, 3049–3060.

Boutros, M., Paricio, N., Strutt, D.I., Mlodzik, M. 1998. Dishevelled activates JNK and discriminates between JNK pathways in planar polarity and wingless signaling. Cell 94, 109–118.

Bouwmeester, T. 2001. The Spemann-Mangold organizer: The control of fate specification and morphogenetic rearrangements during gastrulation in *Xenopus*. Int. J. Dev. Biol. 45, 251–258.

Bradley, R.S., Brown, A.M. 1990. The proto-oncogene int-1 encodes a secreted protein associated with the extracellular matrix. EMBO J. 9, 1569–1575.

Burstyn-Cohen, T., Kalcheim, C. 2002. Association between the cell cycle and neural crest delamination through specific regulation of G1/S transition. Dev. Cell 3, 383–395.

Burstyn-Cohen, T., Stanleigh, J., Sela-Donenfeld, D., Kalcheim, C. 2004. Canonical Wnt activity regulates trunk neural crest delamination linking BMP/noggin signaling with G1/S transition. Development 131, 5327–5339.

Cameron, R.S., Ruffin, J.W., Cho, N.K., Cameron, P.L., Rakic, P. 1997. Developmental expression, pattern of distribution, and effect on cell aggregation implicate a neuron-glial junctional domain protein in neuronal migration. J. Comp. Neurol. 387, 467–488.

Carl, T.F., Dufton, C., Hanken, J., Klymkowsky, M.W. 1999. Inhibition of neural crest migration in *Xenopus* using antisense slug RNA. Dev. Biol. 213, 101–115.

Carreira-Barbosa, F., Concha, M.L., Takeuchi, M., Ueno, N., Wilson, S.W., Tada, M. 2003. Prickle 1 regulates cell movements during gastrulation and neuronal migration in zebrafish. Development 130, 4037–4046.

Carron, C., Bourdelas, A., Li, H.Y., Boucaut, J.C., Shi, D.L. 2005. Antagonistic interaction between IGF and Wnt/JNK signaling in convergent extension in *Xenopus* embryo. Mech. Dev. 122, 1234–1247.

Cavodeassi, F., Carreira-Barbosa, F., Young, R.M., Concha, M.L., Allende, M.L., Houart, C., Tada, M., Wilson, S.W. 2005. Early stages of zebrafish eye formation require the coordinated activity of Wnt11, Fz5, and the Wnt/beta-catenin pathway. Neuron 47, 43–56.

Chen, Y., Stump, R.J., Lovicu, F.J., McAvoy, J.W. 2004. Expression of Frizzleds and secreted frizzled-related proteins (Sfrps) during mammalian lens development. Int. J. Dev. Biol. 48, 867–877.

Choi, S.C., Han, J.K. 2002. *Xenopus* Cdc42 regulates convergent extension movements during gastrulation through Wnt/Ca^{2+} signaling pathway. Dev. Biol. 244, 342–357.

Chuang, J.C., Raymond, P.A. 2002. Embryonic origin of the eyes in teleost fish. Bioessays 24, 519–529.

Ciruna, B., Jenny, A., Lee, D., Mlodzik, M., Schier, A.F. 2006. Planar cell polarity signalling couples cell division and morphogenesis during neurulation. Nature 439, 220–224.

Darken, R.S., Scola, A.M., Rakeman, A.S., Das, G., Mlodzik, M., Wilson, P.A. 2002. The planar polarity gene strabismus regulates convergent extension movements in *Xenopus*. EMBO J. 21, 976–985.

Davidson, L.A., Hoffstrom, B.G., Keller, R., DeSimone, D.W. 2002. Mesendoderm extension and mantle closure in *Xenopus laevis* gastrulation: Combined roles for integrin alpha(5)beta (1), fibronectin, and tissue geometry. Dev. Biol. 242, 109–129.

De Calisto, J., Araya, C., Marchant, L., Riaz, C.F., Mayor, R. 2005. Essential role of non-canonical Wnt signalling in neural crest migration. Development 132, 2587–2597.

De Cat, B., Muyldermans, S.Y., Coomans, C., Degeest, G., Vanderschueren, B., Creemers, J., Biemar, F., Peers, B., David, G. 2003. Processing by proprotein convertases is required for glypican-3 modulation of cell survival, Wnt signaling, and gastrulation movements. J. Cell Biol. 163, 625–635.

de Melker, A.A., Desban, N., Duband, J.L. 2004. Cellular localization and signaling activity of beta-catenin in migrating neural crest cells. Dev. Dyn. 230, 708–726.

Dhoot, G.K., Gustafsson, M.K., Ai, X., Sun, W., Standiford, D.M., Emerson, C.P., Jr. 2001. Regulation of Wnt signaling and embryo patterning by an extracellular sulfatase. Science 293, 1663–1666.

Ding, J., Yang, L., Yan, Y.T., Chen, A., Desai, N., Wynshaw-Boris, A., Shen, M.M. 1998. Cripto is required for correct orientation of the anterior-posterior axis in the mouse embryo. Nature 395, 702–707.

Edmonds, L.D., James, L.M. 1993. Temporal trends in the birth prevalence of selected congenital malformations in the Birth Defects Monitoring Program/Commission on Professional and Hospital Activities, 1979–1989. Teratology 48, 647–649.

Elul, T., Keller, R. 2000. Monopolar protrusive activity: A new morphogenic cell behavior in the neural plate dependent on vertical interactions with the mesoderm in *Xenopus*. Dev. Biol. 224, 3–19.

Fang, X., Ji, H., Kim, S.W., Park, J.I., Vaught, T.G., Anastasiadis, P.Z., Ciesiolka, M., McCrea, P.D. 2004. Vertebrate development requires ARVCF and p120 catenins and their interplay with RhoA and Rac. J. Cell Biol. 165, 87–98.

Fanto, M., McNeill, H. 2004. Planar polarity from flies to vertebrates. J. Cell Sci. 117, 527–533.

Formstone, C.J., Mason, I. 2005. Combinatorial activity of Flamingo proteins directs convergence and extension within the early zebrafish embryo via the planar cell polarity pathway. Dev. Biol. 282, 320–335.

Galceran, J., Miyashita-Lin, E.M., Devaney, E., Rubenstein, J.L., Grosschedl, R. 2000. Hippocampus development and generation of dentate gyrus granule cells is regulated by LEF1. Development 127, 469–482.

Garriock, R.J., D'Agostino, S.L., Pilcher, K.C., Krieg, P.A. 2005. Wnt11-R, a protein closely related to mammalian Wnt11, is required for heart morphogenesis in Xenopus. Dev. Biol. 279, 179–192.

Geis, K., Aberle, H., Kuhl, M., Kemler, R., Wedlich, D. 1998. Expression of the Armadillo family member p120cas1B in Xenopus embryos affects head differentiation but not axis formation. Dev. Genes Evol. 207, 471–481.

Giraldez, A.J., Copley, R.R., Cohen, S.M. 2002. HSPG modification by the secreted enzyme Notum shapes the Wingless morphogen gradient. Dev. Cell 2, 667–676.

Gleeson, J.G., Walsh, C.A. 2000. Neuronal migration disorders: From genetic diseases to developmental mechanisms. Trends Neurosci. 23, 352–359.

Glickman, N.S., Kimmel, C.B., Jones, M.A., Adams, R.J. 2003. Shaping the zebrafish notochord. Development 130, 873–887.

Glinka, A., Wu, W., Delius, H., Monaghan, A.P., Blumenstock, C., Niehrs, C. 1998. Dickkopf-1 is a member of a new family of secreted proteins and functions in head induction. Nature 391, 357–362.

Goto, T., Davidson, L., Asashima, M., Keller, R. 2005. Planar cell polarity genes regulate polarized extracellular matrix deposition during frog gastrulation. Curr. Biol. 15, 787–793.

Habas, R., Dawid, I.B., He, X. 2003. Coactivation of Rac and Rho by Wnt/Frizzled signaling is required for vertebrate gastrulation. Genes Dev. 17, 295–309.

Hall, A. 1998. Rho GTPases and the actin cytoskeleton. Science 279, 509–514.

Hamblet, N.S., Lijam, N., Ruiz-Lozano, P., Wang, J., Yang, Y., Luo, Z., Mei, L., Chien, K.R., Sussman, D.J., Wynshaw-Boris, A. 2002. Dishevelled 2 is essential for cardiac outflow tract development, somite segmentation and neural tube closure. Development 129, 5827–5838.

Hannus, M., Feiguin, F., Heisenberg, C.P., Eaton, S. 2002. Planar cell polarization requires Widerborst, a B' regulatory subunit of protein phosphatase 2A. Development 129, 3493–3503.

Hari, L., Brault, V., Kleber, M., Lee, H.Y., Ille, F., Leimeroth, R., Paratore, C., Suter, U., Kemler, R., Sommer, L. 2002. Lineage-specific requirements of beta-catenin in neural crest development. J. Cell Biol. 159, 867–880.

Harvey, R.P. 2002. Patterning the vertebrate heart. Nat. Rev. Genet. 3, 544–556.

Hatten, M.E. 2002. New directions in neuronal migration. Science 297, 1660–1663.

He, X., Semenov, M., Tamai, K., Zeng, X. 2004. LDL receptor-related proteins 5 and 6 in Wnt/beta-catenin signaling: Arrows point the way. Development 131, 1663–1677.

Heisenberg, C.P., Tada, M., Rauch, G.J., Saúde, L., Concha, M.L., Geisler, R., Stemple, D.L., Smith, J.C., Wilson, S.W. 2000. Silberblick/Wnt11 mediates convergent extension movements during zebrafish gastrulation. Nature 405, 76–81.

Henderson, D.J., Phillips, H.M., Chaudhry, B. 2006. Vang-like 2 and noncanonical Wnt signaling in outflow tract development. Trends Cardiovasc. Med. 16, 38–45.

Hikasa, H., Shibata, M., Hiratani, I., Taira, M. 2002. The Xenopus receptor tyrosine kinase Xror2 modulates morphogenetic movements of the axial mesoderm and neuroectoderm via Wnt signaling. Development 129, 5227–5239.

Honda, T., Tabata, H., Nakajima, K. 2003. Cellular and molecular mechanisms of neuronal migration in neocortical development. Semin. Cell Dev. Biol. 14, 169–174.

Hsieh, J.C. 2004. Specificity of WNT-receptor interactions. Front Biosci. 9, 1333–1338.

Huelsken, J., Vogel, R., Brinkmann, V., Erdmann, B., Birchmeier, C., Birchmeier, W. 2000. Requirement for beta-catenin in anterior-posterior axis formation in mice. J. Cell Biol. 148, 567–578.

Hukriede, N.A., Tsang, T.E., Habas, R., Khoo, P.L., Steiner, K., Weeks, D.L., Tam, P.P., Dawid, I.B. 2003. Conserved requirement of Lim1 function for cell movements during gastrulation. Dev. Cell 4, 83–94.

Hunter, N.L., Hikasa, H., Dymecki, S.M., Sokol, S.Y. 2006. Vertebrate homologues of Frodo are dynamically expressed during embryonic development in tissues undergoing extensive morphogenetic movements. Dev. Dyn. 235, 279–284.

Itoh, K., Sokol, S.Y. 1994. Heparan sulfate proteoglycans are required for mesoderm formation in *Xenopus* embryos. Development 120, 2703–2711.

Janetopoulos, C., Long, Y., Devreotes, P. 2005. Mechanisms of eukaryotic chemotaxis. In: *Cell Migration in Development and Disease* (D. Wedlich, Ed.), Wiley VCH: Weinheim, pp. 33–45.

Jessen, J.R., Topczewski, J., Bingham, S., Sepich, D.S., Marlow, F., Chandrasekhar, A., Solnica-Krezel, L. 2002. Zebrafish trilobite identifies new roles for Strabismus in gastrulation and neuronal movements. Nat. Cell Biol. 4, 610–615.

Jiang, X., Rowitch, D.H., Soriano, P., McMahon, A.P., Sucov, H.M. 2000. Fate of the mammalian cardiac neural crest. Development 127, 1607–1616.

Jopling, C., den Hertog, J. 2005. Fyn/Yes and non-canonical Wnt signalling converge on RhoA in vertebrate gastrulation cell movements. EMBO Rep. 6, 426–431.

Kakinuma, Y., Saito, F., Ohsawa, S., Furuichi, T., Miura, M. 2004a. A sulfatase regulating the migratory potency of oligodendrocyte progenitor cells through tyrosine phosphorylation of beta-catenin. J. Neurosci. Res. 77, 653–661.

Kakinuma, Y., Saito, F., Osawa, S., Miura, M. 2004b. A mechanism of impaired mobility of oligodendrocyte progenitor cells by tenascin C through modification of wnt signaling. FEBS Lett. 568, 60–64.

Kalcheim, C., Burstyn-Cohen, T. 2005. Early stages of neural crest ontogeny: Formation and regulation of cell delamination. Int. J. Dev. Biol. 49, 105–116.

Kane, D.A., Hammerschmidt, M., Mullins, M.C., Maischein, H.M., Brand, M., van Eeden, F.J., Furutani-Seiki, M., Granato, M., Haffter, P., Heisenberg, C.P., Jiang, Y.J., Kelsh, R.N., et al. 1996. The zebrafish epiboly mutants. Development 123, 47–55.

Kane, D.A., McFarland, K.N., Warga, R.M. 2005. Mutations in half baked/E-cadherin block cell behaviors that are necessary for teleost epiboly. Development 132, 1105–1116.

Kawano, Y., Kypta, R. 2003. Secreted antagonists of the Wnt signalling pathway. J. Cell Sci. 116, 2627–2634.

Keller, R. 2002. Shaping the vertebrate body plan by polarized embryonic cell movements. Science 298, 1950–1954.

Keller, R., Davidson, L., Edlund, A., Elul, T., Ezin, M., Shook, D., Skoglund, P. 2000. Mechanisms of convergence and extension by cell intercalation. Phil. Trans. R. Soc. Lond. B 355, 897–922.

Keller, R., Winklbauer, R. 1992. Cellular basis of amphibian gastrulation. Curr. Top. Dev. Biol. 27, 39–89.

Keller, R., Davidson, L.A., Shook, D.R. 2003. How we are shaped: The biomechanics of gastrulation. Differentiation 71, 171–205.

Kibar, Z., Vogan, K.J., Groulx, N., Justice, M.J., Underhill, D.A., Gros, P. 2001. Ltap, a mammalian homolog of *Drosophila* Strabismus/Van Gogh, is altered in the mouse neural tube mutant Loop-tail. Nat. Genet. 28, 251–255.

Kilian, B., Mansukoski, H., Barbosa, F.C., Ulrich, F., Tada, M., Heisenberg, C.P. 2003. The role of Ppt/Wnt5 in regulating cell shape and movement during zebrafish gastrulation. Mech. Dev. 120, 467–476.

Kim, G.H., Han, J.K. 2005. JNK and ROKalpha function in the noncanonical Wnt/RhoA signaling pathway to regulate *Xenopus* convergent extension movements. Dev. Dyn. 232, 958–968.

Kim, S.H., Yamamoto, A., Bouwmeester, T., Agius, E., De Robertis, E.M. 1998. The role of paraxial protocadherin in selective adhesion and cell movements of the mesoderm during *Xenopus* gastrulation. Development 125, 4681–4690.

Kim, S.W., Fang, X., Ji, H., Paulson, A.F., Daniel, J.M., Ciesiolka, M., van Roy, F., McCrea, P.D. 2002. Isolation and characterization of XKaiso, a transcriptional repressor that associates with the catenin Xp120(ctn) in *Xenopus laevis*. J. Biol. Chem. 277, 8202–8208.

Kim, S.W., Park, J.I., Spring, C.M., Sater, A.K., Ji, H., Otchere, A.A., Daniel, J.M., McCrea, P.D. 2004. Non-canonical Wnt signals are modulated by the Kaiso transcriptional repressor and p120-catenin. Nat. Cell Biol. 6, 1212–1220.

Kimura, C., Yoshinaga, K., Tian, E., Suzuki, M., Aizawa, S., Matsuo, I. 2000. Visceral endoderm mediates forebrain development by suppressing posteriorizing signals. Dev. Biol. 225, 304–321.

Kimura, C., Shen, M.M., Takeda, N., Aizawa, S., Matsuo, I. 2001. Complementary functions of Otx2 and Cripto in initial patterning of mouse epiblast. Dev. Biol. 235, 12–32.

Kimura-Yoshida, C., Nakano, H., Okamura, D., Nakao, K., Yonemura, S., Belo, J.A., Aizawa, S., Matsui, Y., Matsuo, I. 2005. Canonical Wnt signaling and its antagonist regulate anterior-posterior axis polarization by guiding cell migration in mouse visceral endoderm. Dev. Cell 9, 639–650.

Kinoshita, N., Iioka, H., Miyakoshi, A., Ueno, N. 2003. PKC delta is essential for Dishevelled function in a noncanonical Wnt pathway that regulates *Xenopus* convergent extension movements. Genes Dev. 17, 1663–1676.

Kirby, M.L., Gale, T.F., Stewart, D.E. 1983. Neural crest cells contribute to normal aorticopulmonary septation. Science 220, 1059–1061.

Kriegstein, A.R., Noctor, S.C. 2004. Patterns of neuronal migration in the embryonic cortex. Trends Neurosci. 27, 392–399.

Kuehl, M., Finnemann, S., Binder, O., Wedlich, D. 1996. Dominant negative expression of a cytoplasmically deleted mutant of XB/U-cadherin disturbs mesoderm migration during gastrulation in *Xenopus laevis*. Mech. Dev. 54, 71–82.

Kuehl, M., Geis, K., Sheldahl, L.C., Pukrop, T., Moon, R.T., Wedlich, D. 2001. Antagonistic regulation of convergent extension movements in *Xenopus* by Wnt/β-catenin and Wnt/Ca^{2+} signaling. Mech. Dev. 106, 61–76.

Kuehl, M., Sheldahl, L.C., Park, M., Miller, J.R., Moon, R.T. 2000a. The Wnt/Ca^{2+} pathway: A new vertebrate Wnt signaling pathway takes shape. Trends Genet. 16, 279–283.

Kuehl, M., Sheldahl, L.C., Malbon, C.C., Moon, R.T. 2000b. Ca(2+)/calmodulin-dependent protein kinase II is stimulated by Wnt and Frizzled homologs and promotes ventral cell fates in *Xenopus*. J. Biol. Chem. 275, 12701–12711.

Kulesa, P.M., Lu, C.C., Fraser, S.E. 2005. Time-lapse analysis reveals a series of events by which cranial neural crest cells reroute around physical barriers. Brain Behav. Evol. 66, 255–265.

LaBonne, C., Bronner-Fraser, M. 1999. Molecular mechanisms of neural crest formation. Annu. Rev. Cell Dev. Biol. 15, 81–112.

LaBonne, C., Bronner-Fraser, M. 2000. Snail-related transcriptional repressors are required in *Xenopus* for both the induction of the neural crest and its subsequent migration. Dev. Biol. 221, 195–205.

Lee, H.H., Frasch, M. 2000. Wingless effects mesoderm patterning and ectoderm segmentation events via induction of its downstream target sloppy paired. Development 127, 5497–5508.

Lee, H.S., Bong, Y.S., Moore, K.B., Soria, K., Moody, S.A., Daar, I.O. 2006. Dishevelled mediates ephrinB1 signalling in the eye field through the planar cell polarity pathway. Nat. Cell Biol. 8, 55–63.

Li, J., Chen, F., Epstein, J.A. 2000. Neural crest expression of Cre recombinase directed by the proximal Pax3 promoter in transgenic mice. Genesis 26, 162–164.

Li, L., Mao, J., Sun, L., Liu, W., Wu, D. 2002. Second cysteine-rich domain of Dickkopf-2 activates canonical Wnt signaling pathway via LRP-6 independently of dishevelled. J. Biol. Chem. 277, 5977–5981.

Lin, X., Perrimon, N. 2002. Developmental roles of heparan sulfate proteoglycans in *Drosophila*. Glycoconj. J. 19, 363–368.

Liu, B.Y., Kim, Y.C., Leatherberry, V., Cowin, P., Alexander, C.M. 2003. Mammary gland development requires syndecan-1 to create a beta-catenin/TCF-responsive mammary epithelial subpopulation. Oncogene 22, 9243–9253.

Liu, J.P., Jessell, T.M. 1998. A role for rhoB in the delamination of neural crest cells from the dorsal neural tube. Development 125, 5055–5067.

Liu, P., Wakamiya, M., Shea, M.J., Albrecht, U., Behringer, R.R., Bradley, A. 1999. Requirement for Wnt3 in vertebrate axis formation. Nat. Genet. 22, 361–365.

Lu, W., Yamamoto, V., Ortega, B., Baltimore, D. 2004a. Mammalian Ryk is a Wnt coreceptor required for stimulation of neurite outgrowth. Cell 119, 97–108.

Lu, X., Borchers, A.G., Jolicoeur, C., Rayburn, H., Baker, J.C., Tessier-Lavigne, M. 2004b. PTK7/CCK-4 is a novel regulator of planar cell polarity in vertebrates. Nature 430, 93–98.

Machon, O., van den Bout, C.J., Backman, M., Kemler, R., Krauss, S. 2003. Role of beta-catenin in the developing cortical and hippocampal neuroepithelium. NeuroScience 122, 129–143.

Mao, J., Wang, J., Liu, B., Pan, W., Farr, G.H., 3rd, Flynn, C., Yuan, H., Takada, S., Kimelman, D., Li, L., Wu, D. 2001. Low-density lipoprotein receptor-related protein-5 binds to Axin and regulates the canonical Wnt signaling pathway. Mol Cell. 7, 801–809.

Marlow, F., Topczewski, J., Sepich, D., Solnica-Krezel, L. 2002. Zebrafish rho kinase 2 acts downstream of wnt11 to mediate cell polarity and effective convergence and extension movements. Curr. Biol. 12, 876–884.

Marsden, M., DeSimone, D.W. 2001. Regulation of cell polarity, radial intercalation and epiboly in *Xenopus*: Novel roles for integrin and fibronectin. Development 128, 3635–3647.

Marsden, M., DeSimone, D.W. 2003. Integrin-ECM interactions regulate cadherin-dependent cell adhesion and are required for convergent extension in *Xenopus*. Curr. Biol. 13, 1182–1191.

Martinez-Barbera, J.P., Beddington, R.S. 2001. Getting your head around Hex and Hesx1: Forebrain formation in mouse. Int. J. Dev. Biol. 45, 327–336.

Marvin, M.J., Di Rocco, G., Gardiner, A., Bush, S.M., Lassar, A.B. 2001. Inhibition of Wnt activity induces heart formation from posterior mesoderm. Genes Dev. 15, 316–327.

Matsui, T., Raya, A., Kawakami, Y., Callol-Massot, C., Capdevila, J., Rodriguez-Esteban, C., Izpisua Belmonte, J.C. 2005. Noncanonical Wnt signaling regulates midline convergence of organ primordia during zebrafish development. Genes Dev. 19, 164–175.

McFarland, K.N., Warga, R.M., Kane, D.A. 2005. Genetic locus half baked is necessary for morphogenesis of the ectoderm. Dev. Dyn. 233, 390–406.

McLaughlin, T., Hindges, R., Yates, P.A., O'Leary, D.D. 2003. Bifunctional action of ephrin-B1 as a repellent and attractant to control bidirectional branch extension in dorsal-ventral retinotopic mapping. Development 130, 2407–2418.

Medina, A., Swain, R.K., Kuerner, K.M., Steinbeisser, H. 2004. *Xenopus* paraxial protocadherin has signaling functions and is involved in tissue separation. EMBO J. 23, 3249–3258.

Mohun, T., Orford, R., Shang, C. 2003. The origins of cardiac tissue in the amphibian, *Xenopus laevis*. Trends Cardiovasc. Med. 13, 244–248.

Mohun, T.J., Leong, L.M., Weninger, W.J., Sparrow, D.B. 2000. The morphology of heart development in *Xenopus laevis*. Dev. Biol. 218, 74–88.

Morgan, R., El-Kadi, A.M., Theokli, C. 2003. Flamingo, a cadherin-type receptor involved in the *Drosophila* planar polarity pathway, can block signaling via the canonical wnt pathway in *Xenopus laevis*. Int. J. Dev. Biol. 47, 245–252.

Moriguchi, T., Kawachi, K., Kamakura, S., Masuyama, N., Yamanaka, H., Matsumoto, K., Kikuchi, A., Nishida, E. 1999. Distinct domains of mouse Dishevelled are responsible for the c-Jun N-terminal Kinase/Stress-activated Protein Kinase activation and the axis formation in vertebrates. J. Biol. Chem. 274, 30957–30962.

Morkel, M., Huelsken, J., Wakamiya, M., Ding, J., van de Wetering, M., Clevers, H., Taketo, M.M., Behringer, R.R., Shen, M.M., Birchmeier, W. 2003. Beta-catenin regulates Cripto- and Wnt3-dependent gene expression programs in mouse axis and mesoderm formation. Development 130, 6283–6294.

Munoz, R., Moreno, M., Oliva, C., Orbenes, C., Larrain, J. 2006. Syndecan-4 regulates non-canonical Wnt signalling and is essential for convergent and extension movements in *Xenopus* embryos. Nat. Cell Biol. 8, 492–500.

Murdoch, J.N., Doudney, K., Paternotte, C., Copp, A.J., Stanier, P. 2001. Severe neural tube defects in the loop-tail mouse result from mutation of Lpp1, a novel gene involved in floor plate specification. Hum. Mol. Genet. 10, 2593–2601.

Nadarajah, B., Brunstrom, J.E., Grutzendler, J., Wong, R.O., Pearlman, A.L. 2001. Two modes of radial migration in early development of the cerebral cortex. Nat. Neurosci. 4, 143–150.

Nichols, D.H., Bruce, L.L. 2006. Migratory routes and fates of cells transcribing the Wnt-1 gene in the murine hindbrain. Dev. Dyn. 235, 285–300.

Niehrs, C., Kazanskaya, O., Wu, W., Glinka, A. 2001. Dickkopf1 and the Spemann-Mangold head organizer. Int. J. Dev. Biol. 45, 237–240.

Nieto, M.A., Sargent, M.G., Wilkinson, D.G., Cooke, J. 1994. Control of cell behavior during vertebrate development by Slug, a zinc finger gene. Science 264, 835–839.

Ohkawara, B., Yamamoto, T.S., Tada, M., Ueno, N. 2003. Role of glypican 4 in the regulation of convergent extension movements during gastrulation in *Xenopus laevis*. Development 130, 2129–2138.

Ouchi, Y., Tabata, Y., Arai, K., Watanabe, S. 2005. Negative regulation of retinal-neurite extension by beta-catenin signaling pathway. J. Cell Sci. 118, 4473–4483.

Pandur, P., Lasche, M., Eisenberg, L.M., Kuhl, M. 2002. Wnt-11 activation of a non-canonical Wnt signalling pathway is required for cardiogenesis. Nature 418, 636–641.

Park, M., Moon, R.T. 2002. The planar cell-polarity gene stbm regulates cell behaviour and cell fate in vertebrate embryos. Nat. Cell Biol. 4, 20–25.

Park, T.J., Gray, R.S., Sato, A., Habas, R., Wallingford, J.B. 2005. Subcellular localization and signaling properties of dishevelled in developing vertebrate embryos. Curr. Biol. 15, 1039–1044.

Paulson, A.F., Fang, X., Ji, H., Reynolds, A.B., McCrea, P.D. 1999. Misexpression of the Catenin p120^{ctn}1A perturbs *Xenopus* gastrulation but does not elicit Wnt-directed axis specification. Dev. Biol. 207, 350–363.

Penzo-Mendez, A., Umbhauer, M., Djiane, A., Boucaut, J.C., Riou, J.F. 2003. Activation of Gbetagamma signaling downstream of Wnt-11/Xfz7 regulates Cdc42 activity during *Xenopus* gastrulation. Dev. Biol. 257, 302–314.

Pera, E.M., De Robertis, E.M. 2000. A direct screen for secreted proteins in *Xenopus* embryos identifies distinct activities for the Wnt antagonists Crescent and Frzb-1. Mech. Dev. 96, 183–195.

Perea-Gomez, A., Lawson, K.A., Rhinn, M., Zakin, L., Brulet, P., Mazan, S., Ang, S.L. 2001a. Otx2 is required for visceral endoderm movement and for the restriction of posterior signals in the epiblast of the mouse embryo. Development 128, 753–765.

Perea-Gomez, A., Rhinn, M., Ang, S.L. 2001b. Role of the anterior visceral endoderm in restricting posterior signals in the mouse embryo. Int. J. Dev. Biol. 45, 311–320.

Person, A.D., Garriock, R.J., Krieg, P.A., Runyan, R.B., Klewer, S.E. 2005. Frzb modulates Wnt-9a-mediated beta-catenin signaling during avian atrioventricular cardiac cushion development. Dev. Biol. 278, 35–48.

Phillips, H.M., Murdoch, J.N., Chaudhry, B., Copp, A.J., Henderson, D.J. 2005. Vangl2 acts via RhoA signaling to regulate polarized cell movements during development of the proximal outflow tract. Circ. Res. 96, 292–299.

Pla, P., Moore, R., Morali, O.G., Grille, S., Martinozzi, S., Delmas, V., Larue, L. 2001. Cadherins in neural crest cell development and transformation. J. Cell. Physiol. 189, 121–132.

Pohl, B.S., Knoechel, W. 2001. Overexpression of the transcriptional repressor FoxD3 prevents neural crest formation in *Xenopus* embryos. Mech. Dev. 103, 93–106.

Raible, D.W., Ragland, J.W. 2005. Reiterated Wnt and BMP signals in neural crest development. Semin. Cell Dev. Biol. 16, 673–682.

Rakic, P. 2003. Developmental and evolutionary adaptations of cortical radial glia. Cereb. Cortex 13, 541–549.

Ren, R., Nagel, M., Tahinci, E., Winklbauer, R., Symes, K. 2006. Migrating anterior mesoderm cells and intercalating trunk mesoderm cells have distinct responses to Rho and Rac during *Xenopus* gastrulation. Dev. Dyn. 235, 1090–1099.

Robb, L., Tam, P.P. 2004. Gastrula organiser and embryonic patterning in the mouse. Semin. Cell Dev. Biol. 15, 543–554.

Rodriguez, J., Esteve, P., Weinl, C., Ruiz, J.M., Fermin, Y., Trousse, F., Dwivedy, A., Holt, C., Bovolenta, P. 2005. SFRP1 regulates the growth of retinal ganglion cell axons through the Fz2 receptor. Nat. Neurosci. 8, 1301–1309.

Ross, M.E., Walsh, C.A. 2001. Human brain malformations and their lessons for neuronal migration. Annu. Rev. Neurosci. 24, 1041–1070.

Ruan, G., Wedlich, D., Koehler, A. 2003. How cell-cell adhesion contributes to early embryonic development. In: *The Vertebrate Organizer* (H. Grunz, Ed.), Berlin: Springer, pp. 201–218.

Saneyoshi, T., Kume, S., Amasaki, Y., Mikoshiba, K. 2002. The Wnt/calcium pathway activates NF-AT and promotes ventral cell fate in *Xenopus* embryos. Nature 417, 295–299.

Schier, A.F., Talbot, W.S. 2005. Molecular genetics of axis formation in zebrafish. Annu. Rev. Genet. 39, 561–613.

Schmitt, A.M., Shi, J., Wolf, A.M., Lu, C.C., King, L.A., Zou, Y. 2006. Wnt-Ryk signalling mediates medial-lateral retinotectal topographic mapping. Nature 439, 31–37.

Schneider, V.A., Mercola, M. 2001. Wnt antagonism initiates cardiogenesis in *Xenopus laevis*. Genes Dev. 15, 304–315.

Schoenwolf, G.C., Garcia-Martinez, V. 1995. Primitive-streak origin and state of commitment of cells of the cardiovascular system in avian and mammalian embryos. Cell. Mol. Biol. Res. 41, 233–240.

Schwarz-Romond, T., Asbrand, C., Bakkers, J., Kuhl, M., Schaeffer, H.J., Huelsken, J., Behrens, J., Hammerschmidt, M., Birchmeier, W. 2002. The ankyrin repeat protein Diversin recruits Casein kinase Iepsilon to the beta-catenin degradation complex and acts in both canonical Wnt and Wnt/JNK signaling. Genes Dev. 16, 2073–2084.

Sepich, D.S., Myers, D.C., Short, R., Topczewski, J., Marlow, F., Solnica-Krezel, L. 2000. Role of the zebrafish trilobite locus in gastrulation movements of convergence and extension. Genesis 27, 159–173.

Shariatmadari, M., Peyronnet, J., Papachristou, P., Horn, Z., Sousa, K.M., Arenas, E., Ringstedt, T. 2005. Increased Wnt levels in the neural tube impair the function of adherens junctions during neurulation. Mol. Cell. Neurosci. 30, 437–451.

Sheldahl, L.C., Park, M., Malbon, C.C., Moon, R.T. 1999. Protein kinase C is differentially stimulated by Wnt and Frizzled homologs in a G-protein-dependent manner. Curr. Biol. 9, 695–698.

Sheldahl, L.C., Slusarski, D.C., Pandur, P., Miller, J.R., Kuhl, M., Moon, R.T. 2003. Dishevelled activates Ca^{2+} flux, PKC, and CamKII in vertebrate embryos. J. Cell Biol. 161, 769–777.

Shibata, M., Itoh, M., Hikasa, H., Taira, S., Taira, M. 2005. Role of crescent in convergent extension movements by modulating Wnt signaling in early *Xenopus* embryogenesis. Mech. Dev. 122, 1322–1339.

Simons, M., Gloy, J., Ganner, A., Bullerkotte, A., Bashkurov, M., Kronig, C., Schermer, B., Benzing, T., Cabello, O.A., Jenny, A., Mlodzik, M., Polok, B., et al. 2005. Inversin, the gene product mutated in nephronophthisis type II, functions as a molecular switch between Wnt signaling pathways. Nat. Genet. 37, 537–543.

Srinivas, S., Rodriguez, T., Clements, M., Smith, J.C., Beddington, R.S. 2004. Active cell migration drives the unilateral movements of the anterior visceral endoderm. Development 131, 1157–1164.

Stainier, D.Y. 2001. Zebrafish genetics and vertebrate heart formation. Nat. Rev. Genet. 2, 39–48.

Steventon, B., Carmona-Fontaine, C., Mayor, R. 2005. Genetic network during neural crest induction: From cell specification to cell survival. Semin. Cell Dev. Biol. 16, 647–654.

Stoller, J.Z., Epstein, J.A. 2005. Cardiac neural crest. Semin. Cell Dev. Biol. 16, 704–715.

Tada, M., Smith, J.C. 2000. Xwnt11 is a target of *Xenopus* Brachyury: Regulation of gastrulation movements via Dishevelled, but not through the canonical Wnt pathway. Development 127, 2227–2238.

Tahinci, E., Symes, K. 2003. Distinct functions of Rho and Rac are required for convergent extension during *Xenopus* gastrulation. Dev. Biol. 259, 318–335.

Takahashi, M., Osumi, N. 2005. Identification of a novel type II classical cadherin: Rat cadherin19 is expressed in the cranial ganglia and Schwann cell precursors during development. Dev. Dyn. 232, 200–208.

Takeuchi, M., Nakabayashi, J., Sakaguchi, T., Yamamoto, T.S., Takahashi, H., Takeda, H., Ueno, N. 2003. The prickle-related gene in vertebrates is essential for gastrulation cell movements. Curr. Biol. 13, 674–679.

Taneyhill, L.A., Bronner-Fraser, M. 2005. Dynamic alterations in gene expression after Wnt-mediated induction of avian neural crest. Mol. Biol. Cell 16, 5283–5293.

Thomas, P., Beddington, R. 1996. Anterior primitive endoderm may be responsible for patterning the anterior neural plate in the mouse embryo. Curr. Biol. 6, 1487–1496.

Thomas, P.Q., Brown, A., Beddington, R.S. 1998. Hex: A homeobox gene revealing peri-implantation asymmetry in the mouse embryo and an early transient marker of endothelial cell precursors. Development 125, 85–94.

Tilghman, R.W., Slack-Davis, J.K., Sergina, N., Martin, K.H., Iwanicki, M., Hershey, E.D., Beggs, H.E., Reichardt, L.F., Parsons, J.T. 2005. Focal adhesion kinase is required for the spatial organization of the leading edge in migrating cells. J. Cell Sci. 118, 2613–2623.

Topczewski, J., Sepich, D.S., Myers, D.C., Walker, C., Amores, A., Lele, Z., Hammerschmidt, M., Postlethwait, J., Solnica-Krezel, L. 2001. The zebrafish glypican knypek controls cell polarity during gastrulation movements of convergent extension. Dev. Cell 1, 251–264.

Toyofuku, T., Hong, Z., Kuzuya, T., Tada, M., Hori, M. 2000. Wnt/frizzled-2 signaling induces aggregation and adhesion among cardiac myocytes by increased cadherin-beta-catenin complex. J. Cell Biol. 150, 225–241.

Ulrich, F., Concha, M.L., Heid, P.J., Voss, E., Witzel, S., Roehl, H., Tada, M., Wilson, S.W., Adams, R.J., Soll, D.R., Heisenberg, C.P. 2003. Slb/Wnt11 controls hypoblast cell migration and morphogenesis at the onset of zebrafish gastrulation. Development 130, 5375–5384.

Ulrich, F., Krieg, M., Schotz, E.M., Link, V., Castanon, I., Schnabel, V., Taubenberger, A., Mueller, D., Puech, P.H., Heisenberg, C.P. 2005. Wnt11 functions in gastrulation by controlling cell cohesion through Rab5c and E-cadherin. Dev. Cell 9, 555–564.

Unterseher, F., Hefele, J.A., Giehl, K., De Robertis, E.M., Wedlich, D., Schambony, A. 2004. Paraxial protocadherin coordinates cell polarity during convergent extension via Rho A and JNK. EMBO J. 23, 3259–3269.

van den Hoff, M.J., Kruithof, B.P., Moorman, A.F. 2004. Making more heart muscle. Bioessays 26, 248–261.

Veeman, M.T., Slusarski, D.C., Kaykas, A., Louie, S.H., Moon, R.T. 2003. Zebrafish prickle, a modulator of noncanonical wnt/fz signaling, regulates gastrulation movements. Curr. Biol. 13, 680–685.

Waldo, K.L., Lo, C.W., Kirby, M.L. 1999. Connexin 43 expression reflects neural crest patterns during cardiovascular development. Dev. Biol. 208, 307–323.

Wallingford, J.B., Rowning, B.A., Vogeli, K.M., Rothbächer, U., Fraser, S.E., Harland, R.M. 2000. Dishevelled controls cell polarity during *Xenopus* gastrulation. Nature 405, 81–85.

Weber, R.J., Pedersen, R.A., Wianny, F., Evans, M.J., Zernicka-Goetz, M. 1999. Polarity of the mouse embryo is anticipated before implantation. Development 126, 5591–5598.

Wilson, S.W., Houart, C. 2004. Early steps in the development of the forebrain. Dev. Cell 6, 167–181.

Wodarz, A., Nusse, R. 1998. Mechanisms of Wnt signaling in development. Annu. Rev. Cell Dev. Biol. 14, 59–88.

Xu, X., Li, W.E., Huang, G.Y., Meyer, R., Chen, T., Luo, Y., Thomas, M.P., Radice, G.L., Lo, C.W. 2001a. Modulation of mouse neural crest cell motility by N-cadherin and connexin 43 gap junctions. J. Cell Biol. 154, 217–230.

Xu, X., Li, W.E., Huang, G.Y., Meyer, R., Chen, T., Luo, Y., Thomas, M.P., Radice, G.L., Lo, C.W. 2001b. N-cadherin and Cx43alpha1 gap junctions modulates mouse neural crest cell motility via distinct pathways. Cell Commun. Adhes. 8, 321–324.

Yamaguchi, T.P., Bradley, A., McMahon, A.P., Jones, S. 1999. A Wnt5a pathway underlies outgrowth of multiple structures in the vertebrate embryo. Development 126, 1211–1223.

Yamamoto, A., Amacher, S.L., Kim, S.H., Geissert, D., Kimmel, C.B., De Robertis, E.M. 1998. Zebrafish paraxial protocadherin is a downstream target of spadetail involved in morphogenesis of gastrula mesoderm. Development 125, 3389–3397.

Yamanaka, H., Moriguchi, T., Masuyama, N., Kusakabe, M., Hanafusa, H., Takada, R., Takada, S., Nishida, E. 2002. JNK functions in the non-canonical Wnt pathway to regulate convergent extension movements in vertebrates. EMBO Rep. 3, 69–75.

Yang, J., Wu, J., Tan, C., Klein, P.S. 2003. PP2A:B56epsilon is required for Wnt/beta-catenin signaling during embryonic development. Development 130, 5569–5578.

Yokota, C., Kofron, M., Zuck, M., Houston, D.W., Isaacs, H., Asashima, M., Wylie, C.C., Heasman, J. 2003. A novel role for a nodal-related protein; Xnr3 regulates convergent extension movements via the FGF receptor. Development 130, 2199–2212.

Zaffran, S., Frasch, M. 2002. Early signals in cardiac development. Circ. Res. 91, 457–469.

Zhong, Y., Brieher, W.M., Gumbiner, B.M. 1999. Analysis of C-cadherin regulation during tissue morphogenesis with an activating antibody. J. Cell Biol. 144, 351–359.

Multiple roles for Wnt signaling in the development of the vertebrate neural crest

Elizabeth Heeg-Truesdell[1] and Carole LaBonne[1,2]

[1]*Department of Biochemistry, Molecular Biology, and Cell Biology,*
Northwestern University, Evanston, Illinois
[2]*Robert H. Lurie Comprehensive Cancer Center, Northwestern University,*
Evanston, Illinois

Contents

The neural crest is a cell type unique to vertebrates that gives rise to a diverse set of derivatives essential to the vertebrate body plan. The formation and subsequent diversification of neural crest cells is a complex multistep process dependent on the reiterative use of multiple intracellular-signaling pathways. Notable among these is the Wnt-signaling pathway, which plays prominent roles at multiple stages of neural crest development.

Advances in Developmental Biology
Volume 17 ISSN 1574-3349
DOI: 10.1016/S1574-3349(06)17006-4

1. Introduction

Neural crest cells are a population of stem cell-like precursor cells that originate at the lateral margins of the neural plate in vertebrate embryos. Around the time of neural tube closure, these cells delaminate from the neural tube and migrate extensively throughout the embryo. Ultimately, neural crest cells contribute to a diverse set of derivatives that includes much of the peripheral nervous system and the craniofacial skeleton, melanocytes, smooth muscle, and the chromaffin cells of the adrenal medulla (Le Douarin, 1982; Fig. 1).

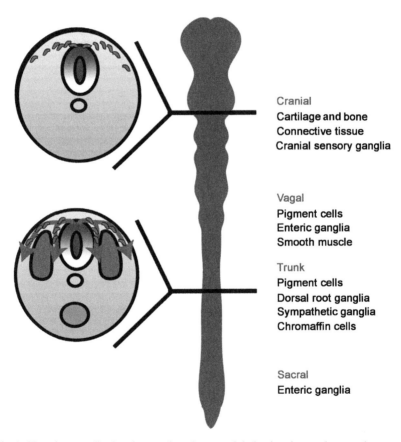

Cranial
Cartilage and bone
Connective tissue
Cranial sensory ganglia

Vagal
Pigment cells
Enteric ganglia
Smooth muscle

Trunk
Pigment cells
Dorsal root ganglia
Sympathetic ganglia
Chromaffin cells

Sacral
Enteric ganglia

Fig. 1. Neural crest cells give rise to a broad range of derivatives in vertebrate embryos, and these derivates differ by axial level. For example, only cranial neural crest gives rise to cartilage and bone whereas the enteric nervous system is derived from vagal and sacral neural crest. Importantly, these differences do not appear to reflect differences in developmental potential. (See Color Insert.)

Neural crest cells have been extensively studied due to their developmental, clinical, and evolutionary relevance. As a consequence of their contribution to such a large number of derivatives, defects in the neural crest are associated with an estimated half of known birth defects (Hall, 1999). These include a broad set of craniofacial defects, as well as syndromes such as Waardenburg's and Di Georges (Wilkie and Morriss-Kay, 2001; Goodman, 2003; Van de Putte et al., 2003). Moreover, a diverse group of cancers are of neural crest origin including melanomas, pheochromocytomas, and neuroblastomas (Nakagawara and Ohira, 2004). Because genes that play essential roles in neural crest development are misregulated in these cancers (Rosivatz et al., 2002; Heeg-Truesdell and LaBonne, 2004; Rothhammer et al., 2004), elucidating the mechanisms that control the normal development of neural crest cells can provide important insights into how these genetic regulatory programs are inappropriately reactivated during tumorigenesis. From an evolutionary perspective, neural crest cells are of central importance as they distinguish the craniates/vertebrates from their close relatives, the nonvertebrate chordates. Indeed, these cells are sometimes considered to be a "fourth germ layer" in this context (Hall, 2000). Thus, an understanding of neural crest development is essential to understanding the evolutionary origins of the vertebrates.

In recent years, studies of neural crest cells carried out in a variety of model organisms have shed important light on our understanding of how this stem cell population first forms, how these cells become migratory, and how they ultimately are instructed to give rise to specific derivatives. A number of signaling molecules, including those of the Wnt, FGF, BMP, and Notch families, have been found to play essential roles in multiple steps during the development of these cells. It has become increasingly clear that neural crest development is a complex multistep process and that elucidating the molecular mechanisms, which mediate distinct aspects of this process, will require understanding the role of each of these signals, both alone and in combination. Moreover, much work remains to be done to determine the mechanisms via which the same signal may elicit distinct responses from neural crest cells at different times in their development. Changes in the response of neural crest cells to the same signal over time may be mediated, in part, by the ever changing cocktail of transcription factors expressed within these cells, to changes in components of the signaling pathways that can in turn influence signal outcome, or by some combination of these mechanisms. Here we discuss the current view of the multiple roles that Wnt signals play during distinct steps in neural crest development.

2. Wnt signals and neural crest induction

Work in a number of model organisms, including frog, chick, and fish, has convincingly implicated the canonical Wnt-signaling pathway in the process by which neural crest cells first become specified. The canonical, or Wnt/β-catenin, pathway centers on the stability of β-catenin, a factor that can serve as a transcriptional coactivator for Wnt target genes (Pandur et al., 2002). Wnt/β-catenin signaling is initiated when an appropriate Wnt ligand binds to a receptor complex consisting of both a Frizzled family serpentine receptor and a single-span transmembrane receptor of the LDL receptor related protein (LRP) family (LRP5/6/Arrow) (reviewed by He, 2003; Yanfeng et al., 2003). In unstimulated cells, cytosolic β-catenin is assembled into a complex scaffolded by Axin, where it is sequentially phosphorylated by CK1 and GSK3 and targeted for ubiquitin-mediated degradation by the proteasome (Amit et al., 2002; Liu et al., 2002; Yanagawa et al., 2002). In response to ligand binding, GSK-3-mediated phosphorylation of β-catenin is inhibited, although the mechanism by which this is accomplished remains poorly understood. As a consequence of this inhibition, β-catenin is stabilized and can translocate to the nucleus where it interacts with DNA-binding proteins of the Lef/Tcf family to activate the transcription of Wnt target genes (Hurlstone and Clevers, 2002).

3. Expression of Wnt ligands and receptors

A number of Wnt ligands and receptors have expression patterns consistent with a role in the induction and/or maintenance of neural crest. It is important to consider both induction and maintenance when evaluating putative roles for these factors, as neural crest specification takes place over a prolonged period, and single cell lineage analysis has indicated that cells do not commit to a neural crest fate until around the time they delaminate from the neuroepithelium (Collazo et al., 1993; Selleck and Bronner-Fraser, 1995; Bronner-Fraser and Fraser, 1988, 1989; Morales et al., 2005). In both *Xenopus* and zebrafish, *Wnt8* is expressed in the paraxial mesoderm underlying the neural plate border and is therefore a reasonable candidate for a Wnt ligand involved in the earliest steps of neural crest formation (Bang et al., 1999). *Xwnt7b* is expressed in the neural and nonneural ectoderm at early neurula stages, also consistent with an early role in neural crest formation. Moreover, following neural tube closure, expression of *Xwnt7b* becomes restricted to the dorsal neural tube, where it could then be involved in the maintenance of neural crest precursors (Chang and Hemmati-Brivanlou, 1998). *Wnt6* has been suggested as a candidate neural crest inducer in avian embryos, where it is expressed in the nonneural ectoderm that abuts the neural plate (Garcia-Castro et al., 2002). *Wnt6* is also expressed in mouse and *Xenopus* early in development (Gavin et al., 1990; Wolda et al., 1993). *Wnt1* and *Wnt3a* are expressed in the lateral neural plate and neural folds in *Xenopus*, and then subsequently in the dorsal neural tube (Saint-Jeannet et al., 1997). These two ligands are also expressed in the dorsal neural tube in fish, chick, and mouse, and therefore could play an evolutionarily conserved role in neural crest cell maintenance.

In order to respond to any of these ligands, cells in the early ectoderm that are competent to give rise to neural crest precursors must express appropriate Frizzled receptors. Two Frizzled receptors, *Xfz3* and *Xfz7*, are expressed at a time and place consistent with a role in neural crest induction in *Xenopus* (Deardorff et al., 2001; Abu-Elmagd et al., 2006). However, *Xfz7* expression appears to be the most consistent with an early role in induction (Abu-Elmagd et al., 2006). *Fz7* is also expressed in the neural folds in avian embryos, as is a third member of this family, *Fz9*, which is also expressed in this region in fish and mice. More support for a role for canonical Wnt signaling in the induction and/or maintenance of neural crest cells derives from a report by Garcia-Castro et al. (2002), who found that β-catenin accumulates in the nucleus of presumptive neural crest cells during neurulation. This finding provides important evidence that these cells have received and responded to a canonical Wnt signal (Garcia-Castro et al., 2002).

4. Gain- and loss-of-function studies

Both gain- and loss-of-function experiments have been used to establish the role of canonical Wnt signaling in the process of neural crest induction. Initial evidence that these signals were important came from misexpression experiments in *Xenopus* embryos and explants. A number of Wnt ligands, including Wnt1, Wnt3a, Wnt8, and Wnt7b, have been shown to induce the expression of early neural crest markers when mis- or overexpressed in early *Xenopus* embryos or in neuralized animal caps (Saint-Jeannet et al., 1997; Chang and Hemmati-Brivanlou, 1998; LaBonne and Bronner-Fraser, 1998; Bang et al., 1999). Other Wnt-signaling components have been shown to possess similar activities, including β-catenin and the Wnt receptors Xfz3, Xfz7, and LRP6 (LaBonne and Bronner-Fraser, 1998; Tamai et al., 2000; Deardorff et al., 2001; Abu-Elmagd et al., 2006). Wnts have also been shown to induce neural crest markers in chick intermediate neural plate explants (Garcia-Castro et al., 2002; Taneyhill and Bronner-Fraser, 2005). Because Wnts can also regulate cell proliferation in some contexts, it was important to determine if the effects this signaling pathway has on neural crest cells were distinct from its ability to influence cell-cycle progression. Accordingly, Saint-Jeannet et al. (1997) demonstrated that Wnt-1 and Wnt-3a retain their ability to induce neural crest cells even in *Xenopus* embryos treated with the DNA synthesis inhibitor HUA, indicating that this activity is independent of cell proliferation.

A role for the canonical Wnt pathway in neural crest formation has been confirmed by loss-of-function experiments. In *Xenopus* embryos, dominant negative forms of Xwnt8 (LaBonne and Bronner-Fraser, 1998), LRP6 (Tamai et al., 2000), Xfz3 (Deardorff et al., 2001), or Xfz7 (Abu-Elmagd et al., 2006) have all been shown to inhibit the expression of neural crest markers. Similarly, morpholino-mediated depletion of Xfz3 (Deardorff et al., 2001), Xfz7 (Abu-Elmagd et al., 2006), or β-catenin (Wu et al., 2005) also prevents neural crest formation. Interestingly, Wnt-unresponsive mutants of Tcf-3, but not Lef-1, can block neural crest formation (Heeg-Truesdell and LaBonne, 2006), supporting earlier studies indicating that there is functional diversity between members of this family of Wnt-responsive transcription factors (Roel et al., 2002; Liu et al., 2005). In mice, compound-mutant embryos lacking both Wnt1 and Wnt3a show major deficiencies in neural crest cells (Ikeya et al., 1997), as do conditional knockouts of β-catenin using Cre-recombinase under the control of the Wnt1 promoter (Brault et al., 2001; Hari et al., 2002). In avian embryos, injection of cells expressing a dominant negative Wnt1 adjacent to the neural fold has been found to block *Slug* expression and the subsequent generation of HNK1-expressing migratory neural crest cells (Garcia-Castro et al., 2002).

Elegant work in zebrafish supports an early role for Wnt/β-catenin signaling in neural crest formation in this system as well. Lewis et al. (2004) used transgenic zebrafish embryos expressing a dominant negative Tcf3 (ΔTcf3) under the control of a heat shock promoter to show that Wnt signaling is required for endogenous neural crest induction, and also demonstrated that in this system, Wnt 8.1 is at least one of the necessary ligands. The requirement for Tcf3-mediated Wnt responses is cell autonomous, as wild-type neural crest cells grafted into embryos in which Tcf3 function had been inhibited continued to express neural crest markers, while ΔTcf3-expressing cells failed to express neural crest markers when grafted into wild-type embryos. Using morpholino-mediated depletion, the authors determined that Wnt 8.1, which is expressed in the paraxial mesoderm, is required for the initial expression of neural crest markers. However, neural crest formation later recovered in these embryos, indicating that other Wnt ligands can ultimately compensate for the loss of Wnt 8.1 (Lewis et al., 2004).

5. Wnt signals and a hierarchy of regulatory steps

Taken together, the above studies provide overwhelming evidence that canonical Wnt signals are required for neural crest formation. Complicating the interpretation of these studies, however, is the emerging realization that Wnt signaling may play distinct roles at several times throughout the induction process. Appreciation of this fact requires a more complete overview of what these different steps may be.

During gastrulation and neural induction, distinct regions of the embryonic ectoderm become progressively specified to form epidermis, CNS progenitors, neural crest, or cranial placodes. The neural crest and placodes arise at the border between cells of the presumptive CNS and epidermal cells (Fig. 2). Preceding the expression of definitive markers of these two cell types is a group of genes whose expression spans the neural crest and placodal domains, as well as dorsal CNS precursors (Heeg-Truesdell and LaBonne, 2004; Meulemans and Bronner-Fraser, 2004). Collectively, these factors have been termed "neural plate border specifiers" (Meulemans and Bronner-Fraser, 2004). Canonical Wnt signaling has been shown to be required for the expression of a subset of these neural plate border specifiers, including Pax3/7 and Msx1/2 (Bang et al., 1999). Other neural plate border specifiers appear to be regulated by FGF or BMP signaling (Nakata et al., 1997; Luo et al., 2001; Aruga et al., 2002; Monsoro-Burq et al., 2003; Tribulo et al., 2003). Interestingly, while two of these factors, Pax3 and ZicR1, can cooperate to induce the formation of ectopic neural crest cells outside the endogenous neural crest domain, their ability to do so remains dependent upon a Wnt signal (Monsoro-Burq et al., 2005). This suggests that a Wnt signal is also required for at least one subsequent step in the induction process.

Also functioning upstream of the expression of definitive neural crest markers is the proto-oncogene c-myc. This bHLH transcription factor is expressed in a more restricted domain than the above-mentioned neural plate border specifiers, but its expression still spans both neural crest and cranial placode precursors, suggesting that these two cell types share a common, c-myc expressing, precursor (Bellmeyer et al., 2003). c-myc has been shown to be a direct target of the Wnt-signaling pathway in a number of cell types and cancers (He et al., 1998; Willert et al., 2002), and consistent with this, expression of *c-myc* at the neural plate border is blocked in

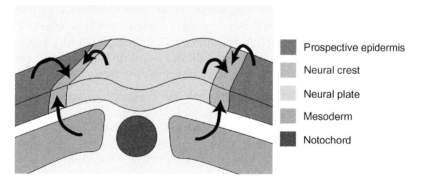

Fig. 2. Schematic summarizing the tissue interactions implicated in neural crest induction. A gradient of BMP signaling is believed to pattern the early ectoderm, with high levels of BMP signaling specifying epidermal fates. The neural plate forms in regions of low or absent BMP signaling, while intermediate levels of BMP signaling are permissive for neural crest induction. Interactions between the neural plate and prospective epidermis are implicated in neural crest formation. A Wnt signal originating from either the prospective epidermis or the underlying paraxial mesoderm is required for the induction of neural crest precursor cells. (See Color Insert.)

Xenopus embryos injected with mRNA encoding a dominant negative form of Wnt8 (Bellmeyer et al., 2003). Blocking Wnt signals was found to prevent expression of *c-myc* in neural crest forming regions at gastrula stages, placing Wnt-dependent *c-myc* expression near the top of the known hierarchy of regulatory events leading to neural crest formation. Interestingly, over-expression of dominant negative Wnt8 was also found to block *c-myc* expression in the anterior/transverse neural fold, providing evidence that the role of Wnt signals in patterning the neural plate border is not restricted to posterior cell types. While c-myc is known to regulate the expression of many target genes, more recent work has suggested that the small HLH protein Id3 is the major Myc target required for neural crest formation (Kee and Bronner-Fraser, 2005; Light et al., 2005) and that together c-myc and Id3 play a central role in the establishment and maintenance of a stem cell-like state in neural crest cells (Light et al., 2005).

c-myc, Id3, and the neural plate border specifiers lie upstream of, and are required for, the expression of genes that mark definitive neural crest. These genes include the transcription factors Slug/Snail, Sox 8/9/10, and FoxD3, which collectively have been referred to as "neural crest specifiers" (reviewed in Meulemans and Bronner-Fraser, 2004). While the expression of each of the aforementioned neural crest specifiers has been shown to be Wnt dependent in a number of model systems (reviewed in Heeg-Truesdell and LaBonne, 2004), this could in principle be indirect and due to the earlier requirement for Wnt signaling during induction of the upstream neural plate border specifiers and c-myc. Arguing for a direct role for Wnt signaling in the induction of at least a subset of neural crest specifiers is the presence of a conserved Lef/Tcf-binding site in the Slug promoter that is required for promoter function (Vallin et al., 2001; Sakai et al., 2005). Determining which additional neural plate border and neural crest specifiers are directly dependent upon Wnt responsive *cis*-regulatory elements will be essential to revealing the full picture of the roles that this signaling pathway plays in neural crest precursor formation and will also aid in ordering other aspects of the regulatory interactions that control this multistep process.

6. Wnt signals in delamination and the onset of migration

Subsequent to their roles in the establishment and maintenance of neural crest precursor cells, Wnts also play a role in controlling the onset of neural crest migration. This is essential, as it is their delamination from the neuroepithelium that ultimately distinguishes definitive neural crest cells from the precursor population that contributes to both these cell and the CNS cells of the roof plate (Collazo et al., 1993; Selleck and Bronner-Fraser, 1995; Bronner-Fraser and Fraser, 1988, 1989). The process of delamination requires an epithelial-mesenchymal transition (EMT), and alterations in cell surface properties, cytoskeleton, and cellular morphology that accompany such transitions (Hay, 1995; Nieto, 2001; Savagner, 2001). The EMT is an important event to understand, not only from a neural crest perspective, but also from a clinical perspective, as EMTs are essential to the formation of metastases from epithelial tumors (reviewed in Thiery, 2003). Following their EMT, neural crest cells migrate extensively along characteristic pathways to target sites, where they will differentiate into one of their many possible derivatives.

The signals involved in inducing delamination and migration of neural crest cells are beginning to be elucidated. Both intrinsic properties of the neural crest precursor cells themselves as well as cues from the environment help determine when and if these cells will delaminate and commence migration. Interestingly, in the spinal cord of avian embryos, neural crest cells have been shown to delaminate synchronously from the neural tube when they enter the S phase of the cell cycle (Burstyn-Cohen et al., 2004). As with other progenitor cells within the neuroepithelium, neural crest precursors undergo interkinetic apicobasal nuclear migration as they cycle, with DNA synthesis and delamination occurring at the basal surface, and mitosis occurring at the apical surface. The Wnt pathway appears to influence EMT by controlling exit from the cell cycle during S phase (Burstyn-Cohen et al., 2004).

It has been proposed that a balance between BMP and its inhibitor noggin regulates neural crest delamination in the trunk of avian embryos (Sela-Donenfeld and Kalcheim, 1999). According to this model, at axial levels where *noggin* expression is downregulated, BMP signaling becomes activated at high levels and as a result triggers neural crest cell delamination. Interestingly, at least in spinal cord regions, neural crest cells were found to emigrate synchronously at S phase of the cell cycle (Burstyn-Cohen and Kalcheim, 2002). Moreover, pharmacological agents, such as mimosine, that inhibit the G1/S transition were found to block delamination of neural crest cells. By contrast, treatment with drugs that inhibit the cell cycle at other phases had no effect on neural crest cell emigration. These results provided strong evidence that at least in the avian spinal cord, the transition between G1 and S phase is essential to the process leading to neural crest cell migration (Burstyn-Cohen and Kalcheim, 2002).

This same group subsequently demonstrated that BMP signaling regulates G1/S transition and neural crest delamination indirectly, through canonical Wnt signaling. In order to demonstrate which signals were responsible for regulating EMT, the neural tube was exposed to Noggin-expressing CHO cells or received Noggin via electroporation. Both experimental manipulations decreased the number of BrdU incorporating cells as well as delamination (Burstyn-Cohen et al., 2004). However, it was clear that BMP was not directly responsible for these effects, as activation of this pathway could not rescue mimosine-mediated inhibition of the G1/S transition. Due to

the temporal and spatial expression of Wnt1, and its ability to be regulated
by BMP signaling, it seemed possible that Wnt signaling might mediate the
BMP effects. Using a mutant Disheveled that blocks both canonical and
noncanonical Wnt signaling (Xdd1), Burstyn-Cohen and colleagues demonstrated that Wnt signals are necessary for cells of the dorsal neural tube to
progress from G1 to S. In addition to blocking the cell-cycle transition, the
majority of presumptive neural crest cells expressing Xdd1 failed to emigrate
from the neural tube. Importantly, the authors confirmed that it was the
Wnt/β-catenin pathway that influenced EMT and emigration, by demonstrating that both processes could be blocked using either a dominant negative
form of Lef1 or with a constitutive repressor form of β-catenin (fused with the
Engrailed repressor domain). By contrast, when these authors expressed a
mutant form of Disheveled that is thought to selectively interfere only with
noncanonical Wnt signals, they found that the noncanonical pathway has
little or no effect on neural crest cell emigration.

Consistent with the findings of Burstyn-Cohen et al. (2004), a second
group has reported that β-catenin and Lef-1 transiently localize to the
nucleus of neural crest cells at the time of their delamination (de Melker
et al., 2004). However, in contrast to the findings of the aforementioned
in vivo studies, de Melker et al. (2004), found that overstimulation of Wnt/
β-catenin signals at the time of neural crest emigration actually inhibited
neural crest cell delamination and migration (de Melker et al., 2004). Because
this second study was carried out on explants in culture, and not in
intact embryos, it is difficult to directly compare its findings to those of
Burstyn-Cohen et al. (2004).

Also surprisingly at odds with the work in avian embryos, a recent report
in *Xenopus* has suggested that it is noncanonical Wnt signaling, rather than
Wnt/β-catenin signaling, that controls the onset of neural crest migration
(De Calisto et al., 2005). This study also utilized Disheveled mutants thought
to distinguish between canonical and noncanonical pathways. The authors
found that while blocking noncanonical Wnt signaling did not affect *Slug*
expression, and thus presumably the specification of neural crest progenitors, it did interfere with neural crest migration. Grafting experiments
further suggested that the effects of noncanonical Wnt signaling on neural
crest migration were cell autonomous. This study went on to show that
Wnt11 is expressed, temporally and spatially, in a manner consistent with
a role in influencing neural crest cell migration, and that misexpression of
either Wnt11 or a dominant negative version of this ligand, interferes neural
crest cell migration.

It seems highly unlikely that the differences between the findings of the
Xenopus (De Calisto et al., 2005) and avian (Burstyn-Cohen et al., 2004)
studies discussed above are due to fundamental differences in how the onset
of migration is controlled in the two model organisms. Instead, the contrasting findings can likely be explained by differences related to the axial level

(cranial in *Xenopus*, spinal cord in chick) or the relative developmental time at which the signal perturbations were carried out. Careful future studies will be necessary to clarify the contributions that canonical and noncanonical Wnt signals make to the onset of neural crest migration in the two systems. It will also be important to examine the extent to which cell-cycle progression plays a role in this process in *Xenopus* and in cranial regions in the chick.

Beyond cell-cycle regulation, an additional way that Wnt signaling might regulate neural crest cell emigration is by directly altering the function of cell adherence junctions and/or by regulating the expression and activity of proteins that themselves regulate cell adhesion. In addition to its function as a downstream mediator of canonical Wnt signals, β-catenin serves as an important component of adherence junctions, where it bridges cadherin receptors and the actin cytoskeleton by interacting with the actin-binding protein, α-catenin (Rimm et al., 1995; Perez-Moreno et al., 2003; Gottardi and Gumbiner, 2004). Recent work has suggested that Wnt stimulation not only stabilizes the cytoplasmic pool of β-catenin but also directly modulates the adherence junction function of this protein (Gottardi and Gumbiner, 2004). This study found that Wnt signaling is able to alter the conformation of β-catenin so that it preferentially associates with Tcf in the nucleus as opposed to cadherins at cellular junctions. Using *in vitro* GST pull down assays, these authors demonstrated that β-catenin from cells expressing Wnt1, but not from untreated cells, preferentially bound to Tcf compared to cadherin (Gottardi and Gumbiner, 2004). Thus, a Wnt signal can cause a decrease in β-catenin association with adherence junctions, and therefore a decrease in cellular adhesiveness.

Canonical Wnt signals could also mediate such effects indirectly, by altering the expression and/or function of proteins that themselves regulate cell adhesion. One set of Wnt targets known to regulate cell adhesion are Slug/Snail factors. As previously mentioned, the Slug promoter contains a Wnt-dependent *cis*-regulatory element that is essential for its function (Vallin et al., 2001; Sakai et al., 2005). Moreover, Slug/Snail family repressors can directly induce both physiological and pathological EMTs, in part by directly downregulating the expression of E-cadherin (Batlle et al., 2000; Cano et al., 2000). Interestingly, Slug/Snail family proteins are regulated at the level of protein turnover, and at least in tumor cells, Wnt signaling may play a role in determining if and when Snail is targeted for degradation. The evidence for this is derived from two studies of human Snail carried out in tissue culture cells, both reporting that Snail stability is regulated by GSK3β phosphorylation and β-trcp-mediated ubiquitination (Zhou et al., 2004; Yook et al., 2005). Importantly, however, the well-defined β-trcp destruction motif (DSGX2 + nS), which is also found in β-catenin, is absent in Slug, and is not conserved in nonmammalian Snail proteins. This makes it unlikely that Wnt signaling is a widely deployed means of regulating the stability and thus activity of these factors. Indeed, a recent study has demonstrated that in *Xenopus* embryos, Slug stability is unaffected by this pathway (Vernon and LaBonne, 2006).

7. Wnt signals and neural crest cell fate diversification

The importance of the neural crest to the vertebrate body plan is a consequence of the vast variety of derivatives to which these cells contribute. Most of the peripheral nervous system is formed by these cells, including the entire autonomic nervous system (and its sympathetic, parasympathetic, and enteric nervous system components), the neurons of the dorsal root ganglia, and all of the glia and most of the neurons of the cranial sensory ganglia. Beyond the cell types that comprise the peripheral nervous system, neural crest cells give rise to all nonretinal pigment cells or melanocytes, the chromaffin cells of the andrenomedullary (as well as other endocrine and paraendocrine cells), and most of the craniofacial skeleton, including cartilage, bone, and connective tissue. This is a vast and diverse array of cell types, and it is therefore not surprising that a number of different signaling pathways, including the Wnt pathway, have been implicated in the process by which individual pluripotent neural crest progenitor cells are instructed to form specific derivatives (Fig. 3). Wnt signaling has been most convincingly implicated in the genesis of melanocytes and sensory neurons, and the evidence supporting these roles is discussed below.

In zebrafish, Lewis et al. (2004) were able to distinguish early versus late roles for Wnt/β-catenin signaling in neural crest development by utilizing a transgenic line expressing an inhibitory form of Tcf-3 (missing the β-catenin-binding domain) under the inducible control of the heat shock promoter. This allowed Wnt/β-catenin signals to be blocked at time points after neural crest progenitor cells had already been established, and these experiments revealed a later requirement for this signaling pathway in the formation of melanocytes (Lewis et al., 2004). This finding is consistent with the observation that melanocyte precursors arise close to the Wnt1/Wnt3a-expressing region of the dorsal neural tube. In zebrafish there is limited proliferation of neural crest progenitors, and fate mapping has suggested that more medial neural crest cells will give rise to melanocytes, whereas cells at more lateral positions will form neurons (Schilling and Kimmel, 1994). Moreover, the findings of Lewis et al. (2004) also built nicely on earlier work from the same group demonstrating that activation of β-catenin in individual lateral neural crest cells leads to melanocyte formation at the expense of neurons and glia (Dorsky et al., 1998). By contrast, blocking Wnt signaling in medial neural crest cells was found to inhibit melanocyte formation and promote neuronal and glial differentiation (Dorsky et al., 1998).

Studies in mice also support a role for Wnt/β-catenin signaling in instructing melanocyte formation. Compound mutant mice deficient for both Wnt1

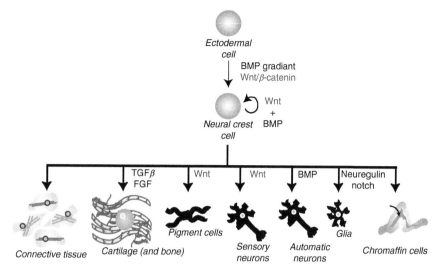

Fig. 3. Many signaling pathways are used reiteratively during neural crest development. Both BMP and Wnt signaling are implicated in the induction and maintenance of neural crest stem cells. These two signaling pathways also play a role in the subsequent diversification of neural crest fates. Wnts play instructive roles in the formation of both pigment cells and sensory neurons. The mechanisms that allow commonly deployed signaling pathways to direct such distinct cellular outcomes remain poorly understood. (See Color Insert.)

and Wnt3a show a dramatic loss of melanocytes (Ikeya et al., 1997). A more recent study has suggested that in this system, the primary role of Wnt1 may be the maintenance of neural crest stem cells, while Wnt3a functions primarily in neural crest cell fate determination (Dunn et al., 2005). However, the mechanisms that might underlie the proposed differences in the activity of these ligands remain unclear. In the case of Wnt3a, it appears to be activating the canonical-signaling pathway, as virally mediated upregulation of β-catenin in neural tube explants was also found to promote melanocyte differentiation at the expense of smooth muscle (Dunn et al., 2005). Consistent with this finding, a Cre-mediated conditional knockout of β-catenin in the neural crest results in deficiencies in melanocyte formation (Hari et al., 2002; Lee et al., 2004).

A role for Wnts in directing melanocyte fates is not unexpected. MITF, a transcription factor that is essential to the survival and terminal differentiation of melanocytes, has been shown to be a direct transcriptional target of Wnt/β-catenin signals (Dorsky et al., 2000; Goding, 2000; Takeda et al., 2000; Saito et al., 2002; Elworthy et al., 2003). Moreover, the expression of a downstream transcriptional target of MITF, the dopachrome tautomerase (DCT) gene, is also directly dependent on an Lef-1-binding site in its promoter (Saito et al., 2002; Yasumoto et al., 2002; Jiao et al., 2004; Ludwig et al., 2004). These findings indicate both that Wnt signaling is directly required for multiple steps in melanocyte differentiation, and, importantly, that is it is also required for the formation of this lineage in mammals.

While studies of MITF regulation had predicted a requirement for Wnt/β-catenin signals in the development of the melanocyte lineage, less expected was a role for this pathway in directing sensory neurogenesis. Nevertheless, mice in which β-catenin has been conditionally inactivated in the neural crest also show deficits in the sensory lineage (Hari et al., 2002; Lee et al., 2004). To further investigate this finding, mice were engineered in which Wnt/β-catenin signals were constitutively active in neural crest progenitor cells (Lee et al., 2004). These mice showed a dramatic increase in the sensory neuron population that was not due to effects on cell proliferation, indicating that the sensory lineage had been promoted at the expense of other neural crest-derived cell types such as sympathetic neurons and melanocytes (Lee et al., 2004). Future work must reconcile these findings with those that indicate that Wnt/β-catenin signaling promotes melanocyte differentiation. Given that the murine loss-of-function studies indicate that Wnt signals are necessary for the formation of both derivatives, it seems likely that the apparent difference in the results of gain-of-function studies in the mouse and fish will prove to be based in timing. For example, activation of Wnt/β-catenin signals in neural crest progenitors at early time points may promote sensory neuron development, whereas later activation of these same signals may lead to melanocyte formation. The challenge for the future will thus be to elucidate how the same signal can elicit distinct responses at different stages of development.

Given that constitutive activation of β-catenin induces almost all neural crest progenitors in the dorsal neural tube to form sensory neurons but that these progenitors themselves arise in Wnt-expressing region of the neural tube, what mechanisms limit sensory neurogenesis during normal development? A more recent study from the same group examined the response of neural crest stem cells to combined Wnt/β-catenin and BMP signals, and found that these factors can act synergistically to maintain neural crest cells in a progenitor state (Kleber et al., 2005). This study was carried out on neural crest stem cells in culture, which allows control over the influence of additional signals and eliminates any effects from migration defects. In these cultures, Wnt/β-catenin signaling promotes sensory neuron fates, as it had been shown to in vivo, whereas BMP signaling promotes autonomic neurogenesis. Perhaps most intriguingly, when the two signals are combined, they prove mutually antagonistic, and as a consequence differentiation is suppressed and stem cell potency is maintained.

Using a Wnt-responsive reporter in vivo, this group also found that Wnt signaling is active in regions where neural crest cells are emigrating, but decreases in postmigratory neural crest cells (Kleber et al., 2005). Interestingly, this finding is consistent with work in chick suggesting that Wnt/β-catenin signals regulate the process of neural crest delamination (Burstyn-Cohen et al., 2004).

In the context of lineage diversification, however, the findings of Kleber and colleagues suggest that there may be a finite window of time during neural crest development when these cells are competent to respond to Wnt/β-catenin signals by adopting a sensory neuron fate. The model proposed by this group suggests that a combination of Wnt and BMP signaling maintains neural crest cells in a stem cell state, whereas Wnt signals act upon early emigrating cells to promote sensory fates, and BMP stimulation to promote autonomic fates (Kleber et al., 2005). This study also highlights the importance of understanding the roles of each signaling pathways not only individually, but also in combination.

8. Conclusions and future directions

This chapter reviews the numerous essential roles that Wnt signals play at multiple stages of neural crest development. It would appear from the sum of these studies that Wnt/β-catenin signals are essential to at least five distinct steps in the genesis of specific neural crest derivatives from cells at the neural plate border. Yet it is also true that Wnts are unlikely to act alone at any of these steps, but instead function together with other signaling pathways including the BMP, FGF, and Notch pathways. An important challenge for the future will therefore be to determine the mechanistic basis for such combinatorial signaling.

It will also be important to determine the mechanisms via which a Wnt-regulated G1/S transition leads to the delamination of neural crest cells from the neuroepithelium. Moreover, studies on the role of cell cycle in neural crest emigration need to be extended to other model organisms, as well as to the cranial neural crest in avian embryos. A better understanding of the roles that noncanonical Wnt signals play at all stages of neural crest development, but especially during the onset of neural crest migration, is clearly essential. And finally, a more complete view of the precise ligand–receptor pairs that mediate each of the Wnt-dependent steps in neural crest development discussed above would prove of great value in furthering our understanding these important inductive events.

Acknowledgments

The authors gratefully acknowledge members of the laboratory for valuable discussions. EH-T was supported by the Cellular and Molecular Basis of Disease training program (NIGMS). CL is a Scholar of the GM Cancer Research Foundation. Work in our laboratory referenced within was supported by grants from the American Cancer Society and the NIH to CL.

References

Abu-Elmagd, M., Garcia-Morales, C., Wheeler, G.N. 2006. Frizzled 7 mediates canonical Wnt signalling in neural crest induction. Dev. Biol. 298, 285–298.

Amit, S., Hatzubai, A., Birman, Y., Andersen, J.S., Ben-Shushan, E., Mann, M., Ben-Neriah, Y., Alkalay, I. 2002. Axin-mediated CKI phosphorylation of beta-catenin at Ser 45: A molecular switch for the Wnt pathway. Genes Dev. 16, 1066–1076.

Aruga, J., Tohmonda, T., Homma, S., Mikoshiba, K. 2002. Zic1 promotes the expansion of dorsal neural progenitors in spinal cord by inhibiting neuronal differentiation. Dev. Biol. 244, 329–341.

Bang, A.G., Papalopulu, N., Goulding, M.D., Kintner, C. 1999. Expression of Pax-3 in the lateral neural plate is dependent on a Wnt-mediated signal from posterior nonaxial mesoderm. Dev. Biol. 212, 366–380.

Batlle, E., Sancho, E., Francí, C., Domínguez, D., Monfar, M., Baulida, J., García de Herreros, A. 2000. The transcription factor snail is a repressor of E-cadherin gene expression in epithelial tumour cells. Nat. Cell Biol. 2, 84–89.

Bellmeyer, A., Krase, J., Lindgren, J., LaBonne, C. 2003. The protooncogene c-Myc is an essential regulator of neural crest formation in Xenopus. Dev. Cell 4, 827–839.

Brault, V., Moore, R., Kutsch, S., Ishibashi, M., Rowitch, D.H., McMohan, A.P., Sommer, L., Boussadia, O., Kemlar, R. 2001. Inactivation of the beta-catenin gene by Wnt1-Cre-mediated deletion results in dramatic brain malformation and failure of craniofacial development. Development 128, 1253–1264.

Burstyn-Cohen, T., Kalcheim, C. 2002. Association between the cell cycle and neural crest delamination through specific regulation of G1/S transition. Dev. Cell 3, 383–395.

Burstyn-Cohen, T., Stanleigh, J., Sela-Donenfeld, D., Kalcheim, C. 2004. Canonical Wnt activity regulates trunk neural crest delamination linking BMP/noggin signaling with G1/S transition. Development 131, 5327–5339.

Bronner-Fraser, M., Fraser, S.E. 1988. Cell lineage analysis reveals multipotency of some avian neural crest cells. Nature 335, 161–164.

Bronner-Fraser, M., Fraser, S. 1989. Developmental potential of avian trunk neural crest cells in situ. Neuron 3, 755–766.

Cano, A., Pérez-Moreno, M.A., Rodrigo, I., Locascio, A., Blanco, M.J., del Barrio, M.G., Portillo, F., Nieto, M.A. 2000. The transcription factor snail controls epithelial-mesenchymal transitions by repressing E-cadherin expression. Nat. Cell Biol. 2, 76–83.

Chang, C., Hemmati-Brivanlou, A. 1998. Neural crest induction by Xwnt7B in Xenopus. Dev. Biol. 194, 129–134.

Collazo, A., Bronner-Fraser Fraser, S.E. 1993. Vital dye labelling of Xenopus laevis trunk neural crest reveals multipotency and novel pathways of migration. Development 118, 363–376.

De Calisto, J., Araya, C., Marchant, L., Riaz, C.F., Mayor, R. 2005. Essential role of non-canonical Wnt signalling in neural crest migration. Development 132, 2587–2597.

de Melker, A.A., Desban, N., Duband, J.L. 2004. Cellular localization and signaling activity of beta-catenin in migrating neural crest cells. Dev. Dyn. 230, 708–726.

Deardorff, M.A., Tan, C., Saint-Jeannet, J.P., Klein, P.S. 2001. A role for frizzled 3 in neural crest development. Development 128, 3655–3663.

Dorsky, R.I., Moon, R.T., Raible, D.W. 1998. Control of neural crest cell fate by the Wnt signalling pathway. Nature 396, 370–373.

Dorsky, R.I., Raible, D.W., Moon, R.T. 2000. Direct regulation of nacre, a zebrafish MITF homolog required for pigment cell formation, by the Wnt pathway. Genes Dev. 14, 158–162.

Dunn, K.J., Brady, M., Ochsenbauer-Jambor, C., Snyder, S., Incao, A., Pavan, W.J. 2005. WNT1 and WNT3a promote expansion of melanocytes through distinct modes of action. Pigment Cell Res. 18, 167–180.

Elworthy, S., Lister, J.A., Carney, T.J., Raible, D.W., Kelsh, R.N. 2003. Transcriptional regulation of mitfa accounts for the sox10 requirement in zebrafish melanophore development. Development 130, 2809–2818.

Garcia-Castro, M.I., Marcelle, C., Bronner-Fraser, M. 2002. Ectodermal Wnt function as a neural crest inducer. Science 297, 848–851.

Gavin, B.J, McMahon, J.A, McMahon, A.P. 1990. Expression of multiple novel Wnt-1/int-1-related genes during fetal and adult mouse development. Genes Dev. 4, 2319–2332.

Goding, C.R. 2000. Mitf from neural crest to melanoma: Signal transduction and transcription in the melanocyte lineage. Genes Dev. 14, 1712–1728.

Goodman, F.R. 2003. Congenital abnormalities of body patterning: Embryology revisited. Lancet 362, 651–662.

Gottardi, C.J., Gumbiner, B.M. 2004. Distinct molecular forms of beta-catenin are targeted to adhesive or transcriptional complexes. J. Cell Biol. 167, 339–349.

Hall, B. 1999. *The Neural Crest in Development and Evolution,* New York: Springer Verlag.

Hall, B.K. 2000. Epithelial-mesenchymal interactions. Methods. Mol. Biol. 137, 235–243.

Hari, L., Brault, V., Kleber, M., Lee, H.Y., Ille, F., Leimeroth, R., Paratore, C., Suter, U., Kemler, R., Sommer, L. 2002. Lineage-specific requirements of beta-catenin in neural crest development. J. Cell Biol. 159, 867–880.

Hay, E.D. 1995. An overview of epithelio-mesenchymal transformation. Acta Anat. (Basel) 154, 8–20.

He, X. 2003. A Wnt-Wnt situation. Dev. Cell 4, 791–797.

He, T.C., Sparks, A.B., Rago, C., Hermeking, H., Zawel, L., da Costa, L.T., Morin, P.J., Vogelstein, B., Kinzler, K.W. 1998. Identification of c-MYC as a target of the APC pathway. Science 281(5382), 1509–1512.

Heeg-Truesdell, E., LaBonne, C. 2004. A slug, a fox, a pair of sox: Transcriptional responses to neural crest inducing signals. Birth Defects Res. C Embryo Today 72, 124–139.

Heeg-Truesdell, E., LaBonne, C. 2006. Neural induction in *Xenopus* requires inhibition of Wnt-β-catenin signaling. Dev. Biol. 298, 71–86.

Hurlstone, A., Clevers, H. 2002. T-cell factors: Turn-ons and turn-offs. EMBO J. 21, 2303–2311.

Ikeya, M., Lee, S.M., Johnson, J.E., McMahon, A.P., Takada, S. 1997. Wnt signalling required for expansion of neural crest and CNS progenitors. Nature 389, 966–970.

Jiao, Z., Mollaaghababa, R., Pavan, W.J., Antonellis, A., Green, E.D., Hornyak, T.J. 2004. Direct interaction of Sox10 with the promoter of murine Dopachrome Tautomerase (Dct) and synergistic activation of Dct expression with Mitf. Pigment Cell Res. 17, 352–362.

Kee, Y., Bronner-Fraser, M. 2005. To proliferate or to die: Role of Id3 in cell cycle progression and survival of neural crest progenitors. Genes Dev. 19, 744–755.

Kleber, M., Lee, H.Y., Wurdak, H., Buchstaller, J., Riccomagno, M.M., Ittner, L.M., Suter, U., Epstein, D.J., Sommer, L. 2005. Neural crest stem cell maintenance by combinatorial Wnt and BMP signaling. J. Cell Biol. 169, 309–320.

LaBonne, C., Bronner-Fraser, M. 1998. Neural crest induction in *Xenopus*: Evidence for a two-signal model. Development 125, 2403–2414.

Le Douarin, N.M. 1982. *The Neural Crest.* Cambridge: Cambridge University Press.

Lee, H.Y., Kleber, M., Hari, L., Brault, V., Suter, U., Taketo, M.M., Kemler, R., Sommer, L. 2004. Instructive role of Wnt/beta-catenin in sensory fate specification in neural crest stem cells. Science 303, 1020–1023.

Lewis, J.L., Bonner, J., Modrell, M., Ragland, J.W., Moon, R.T., Dorsky, R.I., Raible, D.W. 2004. Reiterated Wnt signaling during zebrafish neural crest development. Development 131, 1299–1308.

Light, W., Vernon, A.E., Lasorella, A., Iavarone, A., LaBonne, C. 2005. *Xenopus* Id3 is required downstream of Myc for the formation of multipotent neural crest progenitor cells. Development 132, 1831–1841.

Liu, C., Li, Y., Semenov, M., Han, C., Baeg, G.H., Tan, Y., Zhang, Z., Lin, X., He, X. 2002. Control of beta-catenin phosphorylation/degradation by a dual-kinase mechanism. Cell 108, 837–847.

Liu, F., van den Broek, O., Destree, O., Hoppler, S. 2005. Distinct roles for *Xenopus* Tcf/Lef genes in mediating specific responses to Wnt/β-catenin signalling in mesoderm development. Development 132, 5375–5385.

Ludwig, A., Rehberg, S., Wegner, M. 2004. Melanocyte-specific expression of dopachrome tautomerase is dependent on synergistic gene activation by the Sox10 and Mitf transcription factors. FEBS Lett. 556, 236–244.

Luo, T., Matsuo-Takasaki, M., Lim, J.H., Sargent, T.D. 2001. Differential regulation of Dlx gene expression by a BMP morphogenetic gradient. Int. J. Dev. Biol. 45, 681–684.

Meulemans, D., Bronner-Fraser, M. 2004. Gene-regulatory interactions in neural crest evolution and development. Dev. Cell 7, 291–299.

Monsoro-Burq, A.H., Fletcher, R.B., Harland, R.M. 2003. Neural crest induction by paraxial mesoderm in *Xenopus* embryos requires FGF signals. Development 130, 3111–3124.

Monsoro-Burq, A.H., Wang, E., Harland, R. 2005. Msx1 and Pax3 cooperate to mediate FGF8 and WNT signals during *Xenopus* neural crest induction. Dev. Cell 8, 167–178.

Morales, A.V., Barbas, J.A., Nieto, M.A. 2005. How to become neural crest: From segregation to delamination. Semin. Cell Dev. Biol. 16, 655–662.

Nakagawara, A., Ohira, M. 2004. Comprehensive genomics linking between neural development and cancer: Neuroblastoma as a model. Cancer Lett. 204, 213–224.

Nakata, K., Nagai, T., Aruga, J., Mikoshiba, K. 1997. *Xenopus* Zic3, a primary regulator both in neural and neural crest development. Proc. Natl. Acad. Sci. USA 94, 11980–11985.

Nieto, M.A. 2001. The early steps of neural crest development. Mech. Dev. 105, 27–35.

Pandur, P., Lasche, M., Eisenberg, L.M., Kuhl, M. 2002. Wnt-11 activation of a non-canonical Wnt signalling pathway is required for cardiogenesis. Nature 418, 636–641.

Perez-Moreno, M., Jamora, C., Fuchs, E. 2003. Sticky business: Orchestrating cellular signals at adherens junctions. Cell 112, 535–548.

Rimm, D.L., Koslov, E.R., Kebriaei, P, Cianci, C.D., Morrow, J.S. 1995. Alpha 1(E)-catenin is an actin-binding and -bundling protein mediating the attachment of F-actin to the membrane adhesion complex. Proc. Natl. Acad. Sci. USA 92, 8813–8817.

Roel, G., Hamilton, F.S., Gent, Y., Bain, A.A., Destree, O., Hoppler, S. 2002. Lef-1 and Tcf-3 transcription factors mediate tissue-specific Wnt signaling during *Xenopus* development. Curr. Biol. 12, 1941–1945.

Rosivatz, E., Becker, I., Specht, K., Fricke, E., Luber, B., Busch, R., Höfler, H., Becker, K.F. 2002. Differential expression of the epithelial-mesenchymal transition regulators snail, SIP1, and twist in gastric cancer. Am. J. Pathol. 161, 1881–1891.

Rothhammer, T., Hahne, J.C., Florin, A., Poser, I., Soncin, F., Wernert, N., Bosserhoff, A.K. 2004. The Ets-1 transcription factor is involved in the development and invasion of malignant melanoma. Cell Mol. Life Sci. 61, 118–128.

Saint-Jeannet, J.P., He, X., Varmus, H.E., Dawid, IB. 1997. Regulation of dorsal fate in the neuraxis by Wnt-1 and Wnt-3a. Proc. Natl. Acad. Sci. USA 94, 13713–13718.

Saito, H., Yasumoto, K., Takeda, K., Takahashi, K., Fukuzaki, A., Orikasa, S., Shibahara, S. 2002. Melanocyte-specific microphthalmia-associated transcription factor isoform activates its own gene promoter through physical interaction with lymphoid-enhancing factor 1. J. Biol. Chem. 277, 28787–28794.

Sakai, D., Tanaka, Y., Endo, Y., Osumi, N., Okamoto, H., Wakamatsu, Y. 2005. Regulation of Slug transcription in embryonic ectoderm by beta-catenin-Lef/Tcf and BMP-Smad signaling. Dev. Growth Differ. 47(7), 471–482.

Savagner, P. 2001. Leaving the neighborhood: Molecular mechanisms involved during epithelial-mesenchymal transition. Bioessays 23, 912–923.

Sela-Donenfeld, D., Kalcheim, C. 1999. Regulation of the onset of neural crest migration by coordinated activity of BMP4 and noggin in the dorsal neural tube. Development 126, 4749–4762.

Selleck, M.A., Bronner-Fraser, M. 1995. Origins of the avian neural crest: The role of neural plate-epidermal interactions. Development 121, 525–538.

Schilling, T.F., Kimmel, C.B. 1994. Segment and cell type lineage restrictions during pharyngeal arch development in the zebrafish embryo. Development 120, 483–494.

Takeda, K., Yasumoto, K., Takada, R., Takada, S., Watanabe, K., Udono, T., Saito, H., Takahashi, K., Shibahara, S. 2000. Induction of melanocyte-specific microphthalmia-associated transcription factor by Wnt-3a. J. Biol. Chem. 275, 14013–14016.

Tamai, K., Semenov, M., Kato, Y., Spokony, R., Liu, C., Katsuyama, Y., Hess, F., Saint-Jeannet, J.P., He, X. 2000. LDL-receptor-related proteins in Wnt signal transduction. Nature 407, 530–535.

Taneyhill, L.A., Bronner-Fraser, M. 2005. Recycling signals in the neural crest. J. Biol. 4, 10.

Thiery, J.P. 2003. Epithelial-mesenchymal transitions in development and pathologies. Curr. Opin. Cell Biol. 15, 740–746.

Tribulo, C., Aybar, M.J., Nguyen, V.H., Mullins, M.C., Mayor, R. 2003. Regulation of Msx genes by a Bmp gradient is essential for neural crest specification. Development 130, 6441–6452.

Vallin, J., Thuret, R., Giacomello, E., Faraldo, M.M., Thiery, J.P., Broders, F. 2001. Cloning and characterization of three *Xenopus* slug promoters reveal direct regulation by Lef/beta-catenin signaling. J. Biol. Chem. 276, 30350–30358.

Van de Putte, T., Maruhashi, M., Francis, A., Nelles, L., Kondoh, H., Huylebroeck, D., Higashi, Y. 2003. Mice lacking ZFHX1B, the gene that codes for Smad-interacting protein-1, reveal a role for multiple neural crest cell defects in the etiology of Hirschsprung disease-mental retardation syndrome. Am. J. Hum. Genet. 72, 465–470.

Vernon, A.E., LaBonne, C. 2006. Slug stability is dynamically regulated during neural crest development by the F-box protein, Ppa. Development 133(17), 3359–3370.

Wilkie, A.O., Morriss-Kay, G.M. 2001. Genetics of craniofacial development and malformation. Nat. Rev. Genet. 2, 458–468.

Willert, J., Epping, M., Pollack, J.R., Brown, P.O., Nusse, R. 2002. A transcriptional response to Wnt protein in human embryonic carcinoma cells. BMC Dev. Biol. 2, 8.

Wolda, S.L., Moody, C.J., Moon, R.T. 1993. Overlapping expression of Xwnt-3A and Xwnt-1 in neural tissue of *Xenopus laevis* embryos. Dev. Biol. 155, 46–57.

Wu, J., Yang, J., Klein, PS. 2005. Neural crest induction by the canonical Wnt pathway can be dissociated from anterior-posterior neural patterning in *Xenopus*. Dev. Biol. 279, 220–232.

Yanagawa, S., Matsuda, Y., Lee, J.S., Matsubayashi, H., Sese, S., Kadowaki, T., Ishimoto, A. 2002. Casein kinase I phosphorylates the Armadillo protein and induces its degradation in *Drosophila*. EMBO J. 21, 1733–1742.

Yanfeng, W., Saint-Jeannet, J.P., Klein, P.S. 2003. Wnt-frizzled signaling in the induction and differentiation of the neural crest. Bioessays 25, 317–325.

Yasumoto, K., Takeda, K., Saito, H., Watanabe, K., Takahashi, K., Shibahara, S. 2002. Microphthalmia-associated transcription factor interacts with LEF-1, a mediator of Wnt signaling. EMBO J. 21, 2703–2714.

Yook, J.I., Li, X.Y., Ota, I., Fearon, E.R., Weiss, S.J. 2005. Wnt-dependent regulation of the E-cadherin repressor snail. J. Biol. Chem. 280, 11740–11748.

Zhou, B.P., Deng, J., Xia, W., Xu, J., Li, Y.M., Gunduz, M., Hung, M.C. 2004. Dual regulation of Snail by GSK-3beta-mediated phosphorylation in control of epithelial-mesenchymal transition. Nat. Cell Biol. 6, 931–940.

Wnt pathways in angiogenesis

T. Néstor H. Masckauchán and Jan Kitajewski

Department of Pathology, OB/GYN and Institute of Cancer Genetics,
Columbia University Medical Center, New York, New York

Contents

Analysis of Wnt- or Frizzled-deficient mice provides genetic evidence that angiogenesis is dependent on Wnt signaling. This evidence involves Wnt2, 4, and 7b, Frizzled 4 and 5, Wnt coreceptor low-density-lipoprotein receptor-related protein 5 (LRP5), and Norrin, a non-Wnt ligand for Frizzled 4. In addition, known Wnt-signal modulators, such as secreted Frizzled related proteins (sFRP), can regulate angiogenesis. In the past few years, genetic disorders involving Frizzled 4, Norrin, and LRP5 have been linked to pathological phenotypes in angiogenesis in both mice and humans. Human hereditary disorders of the eye, such as familial exudative vitreoretinopathy (FEVR) and Norrie disease (ND), are related to aberrant Wnt signaling in pathological retinal angiogenesis. Growing evidence *in vitro* has also linked Wnts and Wnt cascade components with regulation of angiogenesis. Canonical Wnt signaling can modulate endothelial cell

Advances in Developmental Biology
Volume 17 ISSN 1574-3349
DOI: 10.1016/S1574-3349(06)17007-6

function *in vitro*, promoting proliferation, survival, and migration of endothelial cells. Known angiogenic factors, like vascular endothelial growth factor (VEGF), fibroblast growth factor (FGF), and interleukin-8, are direct downstream targets of Wnt signaling and may function in Wnt-mediated angiogenesis. It is now clear that Wnts and Frizzled function in angiogenesis. Future work will help establish the relative contribution of Wnt signaling to the developing vasculature and pathological angiogenesis.

1. Angiogenesis: A brief introduction

Angiogenesis is a physiological process aimed at providing a vascular network for the delivery of oxygen and consists of the generation of new blood vessels from preexisting ones. This process is accomplished by either sprouting or intussusception from existing vessels and occurs in several stages. First, vascular permeability increases across the endothelial cell layer through dissolution of adherens junctions (Pepper, 2001). Basal membrane is then degraded, controlled mainly by matrix metalloproteinases (MMPs), to allow endothelial cells to sprout into the interstitial space. Then, cells proliferate outside the vasculature, interacting with neighboring extracellular matrix components. These extravascular cells are able to form new structures with lumens, while other cells migrate to the site of angiogenesis. Finally, the newly formed vessels mature and stabilize, while mural cells are recruited by factors secreted from endothelial cells.

Angiogenesis mostly occurs during embryonic development in mammals, when the vascular plexus extends and remodels into an organized network. After birth, angiogenesis still occurs concomitant with body growth, but in adulthood, most blood vessels are quiescent and ongoing angiogenesis is mostly restricted to the ovaries and womb during menstrual cycles, placental tissue during pregnancy, and to wounds during healing. A large number of pathological processes also involve reactivation of angiogenesis, including diabetes, endometriosis, and inflammatory diseases. Angiogenesis is critical in tumorigenesis and metastasis, as well as in several inherited vascular retinal disorders, which we will discuss later. Angiogenesis is finely regulated by a wide variety of growth factors working in timely coordination. Promoters of angiogenesis include molecules such as vascular endothelial growth factor (VEGF), hepatocyte growth factor (HGF), epidermal growth factor (EGF), fibroblast growth factor 2 (FGF-2), prostaglandins, hypoxia-induced factor 1 α (HIF1α), and inhibitors like angiostatin and endostatin.

Recent experimental evidence has placed Wnt growth factors among the regulators of angiogenesis. The evidence comes from a wide variety of experimental paradigms, including knockout mouse models, expression analysis, signaling, and *in vitro* angiogenesis assays using cultured endothelial cells. In humans, identification of genetic defects in several hereditary disorders that display similar phenotypes of the retinal vasculature has identified Wnt-signaling components as critical to angiogenesis. These findings make it clear that Wnt signaling influences the angiogenic process. Future research in this area will likely focus on understanding the full scope of Wnt/Frizzled angiogenic activities. The diversity of activities ascribed to Wnts is a theme common to many developmental processes and this is true for vascular development, as illustrated by a discussion of vascular phenotypes in Wnt/Frizzled mutant embryos, discussed next.

2. Wnts and Frizzleds in vascular biology

Analysis of Wnt- or Frizzled-deficient mice has provided genetic evidence that developmental angiogenesis is dependent on Wnt signaling.

Wnt2-deficient mouse embryos showed inability to develop a proper fetal capillary placental network (Monkley et al., 1996). This is in accordance with the observation that Wnt2, a Wnt that activates canonical Wnt signaling (Shimizu et al., 1997), can be detected in fetal vessels of the placenta (Monkley et al., 1996). In addition, a Wnt4 knockout mouse showed that Wnt4 is able to repress mesonephric endothelial and steroidogenic cell migration, preventing the formation of male-specific coelomic blood vessels and the production of steroids (Jeays-Ward et al., 2003).

Wnt7b, which can activate canonical Wnt signaling (Wang et al., 2005), is important for lung vascular development, since Wnt7b-null embryos display defects in smooth musculature of larger pulmonary vessels (Shu et al., 2002). At E18.5, Wnt7b-null embryos have enlarged and branched vessels surrounded by extensive lung hemorrhage. Also, Wnt7b-null neonate mice have increased loss of vascular smooth muscle cells, leading to impaired vascular smooth muscle integrity in the lungs. However, a different Wnt7b knockout mouse model showed no evident defects in the vasculature (Parr et al., 2001).

Wnt7b function is critical for proper regression of hyaloid vessels in the developing retina (Lobov et al., 2005). The murine hyaloid vasculature, which develops inside the vitreous body during embryonic life, normally regresses postnatally within 2 weeks. This vascular regression depends on macrophages which secrete Wnt7b. Similarly, homozygous mice null for Lrp5, a Wnt coreceptor required for canonical Wnt signaling, showed reduced vascular endothelial cell apoptosis, implying that Lrp5 is also required for proper hyaloid vessel regression. Thus, in this particular context, canonical Wnt signaling plays an important paracrine functional role inducing cell death in endothelial cells.

The Wnt receptors, Frizzleds, were also implicated as angiogenic receptors, originally from analysis of knockout mouse models. Homozygous

Frizzled-5-null mice died *in utero* around E10.75 due to defects in yolk sac angiogenesis, as evidenced by staining for PECAM, an endothelial cell marker (Ishikawa et al., 2001). Endothelial cell proliferation was significantly reduced in yolk sacs, as measured at E10.25, with large vitelline vessels poorly extended and disorganized branching in the capillary plexus. Placental vasculogenesis was also impaired, although embryonic vasculature was apparently normal. Wnt5a and Wnt10b were found colocalized with Frizzled-5 in the developing yolk sac by *in situ* hybridization, and there was functional interaction of both Wnts with Frizzled-5 as evaluated by secondary axis formation in *Xenopus* embryos. This observation led investigators to believe that both Wnts might be the natural ligands of Frizzled-5 responsible for endothelial growth in the yolk sac. Furthermore, a Frizzled-4-null mouse model showed severe vascular defects (Xu et al., 2004), thus providing important evidence of a link between Wnts and angiogenesis. We will discuss this animal model in greater depth in subsequent sections.

3. Wnt signaling and pathological angiogenesis

The first breakthrough in the study of Wnt signaling in pathological angiogenesis in humans came in 2002 with the finding that mutations in the FRIZZLED-4 (FZD4) gene were detected in a significant number of individuals affected by familial exudative vitreoretinopathy (FEVR) (Robitaille et al., 2002). FEVR is a human hereditary disorder characterized by a lack of development of peripheral retinal vasculature. This may lead to defective hyperpermeable neovasculature that exudates and bleeds, and sometimes to retinal detachment. Researchers were originally able to directly link FZD4 mutations to FEVR in patients from two unrelated families (Robitaille et al., 2002). Additionally, mutant FZD4 was not capable of activating calcium calmodulin-dependent protein kinase II and protein kinase C, while wild-type control FZD4 was capable (Robitaille et al., 2002). This suggested the possibility that FZD4 would be capable of mediating intracellular calcium signaling and that this pathway could be defective in tissues expressing mutant forms of this receptor.

The FZD4 gene is located at one of the four different loci that have ultimately been associated with autosomal FEVR to date. Another gene identified in the same locus encodes for low-density-lipoprotein receptor-related protein, member 5 (LRP5) (Toomes et al., 2004a), a Wnt coreceptor. Loss of LRP5 function causes osteoporosis-pseudoglioma syndrome (OPPG), a recessive disease that develops with low bone mass (Gong et al., 2001). Many patients also present with severe disruption of ocular structure. Persistent hyperplasia of the primary vitreous in children with milder eye involvement has led to speculation that ocular phenotype is a consequence of a failure in the regression of primary vitreal vasculature during fetal growth (Toomes et al., 2004a). When the LRP5 gene was screened in a group of FEVR patients, LRP5 insertion, deletion, and splicing mutations not present in control individuals were found. Together, mutations in LRP5 and FZD4 genes accounted for about 35% of FEVR cases, indicating that other genes are also involved in this disease (Toomes et al., 2004a).

Fzd4 mutant can oligomerize with the wild-type receptor, and the complexes were shown to be retained in the endoplasmic reticulum, depleting the availability of wild-type Fzd4 (Kaykas et al., 2004). Thus, mutant Fzd4 might function by reducing the ability of the wild-type form to transduce signal, explaining the genetic dominance of mutant Fzd4 in FEVR. These findings directly correlated FEVR with aberrant Wnt signaling, elucidating the need for further understanding the mechanisms of signal transduction through Fzd4 and LRP5. Further evidence showing that Fzd4 can also mediate canonical Wnt signaling was also reported (Xu et al., 2004), as we will discuss later, suggesting that Fzd4 can be involved in both Wnt/Ca^{2+} and canonical Wnt signaling.

In accordance with the phenotype observed in FEVR cases, retinas from neonatal Fzd4$^{-/-}$ mice are missing two deeper intraretinal capillary beds, with enlarged and tortuous arteries and veins in the retina, and reduced arteriolar branching (Xu et al., 2004). The hyaloid vasculature, which normally regresses within 2 weeks after birth, is still present by that time in the Fzd4$^{-/-}$ mouse. This animal model also presents frequent retinal hemorrhages not present in the wild-type animals. Altogether, these observations clearly established a direct link between Wnt signaling and retinal angiogenesis.

In a recent and exciting finding, expression of VHL, a tumor suppressor protein that directly participates in the HIF1α-mediated hypoxia response, was found regulated by Tcf4 (Giles et al., 2006). HIF1α expression is a component of the hypoxia response that leads to increased VEGF production. This response can lead to an angiogenic switch in tumor angiogenesis, when increasing demands in nutrient supplies by tumors must be met by the generation of newly formed blood vessels. VHL expression was found completely absent in proliferative intestinal pockets of Tcf4$^{-/-}$ perinatal mice, while normoxic expression HIF1α was detected. Moreover, expression of VHL was very high in early lesions of intestinal mucosa from patients with familial adenomatous polyposis, a cancer form associated with mutations of Wnt-signaling components conferring longer half-life to β-catenin. In later stages of colorectal cancer, VHL expression was not detected, even with higher levels of nuclear β-catenin. This coincided with the presence of higher levels of HIF1α, suggesting that regulation of VHL expression may uncouple from the Wnt pathway at this stage. A suggestion from the findings in colorectal cancer is that Wnt/β-catenin signaling participates in the angiogenic switch. How this ties in with the fact that canonical Wnt signaling can directly upregulate expression of VEGF in colorectal cell lines (Easwaran et al., 2003) still needs further exploration. However, the fact that the VHL/HIF1α axis is directly associated with the angiogenic switch in tumor angiogenesis opens new hypotheses pertaining to Wnt signaling in the regulation of pathological angiogenesis.

4. Norrin, a new player in the field

Mutations detected in some FEVR patients mapped to a gene known as the Norrie gene or Norrie disease product (NDP) gene (Chen et al., 1993). This gene was found by genetic linkage on the short arm of the X chromosome (Bleeker-Wagemakers et al., 1985) and was later cloned and characterized (Berger et al., 1992; Chen et al., 1992). The Norrie gene encodes for a 133-amino acid protein called Norrin, with characteristic cysteine bridges that place it structurally in the cysteine knot growth factor family (Meitinger et al., 1993). This group is shared with transforming growth factor-α (TGF-α) and nerve growth factor (NGF). As we will discuss later, Norrin has proved to be a highly specific ligand for Fzd4 (Xu et al., 2004).

Norrie disease (ND) is an X-linked congenital retinal dysplasia that involves vascular defects in the retina at birth (MacDonald and Sasi, 1994). The disease evolves with different degrees of visual impairment and, in the most severe cases, can present tumor-resembling structures called pseudogliomas in the first months of life, leading to total blindness. There is also mental retardation in about 50% of the male cases and progressive sensorineural hearing loss in about 40% of these cases. Analysis of the NDP gene in patients has revealed a large variety of missense, nonsense, deletion, insertion, and splice-site mutations (Schuback et al., 1995). The small size of the NDP mRNA and the coding sequence allows for fast and simple mutation detection in ND patients. This has been valuable, as clinical diagnostic of this disorder has often been difficult in spite of the existence of classic signs such as specific ocular symptoms, progressive deafness, and mental retardation.

An interesting observation about NDP mutants is that most point mutations seem to be at or near cysteine residues preserved in the superfamily of proteins to which Norrin belongs, emphasizing their structural importance. Mutations in the NDP gene have also been associated with X-linked cases of FEVR, while autosomal dominant forms of this disorder were linked to Fzd4 and LRP5 genes (Toomes et al., 2004a,b). Two individuals belonging to the same family and affected by the same exact mutation in the NDP gene can present two different phenotypes of X-linked FEVR (Riveiro-Alvarez et al., 2005), indicating other genetic modifiers may be involved in Norrin function.

The NDP gene has also been linked to cases of retinopathy of prematurity (Shastry et al., 1997), a retinal vascular disease with similarities to FEVR. Retinopathy of prematurity occurs in children with short gestational age and can lead to retinal detachment. Mutations in the NDP gene have also been linked to Coats' disease, or retinal telangiectasis (Black et al., 1999), a disease that proceeds with abnormal retinal vascular development with exudative retinal detachment.

Norrin is normally expressed in brain and retina (Berger et al., 1992; Chen et al., 1992), and *in situ* hybridization has detected expression of Ndph, the mouse homologue gene, in the neural layers of the retina, brain, and in the spiral ganglion and stria vascularis of the cochlea (Berger et al., 1996). A Norrin-null mouse model showed retrolental structures in the vitreous body and a disorganized retinal ganglion cell layer as a primary pathogenic event (Berger et al., 1996). Another Norrin-null mouse model revealed that the hearing loss in these mice appears at around 3 months of age, and that the earliest primary site of cellular pathology occurs in the stria vascularis, with abnormal variation in vessel size and decreased number of vessels (Rehm et al., 2002). This suggests that one of the principal functions of Norrin in the ear is to regulate the interaction of the cochlea with its vasculature. Overall, phenotypic findings in Norrin-null-mouse models are consistent with those found in ND patients.

Norrin has been shown to play an important role in the proper regression of the hyaloid vasculature in mice after birth (Richter et al., 1998; Ohlmann et al., 2004). Hyaloid vasculature, which should completely disappear between P16 to P21, is still present in Norrin-null mice at this time, with only the single peripheral interconnecting branches having disappeared (Ohlmann et al., 2004). Lack of regression of hyaloid vasculature is also present in LRP5 (Kato et al., 2002) and Fzd4-null mouse models (Xu et al., 2004). Such lack of regression also appears in mice lacking known regulators of angiogenesis, like collagen XVIII and its proteolytically derived product endostatin (Fukai et al., 2002) and angiopoietin 2 (Hackett et al., 2002). This observation further supports the idea that Wnt-signaling components and Norrin clearly play a role in the regulation of angiogenesis. Norrin-null mice also indicate that the absence of functional Norrin primarily affects endothelial cells while only secondarily affecting astrocytes and pericytes, impairing the sprouting process into the deeper capillary network of the retina (Luhmann et al., 2005a). This leads to inner retinal hypoxia, as revealed by higher levels of $HIF1\alpha$ and VEGFA. Impaired endothelial cell function might explain secondary consequences observed in the retina of Norrin-null mice, like leaky superficial vasculature that could lead to displacement of vessels and delayed regression of the hyaloid vasculature.

Although Norrin has no sequence homology with Wnts, the significant similarities in vascular phenotypes between $Fzd4^{-/-}$ and $Ndph^{-/-}$ suggested that Norrin might be able to bind Fzd4 (Xu et al., 2004). That was in fact the case, as Norrin showed high specificity and affinity *in vitro* for this Wnt receptor, and triggered a strong activation of canonical Wnt signaling through Fzd4 and LRPs in luciferase reporter assays in 293 cells. Interestingly, mutant forms of Norrin were defective for signaling through Fzd4 in these assays. Like Wnts, Norrin binds extracellular matrix and appears to act locally. These findings directly linked a Frizzled ligand with vascular disease, opening an entirely new field in Wnt signaling research in retinal angiogenesis. The observations that Norrin can bind Fzd4 with high affinity and that Fzd4 and its correceptor LRP5 are also mutated in cases of FEVR have put both diseases into the context of one common Wnt-signaling pathway.

It is important to note that Norrin and Fzd4-null mice phenotypes do not exactly match. Fzd4$^{-/-}$ has more severe intraretinal hemorrhages and the development of perpendicular vessels emerging from the superficial network into deeper layers is present. Also, impaired association between astrocytes and endothelial cells appears much later in Norrin-null mice than in Fzd4$^{-/-}$ mice (Xu et al., 2004; Luhmann et al., 2005a). These observations strongly suggest that additional players yet to be elucidated are involved in normal vascular development in the eye.

Norrin function might not be restricted to retinal vascular development, as *in situ* hybridization analysis showed significant levels of transcript outside retinal tissue, such as in the cerebellum, hippocampus, olfactory bulb, and cortex and epithelium of rabbit brain (Hartzer et al., 1999). Another interesting observation is that peripheral venous insufficiency has been reported in association with ND in two separate cases (Rehm et al., 1997; Michaelides et al., 2004), one of them being in several affected individuals of a large family. Homozygous mutation of the murine Norrin gene (Ndph) resembles many of the characteristics of the human disease, but it can also lead to incomplete decidualization during pregnancy and defective vascular development of the decidua, with consequent infertility (Luhmann et al., 2005b). Interestingly, an Fzd4$^{-/-}$ mouse line has also been found to be infertile (Hsieh et al., 2005), with nonfunctional corpora lutea, probably due to impaired angiogenesis. These observations support a role for Norrin in female reproductive angiogenesis. The fact that the Ndph$^{-/-}$ mice have functional corpora lutea, while Fzd4$^{-/-}$ mice also do not indicate the possibility of other molecular players interacting through Fzd4 in ovarian angiogenesis.

Using mice with a targeted disruption of the Ndph gene, which presented complete absence of intraretinal capillaries, one group used a strategy of expressing ectopic Norrin in the eye utilizing a lens-specific promoter (Ohlmann et al., 2005) to attempt to restore function. This experiment demonstrated that Norrin can induce growth of ocular capillaries, and what is even more important, was able to restore retinal vascular network respecting the normal retinal architecture. Retinas of these mice were also restored at a functional level, as evaluated by electroretinography. This might open an interesting possibility for Norrin as a therapeutic agent.

5. Wnt regulation of angiogenesis

As a role of Wnts in angiogenesis became progressively more apparent, it was important to understand which Wnt ligands and signaling components were endogenously expressed by endothelial and mural cells (i.e., pericytes or smooth muscle cells). Human umbilical vein endothelial cells (HUVEC) and smooth muscle cells (SMC) from human pulmonary artery were initially reported to express Wnt5a and Frizzled-3 (Wright et al., 1999). Human microvascular endothelial cells from dermis (HMVECd) show expression of Wnt7a, 10b, and 13, and Frizzled members 4, 5, and 6 (Cheng et al., 2003; Masckauchan et al., 2005). Both HUVEC and HMVEC also express β-catenin-associated transcription factors Tcf-1, Tcf-3, Tcf-4, and Lef-1 (Masckauchan et al., 2005). Evidence that canonical Wnt signaling is activated in endothelium *in vivo* comes from a β-catenin-activated transgenic mouse line (Maretto et al., 2003). In this model, the LacZ gene expressing β-galactosidase reporter gene is under the control of β-catenin/Tcf responsive elements. This system allowed the identification of different sites of expression where canonical Wnt signaling is being activated during development, including endothelial cells of vessels invading the spinal cord from the surrounding mesenchyme at E12.5, identified by expression of the endothelial marker, PECAM.

Studies in cultured endothelial cells indicate Wnt signaling can be mitogenic for endothelium, as with many other cell types. Ectopic expression of Wnt1, which can trigger both canonical (Wright et al., 1999) and non-canonical pathways (Habas et al., 2001), in mouse brain microvascular endothelial cells promotes proliferation (Wright et al., 1999). This was further supported by the observation that mutant, stabilized forms of β-catenin have mitogenic effects in human dermal microvascular endothelial cells (Venkiteswaran et al., 2002). The Wnt/β-catenin axis has also been shown to promote survival in human primary endothelial cells in culture under low serum conditions as well as induce formation of capillary-like networks (Masckauchan et al., 2005).

The gene promoter of VEGF, a central and potent inducer of angiogenesis, contains seven β-catenin/Tcf-binding sites. Hence, it is not surprising that VEGF is upregulated when Wnt/β-catenin signaling is activated by mutations in colon cancer cells (Zhang et al., 2001; Easwaran et al., 2003) and in human endothelial cells (Skurk et al., 2005). Expression of other angiogenic factors has been reported as controlled by the Wnt/β-catenin axis. Members of the fibroblast growth factor (FGF) are important regulators of angiogenesis, and FGF18, endogenously expressed in HUVEC and umbilical artery smooth muscle cells (Antoine et al., 2005), can be regulated by Wnt/β-catenin (Shimokawa et al., 2003). Conversely, basic FGF (FGF2) is well known as a potent inducer of angiogenesis and can induce expression of cyclin D1 via Lef/Tcf-β-catenin signaling (Holnthoner et al., 2002). Other angiogenic target genes of Wnts have been identified in endothelial cells such as interleukin-8 (Masckauchan et al., 2005). Interleukin-8 can induce proliferation and survival of endothelial cells *in vitro*, and upregulate expression of at least two matrix metalloproteinases, MMP-2 and MMP-9, known to play an active role in angiogenesis (Li et al., 2003). Expression of Endothelin-1, a vasoactive and endothelial mitogenic peptide known to play a role in tumor angiogenesis (Knowles et al., 2005), can also be regulated by Wnt/β-catenin (Kim et al., 2005).

PECAM-1, a scaffolding and cell adhesion molecule restrictively expressed in vascular cells, plays an important role in endothelial cell biology as a regulator of junctional, cytoskeletal, adhesive, and signaling pathways (Ilan and Madri, 2003). PECAM-1 is able to bind and modulate tyrosine-phosphorylated β-catenin levels and affect its localization (Ilan et al., 1999). Endothelial cells from PECAM-1 knockout mice showed β-catenin only at cell membranes, concomitant with the finding that PECAM-1 lacking most of the cytoplasmic domain has a much lower ability to induce transcriptional activation of canonical Wnt signaling than the full length molecule (Biswas et al., 2003). Disruption of cell adhesion complexes, where PECAM-1 and catenins function, can mitigate migration of endothelial cells during angiogenesis. Finally, VEGF can also stimulate tyrosine phosphorylation of β-catenin in endothelial cells (Cohen et al., 1999).

Wnt/β-catenin signaling may be important for repair of vascular injury. One study reports that β-catenin/Tcf-4 promote vascular remodeling in rats after balloon injury, inhibiting vascular smooth muscle cell apoptosis (Wang et al., 2002). A role of canonical Wnt signaling in vascular biology may be supported in studies using a mouse model with conditional inactivation of β-catenin in endothelial cells (Cattelino et al., 2003). Endothelial intercellular junctions are affected in mutant embryos, with reduced cell–cell adhesion and increased permeability that leads to hemorrhages and fluid extravasation. Vascular patterning also seems to be seriously affected in the head and placenta. Although β-catenin does not seem necessary for early phases of vasculogenesis and angiogenesis during mouse development, it seems to be required for appropriate vascular patterning and maintenance of vascular integrity in later phases. The open question that remains from these studies is what are the relative roles of the adhesive versus signaling function of β-catenin during vascular development.

Endostatin, a C-terminal fragment of type XVIII collagen, is an antiangiogenic molecule able to inhibit endothelial cell migration and proliferation (O'Reilly et al., 1997). Endostatin can downregulate cyclin D1 expression at a transcriptional level in endothelial cells, a process shown to be critical for endostatin action (Ramchandran et al., 2002). Researchers identified a Lef1 site in the cyclin D1 promoter that is essential for the inhibitory effect of endostatin on cyclin D1 expression, indicating that endostatin action is at least in part mediated by Wnt signaling. This agrees with experimental data showing that endostatin is a modest inhibitor of Wnt signaling in endothelial cells at a high dose (Hanai et al., 2002).

Secreted Frizzled-related proteins (sFRPs), a family of secreted proteins with structural homology to the extracellular cystein-rich domain (CRD) of Frizzled receptors, have also been shown to regulate angiogenesis. FrzA, or sFRP-1, has been shown to reduce proliferation of endothelial cells in culture (Duplaa et al., 1999). This antagonistic role was also replicated in smooth muscle cells *in vitro*, where treatment with FrzA decreased the levels of cyclins and cyclin-dependent kinases (cdks) while upregulating cytosolic levels of phospho-β-catenin, targeted for degradation (Ezan et al., 2004). *In vivo* results supported this idea, as in a unilateral hindlimb muscle ischemia mouse model, adenoviruses coding for FrzA induced a decrease in the number of capillaries, cell proliferation, expression of cyclins E, and D1 and cdk2 activity (Ezan et al., 2004). In contrast to this, a transgenic mouse model overexpressing FrzA was subjected to induced myocardial infarction. This treatment led to higher capillary density and evidence that FrzA reduced infarct size, with lower apoptosis and higher collagen deposition in the scar when compared to control animals (Barandon et al., 2003, 2004). FrzA has also been shown to induce angiogenesis in a chick chorioallantoic membrane model, as evaluated by increasing formation of vessels compared to control (Dufourcq et al., 2002). *In vitro*, FrzA increased migration and tube formation of endothelial cells, independently from other known angiogenic factors (Dufourcq et al., 2002). These apparently contrasting results clearly indicate that the role of FrzA in angiogenesis needs further investigation.

6. Current perspectives

In the last few years, we have seen increasing support for the idea that Wnts are angiogenic factors. Wnt/Frizzled signaling is involved in embryonic vascular development, and Norrin is critical for angiogenesis of the developing retina. There is a need to clarify which of the Wnt pathways are involved in distinct steps of angiogenesis and in which biological contexts Wnts function. Evidence supports a biological role for both canonical and noncanonical Wnt signaling in angiogenesis.

Some hope is slowly becoming apparent for ND and FEVR patients, as progressively more is found about the function of Norrin and the potential of utilizing Norrin as a therapeutic agent. The eventual use of effective therapies requires a much better and thorough understanding of Wnt function in angiogenesis.

The exciting findings we have seen in the last few years and a growing confidence among researchers on the importance of the role of Wnts and Wnt signaling in angiogenesis assures a promising outlook in the field and that we will experience new and fascinating discoveries in the future.

References

Antoine, M., Wirz, W., Tag, C.G., Mavituna, M., Emans, N., Korff, T., Stoldt, V., Gressner, A.M., Kiefer, P. 2005. Expression pattern of fibroblast growth factors (FGFs), their receptors and antagonists in primary endothelial cells and vascular smooth muscle cells. Growth Factors 23, 87–95.

Barandon, L., Couffinhal, T., Ezan, J., Dufourcq, P., Costet, P., Alzieu, P., Leroux, L., Moreau, C., Dare, D., Duplaa, C. 2003. Reduction of infarct size and prevention of cardiac rupture in transgenic mice overexpressing FrzA. Circulation 108, 2282–2289.

Barandon, L., Couffinhal, T., Dufourcq, P., Ezan, J., Costet, P., Daret, D., Deville, C., Duplaa, C. 2004. Frizzled A, a novel angiogenic factor: Promises for cardiac repair. Eur. J. Cardiothorac. Surg. 25, 76–83.

Berger, W., Meindl, A., van de Pol, T.J., Cremers, F.P., Ropers, H.H., Döerner, C., Monaco, A., Bergen, A.A., Lebo, R., Warburg, M., Zergollern, L., Lorenz, B., et al. 1992. Isolation of a candidate gene for Norrie disease by positional cloning. Nat. Genet. 2, 84.

Berger, W., van de Pol, D., Bachner, D., Oerlemans, F., Winkens, H., Hameister, H., Wieringa, B., Hendriks, W., Ropers, H.H. 1996. An animal model for Norrie disease (ND): Gene targeting of the mouse ND gene. Hum. Mol. Genet. 5, 51–59.

Biswas, P., Canosa, S., Schoenfeld, J., Schoenfeld, D., Tucker, A., Madri, J.A. 2003. PECAM-1 promotes beta-catenin accumulation and stimulates endothelial cell proliferation. Biochem. Biophys. Res. Commun. 303, 212–218.

Black, G.C., Perveen, R., Bonshek, R., Cahill, M., Clayton-Smith, J., Lloyd, I.C., McLeod, D. 1999. Coats, disease of the retina (unilateral retinal telangiectasis) caused by somatic mutation in the NDP gene: A role for norrin in retinal angiogenesis. Hum. Mol. Genet. 8, 2031–2035.

Bleeker-Wagemakers, L.M., Friedrich, U., Gal, A., Wienker, T.F., Warburg, M., Ropers, H.H. 1985. Close linkage between Norrie disease, a cloned DNA sequence from the proximal short arm, and the centromere of the X chromosome. Hum. Genet. 71, 211–214.

Cattelino, A., Liebner, S., Gallini, R., Zanetti, A., Balconi, G., Corsi, A., Bianco, P., Wolburg, H., Moore, R., Oreda, B., Kemler, R., Dejana, E. 2003. The conditional inactivation of the beta-catenin gene in endothelial cells causes a defective vascular pattern and increased vascular fragility. J. Cell Biol. 162, 1111–1122.

Chen, Z.Y., Hendriks, R.W., Jobling, M.A., Powell, J.F., Breakefield, X.O., Sims, K.B., Craig, I.W. 1992. Isolation and characterization of a candidate gene for Norrie disease. Nat. Genet. 1, 204–208.

Chen, Z.Y., Battinelli, E.M., Fielder, A., Bundey, S., Sims, K., Breakefield, X.O., Craig, I.W. 1993. A mutation in the Norrie disease gene (NDP) associated with X-linked familial exudative vitreoretinopathy. Nat. Genet. 5, 180–183.

Cheng, C.W., Smith, S.K., Charnock-Jones, D.S. 2003. Wnt-1 signaling inhibits human umbilical vein endothelial cell proliferation and alters cell morphology. Exp. Cell Res. 291, 415–425.

Cohen, A.W., Carbajal, J.M., Schaeffer, R.C., Jr. 1999. VEGF stimulates tyrosine phosphorylation of beta-catenin and small-pore endothelial barrier dysfunction. Am. J. Physiol. 277, H2038–H2049.

Dufourcq, P., Couffinhal, T., Ezan, J., Barandon, L., Moreau, C., Daret, D., Duplaa, C. 2002. FrzA, a secreted frizzled related protein, induced angiogenic response. Circulation 106, 3097–3103.

Duplaa, C., Jaspard, B., Moreau, C., D'Amore, P.A. 1999. Identification and cloning of a secreted protein related to the cysteine-rich domain of frizzled. Evidence for a role in endothelial cell growth control. Circ. Res. 84, 1433–1445.

Easwaran, V., Lee, S.H., Inge, L., Guo, L., Goldbeck, C., Garrett, E., Wiesmann, M., Garcia, P.D., Fuller, J.H., Chan, V., Randazzo, F., Gundel, R., et al. 2003. Beta-Catenin regulates vascular endothelial growth factor expression in colon cancer. Cancer Res. 63, 3145–3153.

Ezan, J., Leroux, L., Barandon, L., Dufourcq, P., Jaspard, B., Moreau, C., Allieres, C., Daret, D., Couffinhal, T., Duplaa, C. 2004. FrzA/sFRP-1, a secreted antagonist of the Wnt-Frizzled pathway, controls vascular cell proliferation *in vitro* and *in vivo*. Cardiovasc. Res. 63, 731–738.

Fukai, N., Eklund, L., Marneros, A.G., Oh, S.P., Keene, D.R., Tamarkin, L., Niemela, M., Ilves, M., Li, E., Pihlajaniemi, T., Olsen, B.R. 2002. Lack of collagen XVIII/endostatin results in eye abnormalities. EMBO J. 21, 1535–1544.

Giles, R.H., Lolkema, M.P., Snijckers, C.M., Belderbos, M., van der Groep, P., Mans, D.A., van Beest, M., van Noort, M., Goldschmeding, R., van Diest, P.J., Clevers, H., Voest, E.E. 2006. Interplay between VHL/HIF1alpha and Wnt/beta-catenin pathways during colorectal tumorigenesis. Oncogene 25, 3065–3070.

Gong, Y., Slee, R.B., Fukai, N., Rawadi, G., Roman-Roman, S., Reginato, A.M., Wang, H., Cundy, T., Glorieux, F.H., Lev, D., Zacharin, M., Oexle, K., et al. 2001. LDL receptor-related protein 5 (LRP5) affects bone accrual and eye development. Cell 107, 513–523.

Habas, R., Kato, Y., He, X. 2001. Wnt/Frizzled activation of Rho regulates vertebrate gastrulation and requires a novel Formin homology protein Daam1. Cell 107, 843–854.

Hackett, S.F., Wiegand, S., Yancopoulos, G., Campochiaro, P.A. 2002. Angiopoietin-2 plays an important role in retinal angiogenesis. J. Cell. Physiol. 192, 182–187.

Hanai, J., Gloy, J., Karumanchi, S.A., Kale, S., Tang, J., Hu, G., Chan, B., Ramchandran, R., Jha, V., Sukhatme, V.P., Sokol, S. 2002. Endostatin is a potential inhibitor of Wnt signaling. J. Cell Biol. 158, 529–539.

Hartzer, M.K., Cheng, M., Liu, X., Shastry, B.S. 1999. Localization of the Norrie disease gene mRNA by *in situ* hybridization. Brain Res. Bull. 49, 355–358.

Holnthoner, W., Pillinger, M., Groger, M., Wolff, K., Ashton, A.W., Albanese, C., Neumeister, P., Pestell, R.G., Petzelbauer, P. 2002. Fibroblast growth factor-2 induces Lef/Tcf- dependent transcription in human endothelial cells. J. Biol. Chem. 277, 45847–45853.

Hsieh, M., Boerboom, D., Shimada, M., Lo, Y., Parlow, A.F., Luhmann, U.F., Berger, W., Richards, J.S. 2005. Mice Null for Frizzled4 (Fzd4–/–) are infertile and exhibit impaired corpora lutea formation and function. Biol. Reprod. 73, 1135–1146.

Ilan, N., Madri, J.A. 2003. PECAM-1: Old friend, new partners. Curr. Opin. Cell Biol. 15, 515–524.

Ilan, N., Mahooti, S., Rimm, D.L., Madri, J.A. 1999. PECAM-1 (CD31) functions as a reservoir for and a modulator of tyrosine-phosphorylated beta-catenin. J. Cell Sci. 112 (Pt. 18), 3005–3014.

Ishikawa, T., Tamai, Y., Zorn, A.M., Yoshida, H., Seldin, M.F., Nishikawa, S., Taketo, M.M. 2001. Mouse Wnt receptor gene Fzd5 is essential for yolk sac and placental angiogenesis. Development 128, 25–33.

Jeays-Ward, K., Hoyle, C., Brennan, J., Dandonneau, M., Alldus, G., Capel, B., Swain, A. 2003. Endothelial and steroidogenic cell migration are regulated by WNT4 in the developing mammalian gonad. Development 130, 3663–3670.

Kato, M., Patel, M.S., Levasseur, R., Lobov, I., Chang, B.H., Glass, D.A., 2nd, Hartmann, C., Li, L., Hwang, T.H., Brayton, C.F., Lang, R.A., Karsenty, G., et al. 2002. Cbfa1-independent decrease in osteoblast proliferation, osteopenia, and persistent embryonic eye vascularization in mice deficient in Lrp5, a Wnt coreceptor. J. Cell Biol. 157, 303–314.

Kaykas, A., Yang-Snyder, J., Heroux, M., Shah, K.V., Bouvier, M., Moon, R.T. 2004. Mutant Frizzled 4 associated with vitreoretinopathy traps wild-type Frizzled in the endoplasmic reticulum by oligomerization. Nat. Cell Biol. 6, 52–58.

Kim, T.H., Xiong, H., Zhang, Z., Ren, B. 2005. beta-Catenin activates the growth factor endothelin-1 in colon cancer cells. Oncogene 24, 597–604.

Knowles, J., Loizidou, M., Taylor, I. 2005. Endothelin-1 and angiogenesis in cancer. Curr. Vasc. Pharmacol. 3, 309–314.

Li, A., Dubey, S., Varney, M.L., Dave, B.J., Singh, R.K. 2003. IL-8 directly enhanced endothelial cell survival, proliferation, and matrix metalloproteinases production and regulated angiogenesis. J. Immunol. 170, 3369–3376.

Lobov, I.B., Rao, S., Carroll, T.J., Vallance, J.E., Ito, M., Ondr, J.K., Kurup, S., Glass, D.A., Patel, M.S., Shu, W., Morrisey, E.E., McMahon, A.P., et al. 2005. WNT7b mediates macrophage-induced programmed cell death in patterning of the vasculature. Nature 437, 417–421.

Luhmann, U.F., Lin, J., Acar, N., Lammel, S., Feil, S., Grimm, C., Seeliger, M.W., Hammes, H.P., Berger, W. 2005a. Role of the Norrie disease pseudoglioma gene in sprouting angiogenesis during development of the retinal vasculature. Invest. Ophthalmol. Vis. Sci. 46, 3372–3382.

Luhmann, U.F., Meunier, D., Shi, W., Luttges, A., Pfarrer, C., Fundele, R., Berger, W. 2005b. Fetal loss in homozygous mutant Norrie disease mice: A new role of Norrin in reproduction. Genesis 42, 253–262.

MacDonald, I.M., Sasi, R. 1994. Molecular genetics of inherited eye disorders. Clin. Invest. Med. 17, 474–498.

Maretto, S., Cordenonsi, M., Dupont, S., Braghetta, P., Broccoli, V., Hassan, A.B., Volpin, D., Bressan, G.M., Piccolo, S. 2003. Mapping Wnt/beta-catenin signaling during mouse development and in colorectal tumors. Proc. Natl. Acad. Sci. USA 100, 3299–3304.

Masckauchan, T.N.H., Shawber, C.J., Funahashi, Y., Li, C.M., Kitajewski, J. 2005. Wnt/beta-catenin signaling induces proliferation, survival and interleukin-8 in human endothelial cells. Angiogenesis 8, 43–51.

Meitinger, T., Meindl, A., Bork, P., Rost, B., Sander, C., Haasemann, M., Murken, J. 1993. Molecular modelling of the Norrie disease protein predicts a cystine knot growth factor tertiary structure. Nat. Genet. 5, 376–380.

Michaelides, M., Luthert, P.J., Cooling, R., Firth, H., Moore, A.T. 2004. Norrie disease and peripheral venous insufficiency. Br. J. Ophthalmol. 88, 1475.

Monkley, S.J., Delaney, S.J., Pennisi, D.J., Christiansen, J.H., Wainwright, B.J. 1996. Targeted disruption of the Wnt2 gene results in placentation defects. Development 122, 3343–3353.

O'Reilly, M.S., Boehm, T., Shing, Y., Fukai, N., Vasios, G., Lane, W.S., Flynn, E., Birkhead, J.R., Olsen, B.R., Folkman, J. 1997. Endostatin: An endogenous inhibitor of angiogenesis and tumor growth. Cell 88, 277–285.

Ohlmann, A., Scholz, M., Goldwich, A., Chauhan, B.K., Hudl, K., Ohlmann, A.V., Zrenner, E., Berger, W., Cvekl, A., Seeliger, M.W., Tamm, E.R. 2005. Ectopic norrin induces growth of ocular capillaries and restores normal retinal angiogenesis in Norrie disease mutant mice. J. Neurosci. 25, 1701–1710.

Ohlmann, A.V., Adamek, E., Ohlmann, A., Lutjen-Drecoll, E. 2004. Norrie gene product is necessary for regression of hyaloid vessels. Invest. Ophthalmol. Vis. Sci. 45, 2384–2390.

Parr, B.A., Cornish, V.A., Cybulsky, M.I., McMahon, A.P. 2001. Wnt7b regulates placental development in mice. Dev. Biol. 237, 324–332.

Pepper, M.S. 2001. Role of the matrix metalloproteinase and plasminogen activator-plasmin systems in angiogenesis. Arterioscler. Thromb. Vasc. Biol. 21, 1104–1117.

Ramchandran, R., Karumanchi, S.A., Hanai, J., Alper, S.L., Sukhatme, V.P. 2002. Cellular actions and signaling by endostatin. Crit. Rev. Eukaryot. Gene Expr. 12, 175–191.

Rehm, H.L., Gutierrez-Espeleta, G.A., Garcia, R., Jimenez, G., Khetarpal, U., Priest, J.M., Sims, K.B., Keats, B.J., Morton, C.C. 1997. Norrie disease gene mutation in a large Costa Rican kindred with a novel phenotype including venous insufficiency. Hum. Mutat. 9, 402–408.

Rehm, H.L., Zhang, D.S., Brown, M.C., Burgess, B., Halpin, C., Berger, W., Morton, C.C., Corey, D.P., Chen, Z.Y. 2002. Vascular defects and sensorineural deafness in a mouse model of Norrie disease. J. Neurosci. 22, 4286–4292.

Richter, M., Gottanka, J., May, C.A., Welge-Lussen, U., Berger, W., Lutjen-Drecoll, E. 1998. Retinal vasculature changes in Norrie disease mice. Invest. Ophthalmol. Vis. Sci. 39, 2450–2457.

Riveiro-Alvarez, R., Trujillo-Tiebas, M.J., Gimenez-Pardo, A., Garcia-Hoyos, M., Cantalapiedra, D., Lorda-Sanchez, I., Rodriguez de Alba, M., Ramos, C., Ayuso, C. 2005. Genotype-phenotype variations in five Spanish families with Norrie disease or X-linked FEVR. Mol. Vis. 11, 705–712.

Robitaille, J., MacDonald, M.L., Kaykas, A., Sheldahl, L.C., Zeisler, J., Dube, M.P., Zhang, L.H., Singaraja, R.R., Guernsey, D.L., Zheng, B., Siebert, L.F., Hoskin-Mott, A., et al. 2002. Mutant frizzled-4 disrupts retinal angiogenesis in familial exudative vitreoretinopathy. Nat. Genet. 32, 326–330.

Schuback, D.E., Chen, Z.Y., Craig, I.W., Breakefield, X.O., Sims, K.B. 1995. Mutations in the Norrie disease gene. Hum. Mutat. 5, 285–292.

Shastry, B.S., Pendergast, S.D., Hartzer, M.K., Liu, X., Trese, M.T. 1997. Identification of missense mutations in the Norrie disease gene associated with advanced retinopathy of prematurity. Arch. Ophthalmol. 115, 651–655.

Shimizu, H., Julius, M.A., Giarre, M., Zheng, Z., Brown, A.M., Kitajewski, J. 1997. Transformation by Wnt family proteins correlates with regulation of beta-catenin. Cell Growth Differ. 8, 1349–1358.

Shimokawa, T., Furukawa, Y., Sakai, M., Li, M., Miwa, N., Lin, Y.M., Nakamura, Y. 2003. Involvement of the FGF18 gene in colorectal carcinogenesis, as a novel downstream target of the beta-catenin/T-cell factor complex. Cancer Res. 63, 6116–6120.

Shu, W., Jiang, Y.Q., Lu, M.M., Morrisey, E.E. 2002. Wnt7b regulates mesenchymal proliferation and vascular development in the lung. Development 129, 4831–4842.

Skurk, C., Maatz, H., Rocnik, E., Bialik, A., Force, T., Walsh, K. 2005. Glycogen-synthase kinase3beta/beta-catenin axis promotes angiogenesis through activation of vascular endothelial growth factor signaling in endothelial cells. Circ. Res. 96, 308–318.

Toomes, C., Bottomley, H.M., Jackson, R.M., Towns, K.V., Scott, S., Mackey, D.A., Craig, J.E., Jiang, L., Yang, Z., Trembath, R., Woodruff, G., Gregory-Evans, C.Y., et al. 2004a. Mutations in LRP5 or FZD4 underlie the common familial exudative vitreoretinopathy locus on chromosome 11q. Am. J. Hum. Genet. 74, 721–730.

Toomes, C., Bottomley, H.M., Scott, S., Mackey, D.A., Craig, J.E., Appukuttan, B., Stout, J.T., Flaxel, C.J., Zhang, K., Black, G.C., Fryer, A., Downey, L.M., et al. 2004b. Spectrum and frequency of FZD4 mutations in familial exudative vitreoretinopathy. Invest. Ophthalmol. Vis. Sci. 45, 2083–2090.

Venkiteswaran, K., Xiao, K., Summers, S., Calkins, C.C., Vincent, P.A., Pumiglia, K., Kowalczyk, A.P. 2002. Regulation of endothelial barrier function and growth by VE-cadherin, plakoglobin, and beta-catenin. Am. J. Physiol. Cell Physiol. 283, C811–C821.

Wang, X., Xiao, Y., Mou, Y., Zhao, Y., Blankesteijn, W.M., Hall, J.L. 2002. A role for the beta-catenin/T-cell factor signaling cascade in vascular remodeling. Circ. Res. 90, 340–347.

Wang, Z., Shu, W., Lu, M.M., Morrisey, E.E. 2005. Wnt7b activates canonical signaling in epithelial and vascular smooth muscle cells through interactions with Fzd1, Fzd10, and LRP5. Mol. Cell Biol. 25, 5022–5030.

Wright, M., Aikawa, M., Szeto, W., Papkoff, J. 1999. Identification of a Wnt-responsive signal transduction pathway in primary endothelial cells. Biochem. Biophys. Res. Commun. 263, 384–388.

Xu, Q., Wang, Y., Dabdoub, A., Smallwood, P.M., Williams, J., Woods, C., Kelley, M.W., Jiang, L., Tasman, W., Zhang, K., Nathans, J. 2004. Vascular development in the retina and inner ear: Control by Norrin and Frizzled-4, a high-affinity ligand-receptor pair. Cell 116, 883–895.

Zhang, X., Gaspard, J.P., Chung, D.C. 2001. Regulation of vascular endothelial growth factor by the Wnt and K-ras pathways in colonic neoplasia. Cancer Res. 61, 6050–6054.

Index

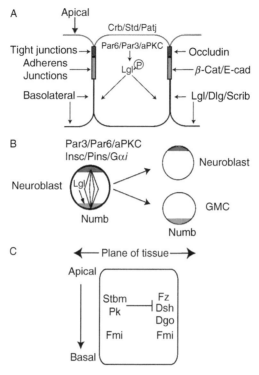

Gretchen L. Dollar and Sergei Y. Sokol, Chapter 2, Figure 1.

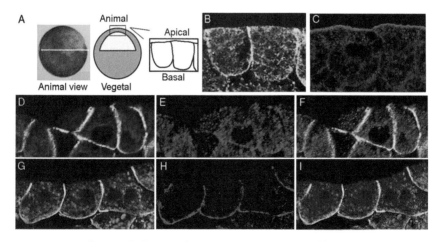

Gretchen L. Dollar and Sergei Y. Sokol, Chapter 2, Figure 2.

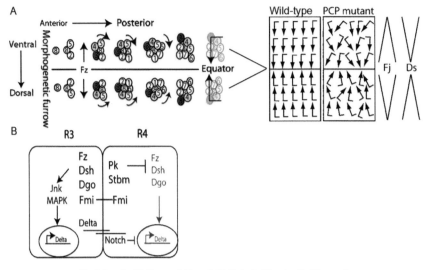

Gretchen L. Dollar and Sergei Y. Sokol, Chapter 2, Figure 4.

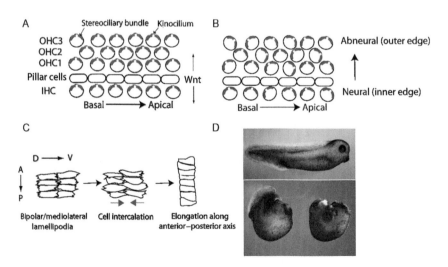

Gretchen L. Dollar and Sergei Y. Sokol, Chapter 2, Figure 5.

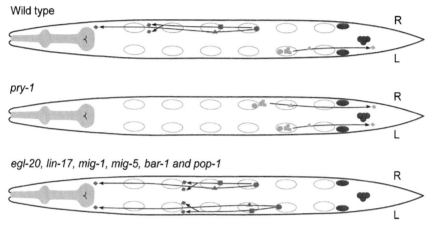

Hendrik C. Korswagen, Chapter 3, Figure 1.

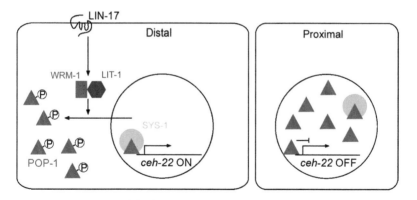

Hendrik C. Korswagen, Chapter 3, Figure 3.

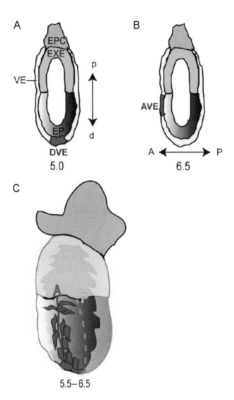

Almut Köhler, Alexandra Schambony and Doris Wedlich, Chapter 5, Figure 2.

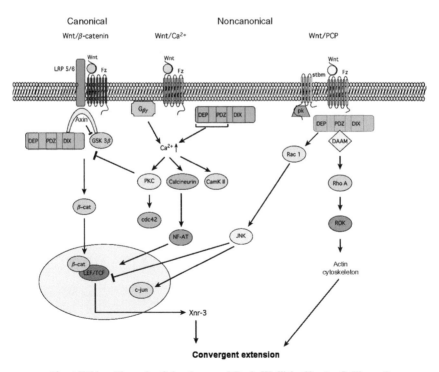

Almut Köhler, Alexandra Schambony and Doris Wedlich, Chapter 5, Figure 3.

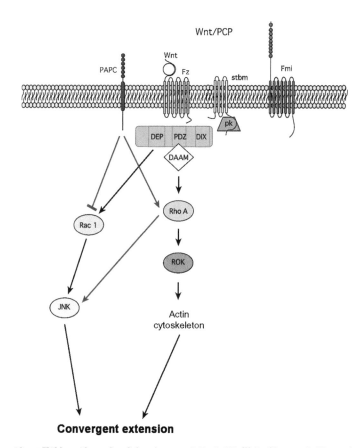

Convergent extension

Almut Köhler, Alexandra Schambony and Doris Wedlich, Chapter 5, Figure 4.

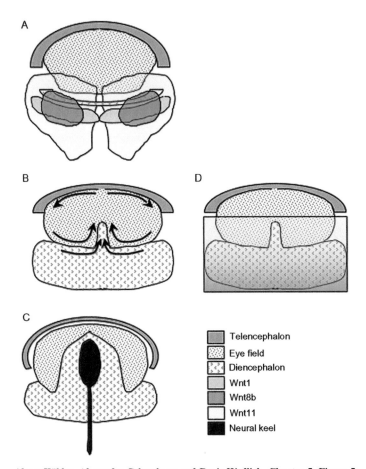

Almut Köhler, Alexandra Schambony and Doris Wedlich, Chapter 5, Figure 5.

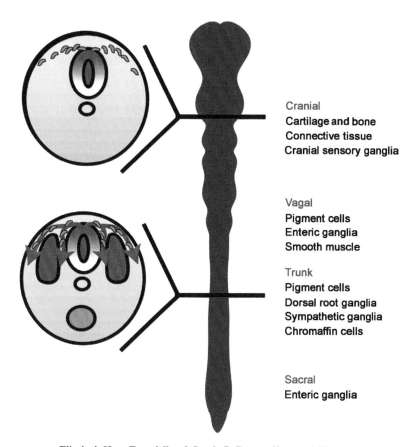

Cranial
Cartilage and bone
Connective tissue
Cranial sensory ganglia

Vagal
Pigment cells
Enteric ganglia
Smooth muscle

Trunk
Pigment cells
Dorsal root ganglia
Sympathetic ganglia
Chromaffin cells

Sacral
Enteric ganglia

Elizabeth Heeg-Truesdell and Carole LaBonne, Chapter 6, Figure 1.

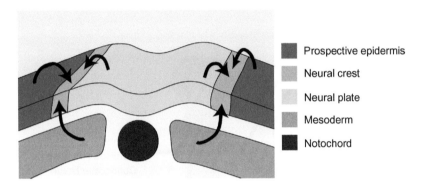

Elizabeth Heeg-Truesdell and Carole LaBonne, Chapter 6, Figure 2.

Prospective epidermis
Neural crest
Neural plate
Mesoderm
Notochord

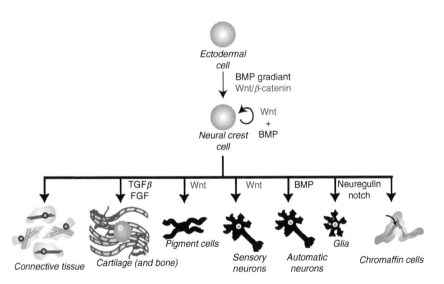

Elizabeth Heeg-Truesdell and Carole LaBonne, Chapter 6, Figure 3.

Printed and bound by CPI Group (UK) Ltd, Croydon, CR0 4YY

08/05/2025

01864966-0007